Plant Anatomy, Morphology and Physiology

Plant Anatomy, Morphology and Physiology

Edited by **Clive Koelling**

SYRAWOOD
PUBLISHING HOUSE
New York

Published by Syrawood Publishing House,
750 Third Avenue, 9th Floor,
New York, NY 10017, USA
www.syrawoodpublishinghouse.com

Plant Anatomy, Morphology and Physiology
Edited by Clive Koelling

International Standard Book Number: 978-1-68286-326-8 (Hardback)

Printed in China.

Contents

Preface

The main aim of this book is to educate learners and enhance their research focus by presenting diverse topics covering this vast field. This is an advanced book which compiles significant studies by distinguished experts. This book addresses successive solutions to the challenges arising in the area of application, along with it; the book provides scope for future developments.

Mankind has been dependent on plants since the early ages. The multiple uses of plants such as in medicine, etc. have raised their economic value as well. This book brings forth some of the most innovative concepts and elucidates the unexplored aspects of botany by exploring a diverse array of topics. Plant cytology and anatomy, taxonomy, plant diversity, ethnobotany, phytopathology, paleobotany, etc., are some of the concepts that have been thoroughly discussed. The aim of this book is to present researches that have transformed this discipline and aided its advancement. It is a ripe text for students and researchers of botany, agriculture, biology, etc.

It was a great honour to edit this book, though there were challenges, as it involved a lot of communication and networking between me and the editorial team. However, the end result was this all-inclusive book covering diverse themes in the field.

Finally, it is important to acknowledge the efforts of the contributors for their excellent chapters, through which a wide variety of issues have been addressed. I would also like to thank my colleagues for their valuable feedback during the making of this book.

Editor

Up until now, it is still not fully understood how GA participates in dissected leaf development. Experimental evidence supports that GA signaling partly mediates the function of KNOXI proteins. On the other hand, KNOXI proteins directly repress the expression of GA20ox and activate the expression of GA2ox in both simple- and compound-leafed species [52,54,56,73]. Thus, KNOXI proteins antagonize the GA activity. Genetic analyses indicate that altered GA homeostasis or signaling modulates *KNOXI* misexpression phenotypes in *A. thaliana*. Moreover, exogenous GA applications or the *pro* mutations suppress the super-compound leaf phenotype in the tomato dominant mutant, *Mouse ears* (*Me*), in which the tomato *KNOXI* gene, *Tkn2*, is misexpressed [56]. Although reducing the GA activity itself is not sufficient to increase the leaflet number in wild-type tomato, it enhances the competence to develop more leaflets of *Me* or *Curl* (*Cu*) mutants [54]. The recessive loss-of-function *clausa* (*clau*) mutant, which fails to delimit *KNOXI* expression domains, displays elevated and broadened *KNOXI* expression and increased leaf complexity. Exogenous GA applications or *pro* mutations suppress the compound leaf phenotype of the *clau* mutant, whereas mutations that lead to GA deficiency act conversely, suggesting that GA signaling acts downstream of the endogenous *KNOXI* pathway in tomato [54]. Thus, KNOXI proteins function in the maintenance of the SAM and promote compound leaf development partly by repressing GA homeostasis and response.

GA signaling is considered to partly mediate the LANCEOLATE (LA) activity in tomato. As discussed earlier, the maintenance of the morphogenetic activity of the leaf marginal blastozone required for the elaboration of leaflets involves the positive regulators, KNOXI proteins, and the negative regulator, LA, in tomato. *LA* promotes determinate growth and is negatively regulated by miR319 [30]. Interestingly, GA-treated wild-type plants resemble the *La-2* mutant in leaf complexity, leaf margin serration, hypocotyl length and anthocyanin content [29]. Exogenous GA applications or constitutive GA responses conditioned by the *pro* mutation suppress the indeterminate leaf growth caused by overexpression of miR319. In addition, the long hypocotyl phenotype of the *La-2* mutant is suppressed by mutations that lead to GA deficiencies or by applications of GA biosynthesis inhibitors, further supporting that the gain-of-function phenotype of *La-2* is mediated by an increased GA activity [29]. Consistent with this, genes involved in the GA metabolism or response is differentially regulated in the *La-2* mutant. Of particular interest among the differentially-regulated genes is *SlGA2ox4*. Overexpression of *SlGA2ox4* partially suppresses the simple-leaf phenotype of *La-2*. The relationship between the GA pathway and LA function appears to be evolutionarily conserved, at least in *A. thaliana* and tomato [29].

CK coordinates a wide range of plant developmental processes [55,74]. During shoot development, CK positively regulates the size and activity of the SAM [75–81]. Interactions between genes related to CK biosynthesis and response and genes required for the SAM maintenance exist at various levels. As discussed, KNOXI proteins activate CK biosynthesis [54,57]. In turn, CK regulates expression and activity of *WUSCHEL*, a key regulator of the meristem [77,82].

Recently, it has been reported that cytokinin regulates dissected leaf morphogenesis in tomato [83]. Manipulation of the endogenous CK level by ectopic expression of the CK metabolic genes, *AtIPT7* and *AtCKX3*, in developing leaf primordia alters leaf complexity in tomato. This has been attributed to CK regulation of the window of morphogenetic activity at leaf margins. Consistent with this, an extended CK response in leaf margins, as indicated by the upregulation of the CK response markers, RRs, has been shown to correlate with an increased leaf complexity [83]. Both local applications of auxin to developing leaf primordia and mutations in *SlIAA9* encoding an auxin response repressor suppress the *AtIPT7*-induced supercompound leaf phenotype, suggesting that CK regulation of dissected leaf morphogenesis requires a localized auxin response. Interestingly, auxin transport and distribution do not appear to be affected by alterations of the CK level. A decrease in the endogenous CK level suppresses and an increase in the CK level compensates for the effect of KNOXI on leaf complexity, supporting the notion that CK acts downstream of KNOXI proteins in dissected leaf development, similar to their interactions in the SAM [83]. Thus, CK positively regulates leaf complexity in tomato.

The antagonistic interactions between GA and CK are dependent on the GA/CK ratio rather than the levels of each hormone, at least in tomato [60]. Intriguingly, the interactions between GA and CK during compound leaf development do not appear to be dependent on the DELLA protein. Together, these results suggest that GA and CK act downstream of the KNOXI and LA proteins to fine-tune the morphogenetic window and, thereby, regulate compound leaf development in tomato.

5. Compound Leaf Development in Legumes

KNOXI proteins maintain the indeterminacy required for compound leaf development in most eudicot species, including tomato and *C. hirsuta*. However, in some leguminous plants (Fabaceae) that belong to the inverted repeat lacking clade (IRLC), *UNIFOLIATA* (*UNI*) in pea and *SINGLE LEAFLET1* (*SGL1*) in *M. truncatula*, encoding legume FLORICAULA (FLO)/LEAFY (LFY)-type transcription factors, have been shown to play a similar role in compound leaf development as *KNOXI* genes in tomato and *C. hirsuta* [9,84]. This is mainly due to the following observations. Firstly, *KNOXI* genes and encoded proteins are not expressed in leaves and leaf primordia and, therefore, do not likely play a role in compound leaf development in pea and *Medicago* [9,13,85,86], although a conflicting report shows that a low level of expression of *KNOXI* genes could be detected in developing leaves in pea and *M. truncatula* [87]. Secondly, pea *uni* and *Medicago sgl1* mutants exhibit simple leaves and *Medicago palmate-like pentafoliata* (*palm1*) mutants, in which *SGL1* is ectopically upregulated, exhibit an increased complexity of compound leaves [88]. Thirdly, a loss-of-function mutant of the *Medicago KNOXI* gene, *BREVIPEDICELLUS* (*MtBP*), does not exhibit compound leaf defects [89].

M. truncatula has been selected as a model species for genetics and genomics studies in legumes, because it has a relatively small and diploid genome (~600 Mb), a short lifecycle and an available genetic transformation system. Recently, its genome has been sequenced [90]. Also available are several comprehensive mutant collections, including the ethyl methane sulfonate (EMS) collection [91], the fast neutron bombardment (FNB) deletion mutant collection [86,92,93] and the insertion mutant collections [94–98]. These mutant collections are excellent resources for forward and reverse genetic studies.

Similar to alfalfa (*M. sativa*) and soybean (*Glycine max*), *M. truncatula* plants exhibit trifoliate compound leaves. *Medicago single leaflet1* (*sgl1*) mutants fail to initiate lateral leaflets and exhibit only simple leaves, similar to pea *uni* mutants, and the mutant phenotypes are consistent with the role of SGL1/UNI in maintaining the morphogenetic activity at leaf margins to elaborate lateral leaflets

(Figure 1) [9,84]. It has been shown that *SGL1* also plays a role in the proximal-distal axis development, because *sgl1* mutants exhibit significantly reduced petioles [9,86].

Figure 1. Compound leaf development in inverted repeat lacking clade (IRLC) legumes. Based on the understanding of compound leaf development in *Medicago truncatula* and pea (*Pisum sativa*) belonging to the IRLC clade of legumes, a genetic model that governs compound leaf development is proposed. The class I homeodomain transcription factor *KNOXI* genes are expressed in the shoot apical meristem (SAM), but their expression is excluded from incipient sites of leaf primordia at the periphery of SAM and subsequent leaf primordia. Loss-of-function mutants of the *FLO/LFY/UNI* ortholog, *SGL1*, exhibit simple leaves with reduced petioles, consistent with the role of *SGL1/UNI* in lateral leaflet initiation and leaf proximodistal axis development. The C2H2 zinc finger transcription factor, PALM1, directly represses the expression of *SGL1* to regulate its spatiotemporal expression domain. *FCL1*, encoding a class M KNOX protein, is required for the leaf proximodistal axis development and leaflet boundary formation. *FCL1* and *SGL1* act in parallel pathways, and both are required for leaf petiole development. *Medicago* auxin efflux PIN10 protein (MtPIN10), the *Arabidopsis* PIN1 ortholog, mediates auxin activity maxima that precede and are required for the initiation of leaf and leaflet primordia and leaflet serration at the distal leaflet margin. The *Medicago NAM/CUC* gene, *MtNAM*, functions in the development of lateral organ boundaries.

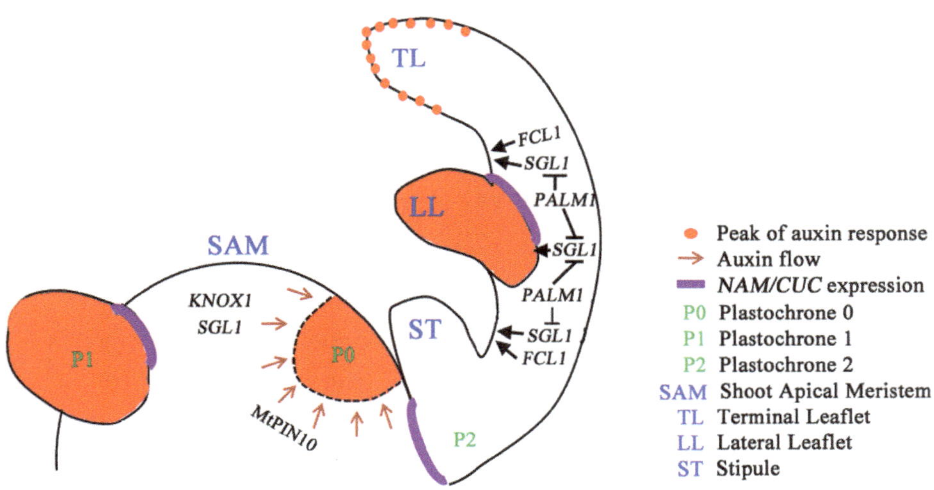

In contrast to the *sgl1* mutants with simple leaves, *Medicago palm1* mutants develop dissected leaves with five leaflets clustered at the leaf tip and increased petioles [88]. The appearance of two extra lateral leaflets occurs at the P3 stage of leaf primordia [88]. *PALM1* encodes a novel transcription factor with an *N*-terminal Cys(2)His(2) zinc finger DNA-binding domain [100] and a *C*-terminal EAR transcription repressor domain [88,100,101]. PALM1 binds to a specific sequence located in the promoter region of *SGL1* and negatively regulates the spatiotemporal expression of *SGL1*. Double mutant analysis shows that *SGL1* is required for the development of lateral leaflets in the *palm1* background. Thus, the intricate interactions between *PALM1* and *SGL1* determine the trifoliate leaf form in *M. truncatula* (Figure 1). Intriguingly, ectopic expression of *PALM1* completely suppresses leaf lobing phenotypes of *A. thaliana* plants that overexpress the *KNOXI* gene, *KNAT1*. This suggests that although KNOXI proteins are not associated with compound leaf development in *M. truncatula*, *PALM1* has the capacity to negatively regulate downstream targets or regulatory networks that are responsive to KNOXI activities. An extension of this hypothesis is that the negative regulation of KNOXI activities by *PALM1* orthologs may be operational in non-IRLC legumes, in which KNOXI proteins may play a role in leaf development. Phylogenetic analysis shows that *PALM1* homologous sequences exist in diverse species, ranging from closely-related legumes to lower land plants [101], suggesting an ancient origin and divergent function of *PALM1* in the development of different leaf forms. Interestingly, loss-of-function of *PALM1* also results in the loss of epicuticular wax crystals on the abaxial leaf surface, suggesting a role of *PALM1* in epicuticular wax biosynthesis and/or distribution [102].

Recently, a class M *KNOX* gene, *FUSED COMPOUND LEAF1* (*FCL1*) has been shown to play a positive role in boundary and proximodistal axis development of compound leaves in *M. truncatula* [86]. Positional cloning and genetic complementation tests reveal that *FCL1* encodes a class M KNOX protein, with a MEINOX domain, but without the DNA-binding homeodomain normally present in the canonical KNOXI proteins. Double mutant analyses show that *fcl1* is epistatic to *palm1*, and on the other hand, *sgl1* is epistatic to *fcl1* in leaf complexity; in terms of petiole development, *SGL1* and *FCL1* act additively. It is not yet clear how *FCL1* regulates compound leaf development in *M. truncatula*. Class M KNOX genes are also found in *Arabidopsis* and tomato [103,104]. It has been shown that the encoded proteins form heterodimers with BEL1-like homeodomain (BELL) proteins that themselves are KNOXI-interacting partners and, thereby, interfere with nuclear targeting and downstream regulatory networks of KNOXI-BELL complexes [103,104]. Although no loss-of-function mutants have been isolated from *A. thaliana* and tomato, ectopic expression of *KNATM-B* results in elongated petioles and narrow, shorter and serrated leaves in *Arabidopsis* [104]. In tomato, a semi-dominant *Petroselinum* (*Pts*) mutant that overexpresses the tomato class M *KNOX* gene, *PTS/TKD1*, due to a promoter mutation, exhibits a proliferation of compound leaves, affecting both primary and secondary leaflets [103]. In *M. truncatula*, the canonical *KNOXI* genes are not expressed in the leaf primordia of wild-type and *fcl1* mutants [86]. It is unlikely that FCL1 interferes with the function of the canonical KNOXI proteins during leaf development. However, FCL1 has been shown to interact in a yeast two-hybrid system with a subset of the *Arabidopsis* BELL homeodomain proteins, albeit with slightly different substrate specificities from KNATM-B [86]. It is plausible that FCL1 plays a role in compound leaf development through interactions with specific BELL proteins in *M. truncatula*. The *fcl1* mutant phenotypes suggest that FCL1 is required to maintain a window of organogenetic activities at leaf margins. Loss-of-function mutations of *FCL1* shorten the window of organogenetic activities and promote precocious leaf maturation, leading to leaflet fusion and clustering and a reduction in petiole length (Figure 1).

MtPIN10, encoding an auxin efflux carrier protein orthologous to the *Arabidopsis* PIN1 protein, has been shown to play a key role in lateral organ development in *M. truncatula* [37,40]. *Medicago pin10* mutants (also named *smooth leaf margin1* or *slm1*) exhibit pleiotropic phenotypes, including an increased number of cotyledons, an increase in compound leaf complexity, smooth leaf margins, altered phyllotaxy and defective flower development. *Trans*-species genetic complementation studies demonstrate that MtPIN10 and *Arabidopsis* PIN1 are functional orthologs. Consistent with previous studies on PIN1 orthologs in compound-leafed species, MtPIN10/SLM1 mediates local auxin activity maxima in incipient sites of leaf primordia and leaflet serration in *M. truncatula* (Figure 1). The various compound leaf phenotypes observed in the *mtpin10* mutants have been attributed to a failure in separating leaf primordia at the flanks of the SAM, which leads to the fusion of leaf primordia at various developmental stages [40].

Boundary genes are required for the separation of leaflets. As discussed earlier, the *NAM/CUC* genes, from the large and evolutionarily conserved family of NAC transcription factor genes, specify organ boundaries in a diverse range of eudicot species [105]. The *M. truncatula NAM/CUC2* ortholog, *NO APICAL MERISTEM* (*MtNAM*), has been shown to exert a similar function as *NAM/CUC* genes in other species [106]. Besides the conserved function in organ separation, *MtNAM* has been shown to play a role in floral organ identity regulation.

6. Perspective and Biological Significance of Leaf Development Studies in *M. truncatula*

Because of the availability of a high-quality genome sequence, orthologous genes of key regulators of compound leaf development from other species can be, in most cases, identified in *M. truncatula*. Conservation and divergence can be further analyzed using protein coding and flanking sequences and synteny information across species. Reverse genetic screening is available to isolate mutant alleles of the orthologous genes, for phenotypic analysis. Genetic interactions between different compound leaf regulators can also be explored by double mutant analysis. These phylogenetic, genetic and functional studies are thus far used to dissect and compare gene regulatory networks controlling compound leaf development in *M. truncatula* and other species. On the other hand, the forward genetic approach has been used to isolate and characterize novel mutants with defects in leaf form, shape or patterning. Collectively, these ongoing studies will likely reveal new insight into novel regulators or the deployment of conserved regulators of compound leaf development. Recently, small RNAs, including microRNA and *trans*-acting small interfering RNA (*ta*-siRNA), known to play a role in plant growth, development and responses to the environment [107,108], have also been shown to regulate dissected leaf development [30,109–111]. It is foreseeable that the small RNA pathways will also be identified in forward or reverse genetic screens of compound leaf mutants in *M. truncatula*.

Leguminous plants, belonging to the third largest family of flowering plants, include many important grain and forage crops, such as soybean (*Glycine max*), lentil (*Lens culinaris*) and alfalfa (*M. sativa*). Various leaf forms, such as pinnate, palmate or higher-ordered compound leaf forms, are found in legume species. However, it is not yet known how these different leaf forms are regulated at the molecular level. Studies of compound leaf development in *M. truncatula* not only expand our knowledge of the regulatory mechanisms that underlie compound leaf development in legumes, but also provide the basis for genetic improvements of the yield and quality of legume crops that are an integral part of sustainable agriculture.

7. Conclusions

Studies in tomato, *C. hirsuta* and legumes with compound leaves and *Arabidopsis* and maize with simple leaves unravel that several conserved genetic frameworks are deployed in different morphogenetic processes, including leaf margin serration and patterning during simple and dissected leaf development in diverse species. One is the auxin efflux PIN protein-mediated auxin concentration gradients, or activity maxima, which precede the initiation of leaves from the flanks of the SAM and leaflets from leaf marginal blastozones and serrations from leaf margins. Another one involves the homeodomain KNOXI proteins that promote indeterminacy and the CIN-like TCP protein that promotes leaf maturation. GA and CK act downstream of KNOXI and TCP to fine-tune the morphogenetic activity for compound leaf development. The NAM/CUC proteins, controlling the organ boundary, also play a role in compound leaf development by suppressing the auxin signal between regions of lamina outgrowth.

Interestingly, SGL1 and PALM1 proteins fine-tune the morphogenetic activity at leaf margins and control compound leaf development in *M. truncatula*, which belongs to the IRLC clade of legumes (Figure 1). In IRLC legumes, KNOXI proteins are not recruited in compound leaf development. However, the processes involving PIN-mediated auxin activity maxima and NAM/CUC proteins are involved in compound leaf development in *M. truncatula*. Intriguingly, a class M KNOX protein with the MEINOX domain, but without the DNA-binding homeodomain, plays a role in trifoliate leaf development. It is foreseeable that future studies in areas of gene regulatory networks, small RNAs and plant growth regulators will expand our knowledge on the origin and evolution of dissected leaves in angiosperm.

Acknowledgments

Funding of this work was provided in part by the Samuel Roberts Noble Foundation, National Science Foundation (IOS-1127155) and Oklahoma Center for The Advancement of Science and Technology (PS12-036).

Conflicts of Interest

The authors declare no conflict of interest.

References

1. Nealson, K.; Conrad, P. Life: Past, present and future. *Phil. Trans. R. Soc. B* **1999**, *354*, 1923–1939.
2. Field, C.; Behrenfeld, M.; Randerson, J.; Falkowski, P. Primary production of the biosphere: Integrating terrestrial and oceanic components. *Science* **1998**, *281*, 237–240.
3. Zimmermann, W. Main results of the telome theory. *Palaeobotanist* **1952**, *1*, 456–470.
4. Champagne, C.; Sinha, N. Compound leaves: Equal to the sum of their parts? *Development* **2004**, *131*, 4401–4412.
5. Goliber, T.; Kessler, S.; Chen, J.J.; Bharathan, G.; Sinha, N. Genetic, molecular, and morphological analysis of compound leaf development. *Curr. Top. Dev. Biol.* **1999**, *43*, 259–290.
6. Allsopp, A. Land and water forms: Physiological aspects. *Handb. Pflanzenphysiol.* **1965**, *15*, 1236–1255.
7. Hay, A.; Tsiantis, M. The genetic basis for differences in leaf form between *Arabidopsis thaliana* and its wild relative *Cardamine hirsuta*. *Nat. Genet.* **2006**, *38*, 942–947.
8. Merrill, E. Heteroblastic seedlings of green ash. I. Predictability of leaf form and primordial length. *Can. J. Bot.* **1986**, *64*, 2645–2649.

9. Wang, H.; Chen, J.; Wen, J.; Tadege, M.; Li, G.; Liu, Y.; Mysore, K.S.; Ratet, P.; Chen, R. Control of compound leaf development by *FLORICAULA/LEAFY* ortholog *SINGLE LEAFLET1* in *Medicago truncatula*. *Plant Physiol.* **2008**, *146*, 1759–1772.
10. Givnish, T.J. Comparative studies of leaf form: Assessing the relative roles of selective pressures and phylogenetic constraints. *New Phytol.* **1987**, *106*, 131–160.
11. Niinemets, Ü. Are compound-leaved woody species inherently shade-intolerant? An analysis of species ecological requirements and foliar support costs. *Plant Ecol.* **1998**, *134*, 1–11.
12. Popma, J.; Bongers, F.; Werger, M. Gap-dependence and leaf characteristics of trees in a tropical lowland rain forest in Mexico. *Oikos* **1992**, *63*, 207–214.
13. Champagne, C.E.; Goliber, T.E.; Wojciechowski, M.F.; Mei, R.W.; Townsley, B.T.; Wang, K.; Paz, M.M.; Geeta, R.; Sinha, N.R. Compound leaf development and evolution in the legumes. *Plant Cell* **2007**, *19*, 3369–3378.
14. Cronk, Q.C. Plant evolution and development in a post-genomic context. *Nat. Rev. Genet.* **2001**, *2*, 607–619.
15. Bharathan, G.; Sinha, N.R. The regulation of compound leaf development. *Plant Physiol.* **2001**, *127*, 1533–1538.
16. Hake, S.; Smith, H.M.; Holtan, H.; Magnani, E.; Mele, G.; Ramirez, J. The role of *knox* genes in plant development. *Annu. Rev. Cell Dev. Biol.* **2004**, *20*, 125–151.
17. Hay, A.; Tsiantis, M. *KNOX* genes: Versatile regulators of plant development and diversity. *Development* **2010**, *137*, 3153–3165.
18. Long, J.A.; Moan, E.I.; Medford, J.I.; Barton, M.K. A member of the KNOTTED class of homeodomain proteins encoded by the *STM* gene of Arabidopsis. *Nature* **1996**, *379*, 66–69.
19. Burko, Y.; Shleizer-Burko, S.; Yanai, O.; Shwartz, I.; Zelnik, I.D.; Jacob-Hirsch, J.; Kela, I.; Eshed-Williams, L.; Ori, N. A role for APETALA1/FRUITFULL transcription factors in tomato leaf development. *Plant Cell* **2013**, 10.1105/tpc.1113.113035.
20. Shani, E.; Burko, Y.; Ben-Yaakov, L.; Berger, Y.; Amsellem, Z.; Goldshmidt, A.; Sharon, E.; Ori, N. Stage-specific regulation of *Solanum lycopersicum* leaf maturation by class 1 KNOTTED1-LIKE HOMEOBOX proteins. *Plant Cell* **2009**, *21*, 3078–3092.
21. Chen, J.J.; Janssen, B.J.; Williams, A.; Sinha, N. A gene fusion at a homeobox locus: Alterations in leaf shape and implications for morphological evolution. *Plant Cell* **1997**, *9*, 1289–1304.
22. Parnis, A.; Cohen, O.; Gutfinger, T.; Hareven, D.; Zamir, D.; Lifschitz, E. The dominant developmental mutants of tomato, *Mouse-ear* and *Curl*, are associated with distinct modes of abnormal transcriptional regulation of a *Knotted* gene. *Plant Cell* **1997**, *9*, 2143–2158.
23. Hareven, D.; Gutfinger, T.; Parnis, A.; Eshed, Y.; Lifschitz, E. The making of a compound leaf: genetic manipulation of leaf architecture in tomato. *Cell* **1996**, *84*, 735–744.
24. Janssen, B.J.; Lund, L.; Sinha, N. Overexpression of a homeobox gene, *LeT6*, reveals indeterminate features in the tomato compound leaf. *Plant Physiol.* **1998**, *117*, 771–786.
25. Efroni, I.; Blum, E.; Goldshmidt, A.; Eshed, Y. A protracted and dynamic maturation schedule underlies Arabidopsis leaf development. *Plant Cell* **2008**, *20*, 2293–2306.
26. Poethig, R.S. Leaf morphogenesis in flowering plants. *Plant Cell* **1997**, *9*, 1077–1087.
27. Hagemann, W.; Gleissberg, S. Organogenetic capacity of leaves: The significance of marginal blastozones in angiosperms. *Plant Syst. Evol.* **1996**, *199*, 121–152.

28. Ben-Gera, H.; Ori, N. Auxin and LANCEOLATE affect leaf shape in tomato via different developmental processes. *Plant Signal Behav.* **2012**, *7*, 1255–1257.
29. Yanai, O.; Shani, E.; Russ, D.; Ori, N. Gibberellin partly mediates LANCEOLATE activity in tomato. *Plant J.* **2011**, *68*, 571–582.
30. Ori, N.; Cohen, A.R.; Etzioni, A.; Brand, A.; Yanai, O.; Shleizer, S.; Menda, N.; Amsellem, Z.; Efroni, I.; Pekker, I.; *et al.* Regulation of *LANCEOLATE* by miR319 is required for compound-leaf development in tomato. *Nat. Genet.* **2007**, *39*, 787–791.
31. Barkoulas, M.; Hay, A.; Kougioumoutzi, E.; Tsiantis, M. A developmental framework for dissected leaf formation in the *Arabidopsis* relative *Cardamine hirsuta*. *Nat. Genet.* **2008**, *40*, 1136–1141.
32. Zoulias, N.; Koenig, D.; Hamidi, A.; McCormick, S.; Kim, M. A role for *PHANTASTICA* in medio-lateral regulation of adaxial domain development in tomato and tobacco leaves. *Ann. Bot.* **2012**, *109*, 407–418.
33. Kim, M.; McCormick, S.; Timmermans, M.; Sinha, N. The expression domain of *PHANTASTICA* determines leaflet placement in compound leaves. *Nature* **2003**, *424*, 438–443.
34. Reinhardt, D.; Pesce, E.-R.; Stieger, P.; Mandel, T.; Baltensperger, K.; Bennett, M.; Traas, J.; Friml, J.; Kuhlemeier, C. Regulation of phyllotaxis by polar auxin transport. *Nature* **2003**, *426*, 255–260.
35. Pinon, V.; Prasad, K.; Grigg, S.P.; Sanchez-Perez, G.F.; Scheres, B. Local auxin biosynthesis regulation by PLETHORA transcription factors controls phyllotaxis in *Arabidopsis*. *Proc. Natl. Acad. Sci. USA* **2012**, *110*, 1–6.
36. Koenig, D.; Bayer, E.; Kang, J.; Kuhlemeier, C.; Sinha, N. Auxin patterns *Solanum lycopersicum* leaf morphogenesis. *Development* **2009**, *136*, 2997–3006.
37. Zhou, C.; Han, L.; Hou, C.; Metelli, A.; Qi, L.; Tadege, M.; Mysore, K.S.; Wang, Z.Y. Developmental analysis of a *Medicago truncatula smooth leaf margin1* mutant reveals context-dependent effects on compound leaf development. *Plant Cell* **2011**, *23*, 2106–2124.
38. DeMason, D.A.; Chawla, R. Roles for auxin during morphogenesis of the compound leaves of pea (*Pisum sativum*). *Planta* **2004**, *218*, 435–448.
39. Avasarala, S.; Yang, J.; Caruso, J.L. Production of phenocopies of the lanceolate mutant in tomato using polar auxin transport inhibitors. *J. Exp. Bot.* **1996**, *47*, 709–712.
40. Peng, J.; Chen, R. Auxin efflux transporter MtPIN10 regulates compound leaf and flower development in *Medicago truncatula*. *Plant Signal Behav.* **2011**, *6*, 1537–1544.
41. Kawamura, E.; Horiguchi, G.; Tsukaya, H. Mechanisms of leaf tooth formation in *Arabidopsis*. *Plant J.* **2010**, *62*, 429–441.
42. Quint, M.; Gray, W.M. Auxin signaling. *Curr. Opin. Plant Biol.* **2006**, *9*, 448–453.
43. Wang, H.; Jones, B.; Li, Z.; Frasse, P.; Delalande, C.; Regad, F.; Chaabouni, S.; Latche, A.; Pech, J.C.; Bouzayen, M. The tomato Aux/IAA transcription factor IAA9 is involved in fruit development and leaf morphogenesis. *Plant Cell* **2005**, *17*, 2676–2692.
44. Zhang, J.; Chen, R.; Xiao, J.; Qian, C.; Wang, T.; Li, H.; Ouyang, B.; Ye, Z. A single-base deletion mutation in *SlIAA9* gene causes tomato (*Solanum lycopersicum*) *entire* mutant. *J. Plant Res.* **2007**, *120*, 671–678.
45. Ben-Gera, H.; Shwartz, I.; Shao, M.R.; Shani, E.; Estelle, M.; Ori, N. *ENTIRE* and *GOBLET* promote leaflet development in tomato by modulating auxin response. *Plant J.* **2012**, *70*, 903–915.

46. Hendelman, A.; Buxdorf, K.; Stav, R.; Kravchik, M.; Arazi, T. Inhibition of lamina outgrowth following *Solanum lycopersicum AUXIN RESPONSE FACTOR 10* (*SlARF10*) derepression. *Plant Mol. Biol.* **2012**, *78*, 561–576.
47. Dharmasiri, N.; Dharmasiri, S.; Estelle, M. The F-box protein TIR1 is an auxin receptor. *Nature* **2005**, *435*, 441–445.
48. Kepinski, S.; Leyser, O. The *Arabidopsis* F-box protein TIR1 is an auxin receptor. *Nature* **2005**, *435*, 446–451.
49. Berger, Y.; Harpaz-Saad, S.; Brand, A.; Melnik, H.; Sirding, N.; Alvarez, J.P.; Zinder, M.; Samach, A.; Eshed, Y.; Ori, N. The NAC-domain transcription factor GOBLET specifies leaflet boundaries in compound tomato leaves. *Development* **2009**, *136*, 823–832.
50. Weiss, D.; Ori, N. Mechanisms of cross talk between gibberellin and other hormones. *Plant Physiol.* **2007**, *144*, 1240–1246.
51. Greenboim-Wainberg, Y.; Maymon, I.; Borochov, R.; Alvarez, J.; Olszewski, N.; Ori, N.; Eshed, Y.; Weiss, D. Cross talk between gibberellin and cytokinin: The *Arabidopsis* GA response inhibitor SPINDLY plays a positive role in cytokinin signaling. *Plant Cell* **2005**, *17*, 92–102.
52. Bolduc, N.; Hake, S. The Maize transcription factor KNOTTED1 directly regulates the gibberellin catabolism gene ga2ox1. *Plant Cell* **2009**, *21*, 1647–1658.
53. Maekawa, T.; Maekawa-Yoshikawa, M.; Takeda, N.; Imaizumi-Anraku, H.; Murooka, Y.; Hayashi, M. Gibberellin controls the nodulation signaling pathway in *Lotus japonicus*. *Plant J.* **2009**, *58*, 183–194.
54. Jasinski, S.; Piazza, P.; Craft, J.; Hay, A.; Woolley, L.; Rieu, I.; Phillips, A.; Hedden, P.; Tsiantis, M. KNOX action in *Arabidopsis* is mediated by coordinate regulation of cytokinin and gibberellin activities. *Curr. Biol.* **2005**, *15*, 1560–1565.
55. Perilli, S.; Moubayidin, L.; Sabatini, S. The molecular basis of cytokinin function. *Curr. Opin. Plant Biol.* **2010**, *13*, 21–26.
56. Hay, A.; Kaur, H.; Phillips, A.; Hedden, P.; Hake, S.; Tsiantis, M. The gibberellin pathway mediates *KNOTTED1*-type homeobox function in plants with different body plans. *Curr. Biol.* **2002**, *12*, 1557–1565.
57. Yanai, O.; Shani, E.; Dolezal, K.; Tarkowski, P.; Sablowski, R.; Sandberg, G.; Samach, A.; Ori, N. *Arabidopsis* KNOXI proteins activate cytokinin biosynthesis. *Curr. Biol.* **2005**, *15*, 1566–1571.
58. Gray, R.A. Alteration of leaf size and shape and other changes caused by gibberellins in plants. *Am. J. Bot.* **1957**, 674–682.
59. Jones, M.G. Gibberellins and the *procera* mutants of tomato. *Planta* **1987**, *172*, 280–284.
60. Fleishon, S.; Shani, E.; Ori, N.; Weiss, D. Negative reciprocal interactions between gibberellin and cytokinin in tomato. *New Phytol.* **2011**, *190*, 609–617.
61. Phillips, A.L.; Ward, D.A.; Uknes, S.; Appleford, N.E.J.; Lange, T.; Huttly, A.K.; Gaskin, P.; Graebe, J.E.; Hedden, P. Isolation and expression of three gibberellin 20-oxidase cDNA clones from *Arabidopsis*. *Plant Physiol.* **1995**, *108*, 1049–1057.
62. Coles, J.P.; Phillips, A.L.; Croker, S.J.; Garcia-Lepe, R.; Lewis, M.J.; Hedden, P. Modification of gibberellin production and plant development in *Arabidopsis* by sense and antisense expression of gibberellin 20-oxidase genes. *Plant J.* **1999**, *17*, 547–556.

63. Thomas, S.G.; Phillips, A.L.; Hedden, P. Molecular cloning and functional expression of gibberellin 2-oxidases, multifunctional enzymes involved in gibberellin deactivation. *Proc. Natl. Acad. Sci. USA* **1999**, *96*, 4698–4703.
64. Yamaguchi, S. Gibberellin metabolism and its regulation. *Annu. Rev. Plant Biol.* **2008**, *59*, 225–251.
65. Silverstone, A.L.; Jung, H.-S.; Dill, A.; Kawaide, H.; Kamiya, Y.; Sun, T.-P. Repressing a repressor: Gibberellin-induced rapid reduction of the RGA protein in *Arabidopsis*. *Plant Cell* **2001**, *13*, 1555–1565.
66. Sun, T.P.; Gubler, F. Molecular mechanism of gibberellin signaling in plants. *Annu. Rev. Plant Biol.* **2004**, *55*, 197–223.
67. Jupe, S.C.; Causton, D.R.; Scott, I.M. Cellular basis of the effects of gibberelin and the PRO gene on stem growth in tomato. *Planta* **1988**, *174*, 106–111.
68. Peng, J.; Carol, P.; Richards, D.E.; King, K.E.; Cowling, R.J.; Murphy, G.P.; Harberd, N.P. The *Arabidopsis GAI* gene defines a signaling pathway that negatively regulates gibberellin responses. *Genes Dev.* **1997**, *11*, 3194–3205.
69. Van Tuinen, A.; Peters, A.H.L.J.; Kendrick, R.E.; Zeevaart, J.A.D.; Koornneef, M. Characterisation of the *procera* mutant of tomato and the interaction of gibberellins with end-of-day far-red light treatments. *Physiol. Plant* **1999**, *106*, 121–128.
70. Marti, C.; Orzaez, D.; Ellul, P.; Moreno, V.; Carbonell, J.; Granell, A. Silencing of DELLA induces facultative parthenocarpy in tomato fruits. *Plant J.* **2007**, *52*, 865–876.
71. Bassel, G.W.; Mullen, R.T.; Bewley, J.D. *procera* is a putative *DELLA* mutant in tomato (*Solanum lycopersicum*): Effects on the seed and vegetative plant. *J. Exp. Bot.* **2008**, *59*, 585–593.
72. Jasinski, S.; Tattersall, A.; Piazza, P.; Hay, A.; Martinez-Garcia, J.F.; Schmitz, G.; Theres, K.; McCormick, S.; Tsiantis, M. *PROCERA* encodes a DELLA protein that mediates control of dissected leaf form in tomato. *Plant J.* **2008**, *56*, 603–612.
73. Sakamoto, T.; Kamiya, N.; Ueguchi-Tanaka, M.; Iwahori, S.; Matsuoka, M. KNOX homeodomain protein directly suppresses the expression of a gibberellin biosynthetic gene in the tobacco shoot apical meristem. *Genes Dev.* **2001**, *15*, 581–590.
74. Werner, T.; Schmuelling, T. Cytokinin action in plant development. *Curr. Opin. Plant. Biol.* **2009**, *12*, 527–538.
75. Werner, T.; Motyka, V.; Strnad, M.; Schmuelling, T. Regulation of plant growth by cytokinin. *Proc. Natl. Acad. Sci. USA* **2001**, *98*, 10487–10492.
76. Giulini, A.; Wang, J.; Jackson, D. Control of phyllotaxy by the cytokinin-inducible response regulator homologue ABPHYL1. *Nature* **2004**, *430*, 1031–1034.
77. Leibfried, A.; To, J.P.C.; Busch, W.; Stehling, S.; Kehle, A.; Demar, M.; Kieber, J.J.; Lohmann, J.U. WUSCHEL controls meristem function by direct regulation of cytokinin-inducible response regulators. *Nature* **2005**, *438*, 1172–1175.
78. Kurakawa, T.; Ueda, N.; Maekawa, M.; Kobayashi, K.; Kojima, M.; Nagato, Y.; Sakakibara, H.; Kyozuka, J. Direct control of shoot meristem activity by a cytokinin-activating enzyme. *Nature* **2007**, *445*, 652–655.
79. Sablowski, R. The dynamic plant stem cell niches. *Curr. Opin. Plant Biol.* **2007**, *10*, 639–644.
80. Zhao, Z.; Andersen, S.U.; Ljung, K.; Dolezal, K.; Miotk, A.; Schultheiss, S.J.; Lohmann, J.U. Hormonal control of the shoot stem-cell niche. *Nature* **2010**, *465*, U1089–U1154.

81. Werner, T.; Motyka, V.; Laucou, V.; Smets, R.; Van Onckelen, H.; Schmulling, T. Cytokinin-deficient transgenic Arabidopsis plants show multiple developmental alterations indicating opposite functions of cytokinins in the regulation of shoot and root meristem activity. *Plant Cell* **2003**, *15*, 2532–2550.
82. Lindsay, D.L.; Sawhney, V.K.; Bonham-Smith, P.C. Cytokinin-induced changes in *CLAVATA1* and *WUSCHEL* expression temporally coincide with altered floral development in *Arabidopsis*. *Plant Sci.* **2006**, *170*, 1111–1117.
83. Shani, E.; Ben-Gera, H.; Shleizer-Burko, S.; Burko, Y.; Weiss, D.; Ori, N. Cytokinin regulates compound leaf development in tomato. *Plant Cell* **2010**, *22*, 3206–3217.
84. Hofer, J.; Turner, L.; Hellens, R.; Ambrose, M.; Matthews, P.; Michael, A.; Ellis, N. *UNIFOLIATA* regulates leaf and flower morphogenesis in pea. *Curr. Biol.* **1997**, *7*, 581–587.
85. Hofer, J.; Gourlay, C.; Michael, A.; Ellis, T.H. Expression of a class 1 *knotted1*-like homeobox gene is down-regulated in pea compound leaf primordia. *Plant Mol. Biol.* **2001**, *45*, 387–398.
86. Peng, J.; Yu, J.; Wang, H.; Guo, Y.; Li, G.; Bai, G.; Chen, R. Regulation of compound leaf development in *Medicago truncatula* by *fused compound leaf1*, a class M *KNOX* gene. *Plant Cell* **2011**, *23*, 3929–3943.
87. Di Giacomo, E.; Sestili, F.; Iannelli, M.A.; Testone, G.; Mariotti, D.; Frugis, G. Characterization of *KNOX* genes in *Medicago truncatula*. *Plant Mol. Biol.* **2008**, *67*, 135–150.
88. Chen, J.; Yu, J.; Ge, L.; Wang, H.; Berbel, A.; Liu, Y.; Chen, Y.; Li, G.; Tadege, M.; Wen, J.; *et al.* Control of dissected leaf morphology by a Cys(2)His(2) zinc finger transcription factor in the model legume *Medicago truncatula*. *Proc. Natl. Acad. Sci. USA* **2010**, *107*, 10754–10759.
89. Ge, L.; Peng, J.; Berbel, A.; Madueno, F.; Chen, R. Regulation of compound leaf development by *PHANTASTICA* in *Medicago truncatula*. *Plant Physiol.* **2013**, doi:10.1104/pp.113.229914.
90. Young, N.D.; Debellé, F.; Oldroyd, G.E.D.; Geurts, R.; Cannon, S.B.; Udvardi, M.K.; Benedito, V.A.; Mayer, K.F.X.; Gouzy, J.; Schoof, H.; *et al.* The *Medicago* genome provides insight into the evolution of rhizobial symbioses. *Nature* **2011**, *480*, 520–524.
91. Benaben, V.; Duc, G.; Lefebvre, V.; Huguet, T. *TE7*, an inefficient symbiotic mutant of *Medicago truncatula* Gaertn. cv. Jemalong. *Plant. Physiol.* **1995**, *107*, 53–62.
92. Rogers, C.; Wen, J.; Chen, R.; Oldroyd, G. Deletion-based reverse genetics in *Medicago truncatula*. *Plant Physiol.* **2009**, *151*, 1077–1086.
93. Wang, H.; Li, G.; Chen, R. Fast neutron bombardment (FNB) induced deletion mutagenesis for forward and reverse genetic studies in plants. In *Floriculture, Ornamental and Plant Biotechnology: Advances and Topical Issues*, 1st ed.; Da Silva, J.T. Ed.; Global Science Books: Isleworth, UK, 2006; pp. 629–639.
94. Scholte, M.; d'Erfurth, I.; Rippa, S.; Mondy, S.; Cosson, V.; Durand, P.; Breda, C.; Trinh, H.; Rodriguez-Llorente, I.; Kondorosi, E.; *et al.* T-DNA tagging in the model legume *Medicago truncatula* allows efficient gene discovery. *Mol. Breeding* **2002**, *10*, 203–215.
95. d'Erfurth, I.; Cosson, V.; Mondy, S.; Brocard, L.; Kondorosi, A.; Ratet, P. The low level of activity of *Arabidopsis thaliana Tag1* transposon correlates with the absence of two minor transcripts in *Medicago truncatula*. *Mol. Breeding* **2006**, *17*, 317–328.

96. d'Erfurth, I.; Cosson, V.; Eschstruth, A.; Lucas, H.; Kondorosi, A.; Ratet, P. Efficient transposition of the *Tnt1* tobacco retrotransposon in the model legume *Medicago truncatula*. *Plant J.* **2003**, *34*, 95–106.
97. d'Erfurth, I.; Cosson, V.; Eschstruth, A.; Rippa, S.; Messinese, E.; Durand, P.; Trinh, H.; Kondorosi, A.; Ratet, P. Rapid inactivation of the maize transposable element *En/Spm* in *Medicago truncatula*. *Mol. Gen. Genet.* **2003**, *269*, 732–745.
98. Tadege, M.; Wen, J.; He, J.; Tu, H.; Kwak, Y.; Eschstruth, A.; Cayrel, A.; Endre, G.; Zhao, P.X.; Chabaud, M.; *et al.* Large-scale insertional mutagenesis using the *Tnt1* retrotransposon in the model legume *Medicago truncatula*. *Plant J.* **2008**, *54*, 335–347.
99. Takatsuji, H. Zinc-finger proteins: The classical zinc finger emerges in contemporary plant science. *Plant Mol. Biol.* **1999**, *39*, 1073–1078.
100. Ohta, M.; Matsui, K.; Hiratsu, K.; Shinshi, H.; Ohme-Takagi, M. Repression domains of class II ERF transcriptional repressors share an essential motif for active repression. *Plant Cell* **2001**, *13*, 1959–1968.
101. Ge, L.; Chen, J.; Chen, R. *Palmate-like pentafoliata1* encodes a novel Cys(2)His(2) zinc finger transcription factor essential for compound leaf morphogenesis in *Medicago truncatula*. *Plant Signal Behav.* **2010**, *5*, 1134–1137.
102. Uppalapati, S.R.; Ishiga, Y.; Doraiswamy, V.; Bedair, M.; Mittal, S.; Chen, J.; Nakashima, J.; Tang, Y.; Tadege, M.; Ratet, P.; *et al.* Loss of abaxial leaf epicuticular wax in *Medicago truncatula irg1/palm1* mutants results in reduced spore differentiation of anthracnose and nonhost rust pathogens. *Plant Cell* **2012**, *24*, 353–370.
103. Kimura, S.; Koenig, D.; Kang, J.; Yoong, F.Y.; Sinha, N. Natural variation in leaf morphology results from mutation of a novel *KNOX* gene. *Curr. Biol.* **2008**, *18*, 672–677.
104. Magnani, E.; Hake, S. *KNOX* lost the *OX*: The *Arabidopsis KNATM* gene defines a novel class of KNOX transcriptional regulators missing the homeodomain. *Plant Cell* **2008**, *20*, 875–887.
105. Blein, T.; Pulido, A.; Vialette-Guiraud, A.; Nikovics, K.; Morin, H.; Hay, A.; Johansen, I.E.; Tsiantis, M.; Laufs, P. A conserved molecular framework for compound leaf development. *Science* **2008**, *322*, 1835–1839.
106. Cheng, X.; Peng, J.; Ma, J.; Tang, Y.; Chen, R.; Mysore, K.S.; Wen, J. *NO APICAL MERISTEM* (*MtNAM*) regulates floral organ identity and lateral organ separation in *Medicago truncatula*. *New Phytol.* **2012**, *195*, 71–84.
107. Chuck, G.; Candela, H.; Hake, S. Big impacts by small RNAs in plant development. *Curr. Opin. Plant Biol.* **2009**, *12*, 81–86.
108. Pulido, A.; Laufs, P. Co-ordination of developmental processes by small RNAs during leaf development. *J. Exp. Bot.* **2010**, *61*, 1277–1291.
109. Chitwood, D.H.; Nogueira, F.T.S.; Howell, M.D.; Montgomery, T.A.; Carrington, J.C.; Timmermans, M.C.P. Pattern formation via small RNA mobility. *Genes Dev.* **2009**, *23*, 549–554.
110. Yan, J.; Cai, X.; Luo, J.; Sato, S.; Jiang, Q.; Yang, J.; Cao, X.; Hu, X.; Tabata, S.; Gresshoff, P.M.; *et al.* The *REDUCED LEAFLET* genes encode key components of the *trans*-acting small interfering RNA pathway and regulate compound leaf and flower development in *Lotus japonicus*. *Plant Physiol.* **2010**, *152*, 797–807.

111. Yifhar, T.; Pekker, I.; Peled, D.; Friedlander, G.; Pistunov, A.; Sabban, M.; Wachsman, G.; Alvarez, J.P.; Amsellem, Z.; Eshed, Y. Failure of the Tomato trans-acting short interfering RNA program to regulate AUXIN RESPONSE FACTOR3 and ARF4 underlies the wiry leaf syndrome. *Plant Cell* **2012**, *24*, 3575–3589.

Diurnal Regulation of Leaf Water Status in High- and Low-Mannitol Olive Cultivars

Riccardo Lo Bianco [1,*] and Giuseppe Avellone [2]

[1] Dipartmento Scienze Agrarie e Forestali, Università degli Studi di Palermo, Viale delle Scienze 11, Palermo 90128, Italy

[2] Dipartimento Scienze e Tecnologie Biologiche Chimiche e Farmaceutiche, Università degli Studi di Palermo, Via Archirafi 32, Palermo 90123, Italy

* Author to whom correspondence should be addressed; E-Mail: riccardo.lobianco@unipa.it

---

Abstract: The role of mannitol and malic acid in the regulation of diurnal leaf water relations was investigated in 'Biancolilla' (high-mannitol) and 'Cerasuola' (low-mannitol) olive trees. Photosynthetic photon flux density (PPFD), vapor pressure deficit (VPD), stomatal conductance (g_s), transpiration rate (T), relative water content (RWC), mannitol and malic acid were measured in 'Biancolilla' and 'Cerasuola' leaves during a dry and hot day of summer in Sicily. In general, leaves of 'Biancolilla' trees exhibited greater mannitol content, higher g_s and T, but lower RWC than leaves of 'Cerasuola' trees. Differences in g_s and T between the two cultivars were evident mainly in mid to late morning. 'Biancolilla' leaves accumulated mannitol at midday and again late in the evening. Stomatal responses to VPD were RWC dependent, and limited somewhat T, only in 'Biancolilla'. Mannitol was directly related to RWC, and may play an osmotic role, in 'Biancolilla' leaves, whereas 'Cerasuola' leaves remained well hydrated by just transpiring less and regardless of mannitol. A day-time accumulation and night-time utilization of mannitol in 'Biancolilla' leaves is proposed as an efficient mechanism to regulate water status and growth.

Keywords: malate; polyol; relative water content; stomatal conductance; transpiration; vapor pressure deficit

---

1. Introduction

Olive (*Olea europaea* L.) is native to and widespread in semi-arid regions where it commonly grows under high temperature and irradiation levels facing long periods of water deficit. Growth and yields of olive largely depend on the resistance to environmental stress [1,2].

During drought periods, olive is able to lower root osmotic potential and reduce leaf hydration, ending growth but keeping some photosynthesis and carbohydrate accumulation [3,4]. This has been mainly attributed to the accumulation of compatible solutes (osmotic adjustment). In particular, mannitol seems to contribute to osmotic adjustment under water deficit [3] and salt stress [5], and it increases in response to low temperatures *in vitro* [6], whereas sorbitol plays a similar role in apple [7], cherry [8], and peach [9]. In olive under severe water deficit, mannitol and glucose may be responsible for reductions of osmotic potential up to 0.32 MPa [4]. Also, in some species of the Oleaceae family, mannitol and malic acid contribute to the osmotic regulation of leaf water status in summer by accumulating around midday, when leaves are exposed to high solar radiation, temperatures and evaporative demand [10].

Sorbitol and mannitol are the most widely distributed sugar alcohols in the plant kingdom [11]. Mannitol is particularly abundant in species of Oleaceae, Apiaceae and Rubiaceae [11–13]. Once synthesized in mature leaves, mannitol may accumulate in source tissues or move to growing sinks [14].

In addition to playing an osmotic role, mannitol and other polyols are strong water-structure formers acting as effective stabilizing/protecting agents at both molecular and whole-cell level [15]. Polyols may also function as scavengers of reactive oxygen species and represent a non-enzymatic mechanism to protect cells from oxidative stress [16]. In transgenic plants exposed to salt or water stress, production of mannitol induces better survival and/or performance compared to wild types [17–19], and this does not seem to be attributable to osmotic protection by mannitol [18,20] but rather to a more specific radical scavenging mechanisms [21]. In olive, mannitol may function as a scavenger of oxygen radicals generated by paraquat applications or excess of light energy [22]. Finally, polyol production may enhance photosynthetic rates (similar to those of C4 species) as well as the efficiency of carbon and energy allocation [23,24].

Malic acid, on the other hand, is the most common organic acid in plant tissues [25,26], and it has been often considered a relevant component of osmotic adjustment during water deficit [27,28].

From previous investigations, we know that olive genotypes may significantly differ for their mannitol content in leaves and other organs [29]. In this study, the role of mannitol and malic acid in the regulation of diurnal leaf water relations was investigated using two olive cultivars with different leaf mannitol content.

2. Results and Discussion

Gas chromatography analyses confirmed that leaves of 'Cerasuola' olive had significantly less (only 62%) mannitol than leaves of 'Biancolilla' olive, whereas malic acid contents were similar in the two cultivars (Table 1). Leaves of 'Biancolilla' olive resulted also less hydrated than leaves of 'Cerasuola' olive, possibly due to greater water consumption by transpiration in the former cultivar (Table 1). Although a direct relationship between RWC and g_s has been documented in other species [30,31], and also observed in this study (discussed later, see Figure 5), the ability of 'Biancolilla' leaves to maintain high g_s and T under low RWC may be due to both greater osmotic adjustment and cell wall thickness (*i.e.*, elasticity), which are known to confer dehydration tolerance to olive leaves [3,4].

Table 1. Average diurnal relative water content (RWC), stomatal conductance (g_s), transpiration rate (T), mannitol content, and malic acid content in leaves of 'Biancolilla' and 'Cerasuola' olive trees on 18 August 2007 near Scillato, Sicily.

	Biancolilla	**Cerasuola**	***p* value** [1]
RWC (%)	78.3	87.1	<0.001
g_s (mmol m^{-2} s^{-1})	110	68.0	<0.001
T (mmol m^{-2} s^{-1})	0.78	0.48	<0.001
Mannitol (mg g^{-1} DW)	770	477	<0.001
Malic acid (mg g^{-1} DW)	16.2	17.2	0.654

[1] from analysis of variance.

As expected, ambient conditions on the sampling day were those of a typical hot and dry summer day in Sicily, with VPD reaching peak values of over 5 kPa around 13:30 HR, during the hours of highest light intensity (Figure 1A). Air and leaf temperatures followed trends similar to VPD, and only 'Cerasuola' leaves exhibited slightly higher temperatures than 'Biancolilla' leaves during evening hours (Figure 1B).

Figure 1. (**a**) Diurnal changes of photosynthetic photon flux density (PPFD) and vapor pressure deficit (VPD); (**b**) air temperature and temperatures of 'Biancolilla' and 'Cerasuola' olive leaves during a typical summer day in Sicily. Error bars represent standard errors of means.

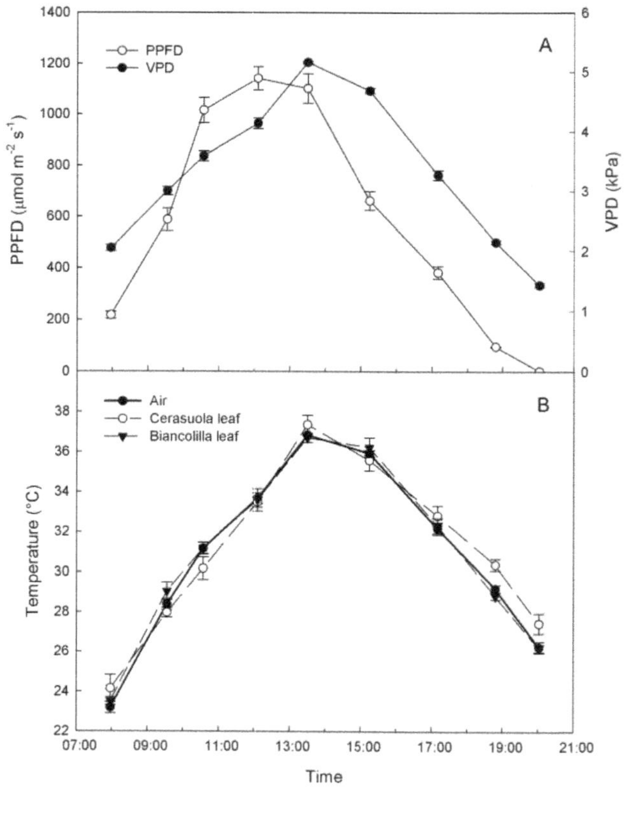

In ‘Cerasuola’ leaves, RWC levels did not exhibit significant diurnal changes, while RWC of ‘Biancolilla’ leaves exhibited low levels in the morning and late afternoon (Figure 2A). Also g_s exhibited different diurnal trends for the two cultivars (Figure 2B). In particular, g_s of ‘Biancolilla’ leaves was constantly higher than g_s of ‘Cerasuola’ leaves during morning hours and until noon; in the afternoon, g_s exhibited similar decreases in both cultivars, with the exception of 19:00 HR when g_s was again higher in ‘Biancolilla’ than in ‘Cerasuola’ leaves. In both cultivars, T followed the typical bell-shaped curve with an earlier peak in ‘Biancolilla’ (12:00 HR) than in ‘Cerasuola’ (13:30 HR) leaves, and major differences between the two cultivars during mid to late morning hours (Figure 2C). Within the general context of olive being indicated as a near-isohydric species [32,33], ‘Cerasuola’ seems to exhibit a more isohydric-like behavior than ‘Biancolilla’.

Figure 2. Diurnal changes of (**a**) relative water content (RWC); (**b**) stomatal conductance (g_s); and (**c**) transpiration rate (T) in leaves of ‘Biancolilla’ and ‘Cerasuola’ olive trees during a typical summer day in Sicily. Error bars represent standard errors of means.

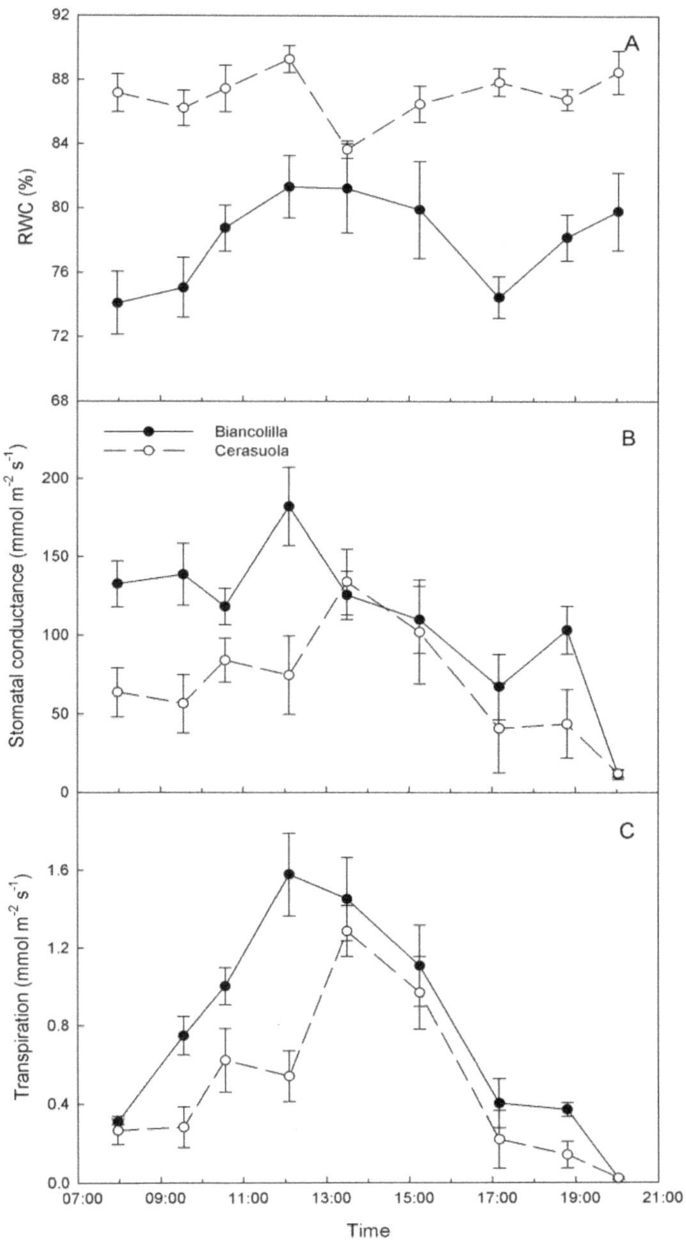

Mannitol content varied significantly during day hours reaching a peak around 13:30 HR in both cultivars (Figure 3A). Mannitol accumulated more in 'Biancolilla' than in 'Cerasuola' leaves during morning hours, decreased in the afternoon, and tended to resume accumulation in the evening only in 'Biancolilla' leaves. Morning accumulation of mannitol in 'Biancolilla' leaves may be the result of concomitant photosynthetic production (open stomates) and reduced translocation to sinks due to unfavorable hydration levels and low sink strength for mannitol. Similarly, accumulation of sorbitol in leaves of drought-stressed peach seedlings is caused by a decrease in sorbitol oxidation rates at the sink site and consequent reduction of its translocation [9].

Figure 3. Diurnal changes of (**a**) mannitol and (**b**) malic acid in leaves of 'Biancolilla' and 'Cerasuola' olive trees during a typical summer day in Sicily. Error bars represent standard errors of means.

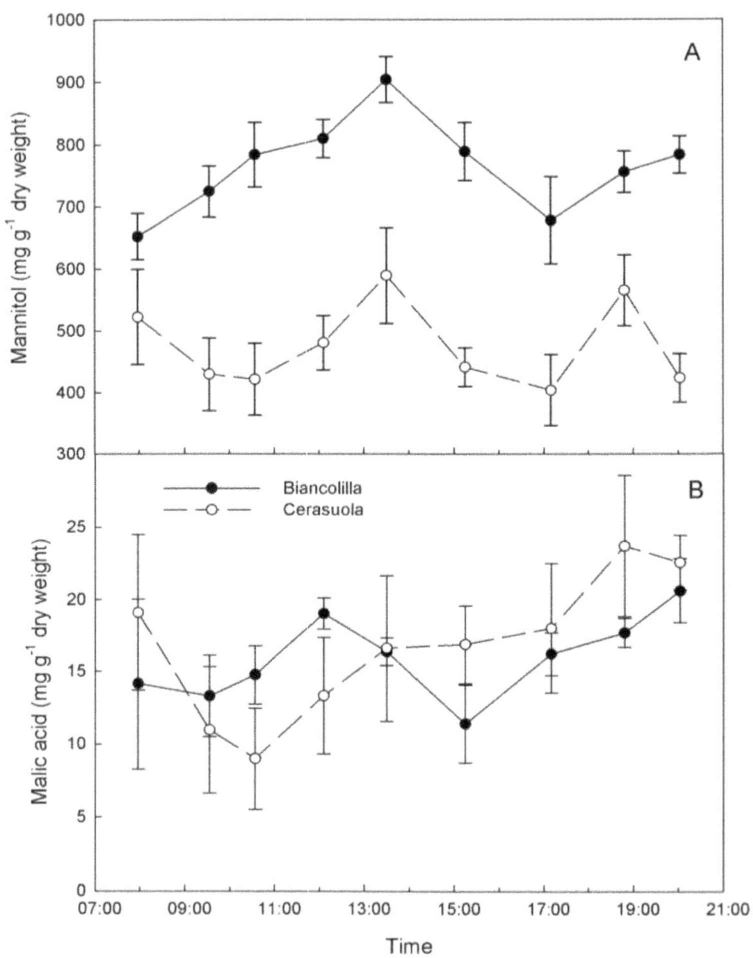

On the other hand, malic acid content varied greatly across leaf samples, and this probably masked any significant change ($p = 0.087$) during day hours (Figure 3B). Formation of calcium chelates during sample freezing and thawing may also have affected malic acid quantification as suggested by Marigo and Peltier [27], although levels of malic acid observed in this study were three-folds higher then those previously reported in olive leaves [34]. In addition to the lack of a significant diurnal trend, malic acid levels were 30 to 50 times lower than mannitol levels, posing some doubts on the real contribution of malic acid to any possible osmotic adjustment in olive.

Figure 4. Relationships between vapor pressure deficit (VPD) and stomatal conductance (g_s) or transpiration rate (T), and between g_s and T in leaves of (**A–C**) 'Biancolilla' and (**D–F**) 'Cerasuola' olive trees during a typical summer day in Sicily. In 'Biancolilla', $T = Ln\ (0.83 + 0.013\ g_s)$; $R^2 = 0.486$, $p < 0.001$; $T = -0.50 + 0.39$ VPD; $R^2 = 0.694$, $p < 0.001$; $g_s = -95 + 120\ VPD - 15.3\ VPD^2$; $R^2 = 0.265$, $p = 0.006$. In 'Cerasuola', $T = -0.05 + 0.008\ g_s$; $R^2 = 0.782$, $p < 0.001$; $T = 0.04 \times 1.93^{VPD}$; $R^2 = 0.706$, $p < 0.001$; $g_s = \exp(2.90 + 0.38VPD)$; $R^2 = 0.353$, $p < 0.001$. Models tested by regression analysis and best equation chosen according to Schwarz's Bayesian Information Criterion.

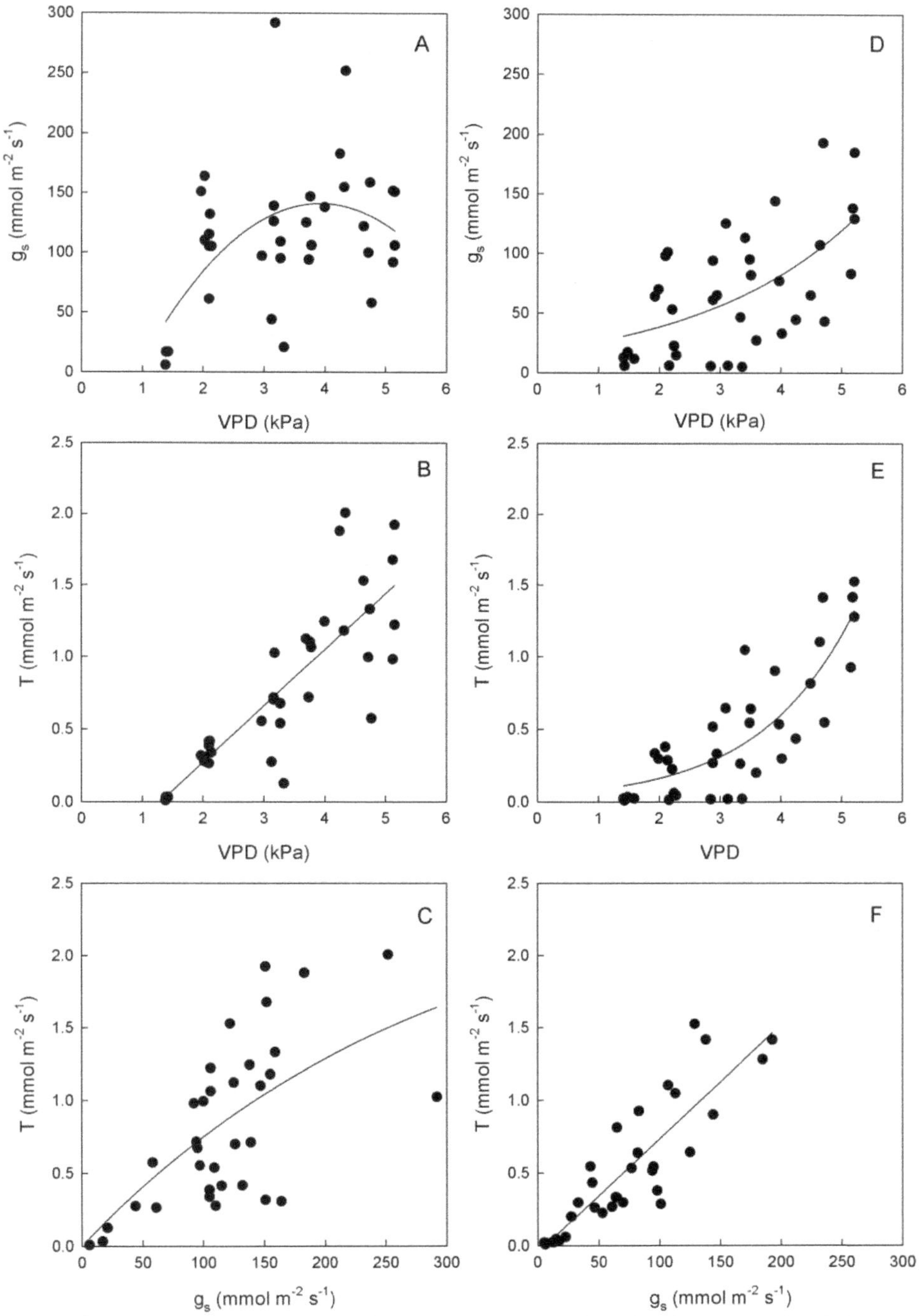

Stomates tended to open as VPD increased in both cultivars, but with different trends (Figure 4). In particular, g_s of 'Biancolilla' leaves increased until VPD reached values of about 4 kPa, after which g_s started decreasing (Figure 4A); on the other hand, g_s of 'Cerasuola' leaves increased exponentially with VPD (Figure 4D) giving no sign of stomatal control on water loss. Also, T responded directly to VPD, according to a linear trend in 'Biancolilla' (Figure 4B) and to an exponential trend in 'Cerasuola' (Figure 4E). Finally, T was related to g_s according to a log-linear model in 'Biancolilla' (Figure 4C), whereas it responded linearly to g_s in 'Cerasuola' (Figure 4F). It is evident that, in the observed ranges, VPD was the main driving force for T, with g_s somewhat limiting T in 'Biancolilla' more than in 'Cerasuola'. A direct association between VPD and T, but also g_s in absence of major dehydration, is expected during daily cycles. Low VPD levels were, in fact, associated to relatively low temperatures and light during early morning and late evening, and light and leaf temperature become the main factors driving T and g_s. The observed differences in the stomatal behavior of the two cultivars may be at least in part due to the differences in leaf hydration level.

In 'Biancolilla', an unexpected increase in RWC was observed in response to VPD increases (Figure 5A); whereas in 'Cerasuola', RWC remained constant until a critical VPD level (around 4 kPa) beyond which RWC started decreasing (Figure 5D). T and g_s responses to RWC were also very different in the two cultivars. Specifically, RWC increases seem to determine exponential increases of both g_s and T in 'Biancolilla' (Figure 5 B,C). On the other hand, there was no relationship between RWC and g_s in 'Cerasuola' (Figure 5E), and rather T losses seem to be responsible for some leaf dehydration (Figure 5F). In this case, 'Biancolilla' leaves exhibited a passive feedback mechanism according to which RWC affects stomatal responses to VPD; conversely, 'Cerasuola' leaves seem to follow more of a feed forward mechanism where chemical signaling could affect stomatal responses to VPD [35].

Figure 5. Relationships between relative water content (RWC) and vapor pressure deficit (VPD), stomatal conductance (g_s), or transpiration rate (T) in leaves of (**A–C**) 'Biancolilla' and (**D–F**) 'Cerasuola' olive trees during a typical summer day in Sicily. In 'Biancolilla', $g_s = \exp(1.95 + 3.64\ \mathrm{RWC})$; $R^2 = 0.234$, $p = 0.005$; $T = 4.29\ \mathrm{RWC}^{6.68}$; $R^2 = 0.500$, $p < 0.001$. In 'Cerasuola', $\mathrm{RWC} = 0.88 + 0.0007\ \mathrm{VPD}^{15}$; $R^2 = 0.394$, $p = 0.003$; $T = 0.88 - 0.014\ \mathrm{RWC}^2$; $R^2 = 0.181$, $p = 0.027$. Models tested by regression analysis and best equation chosen according to Schwarz's Bayesian Information Criterion.

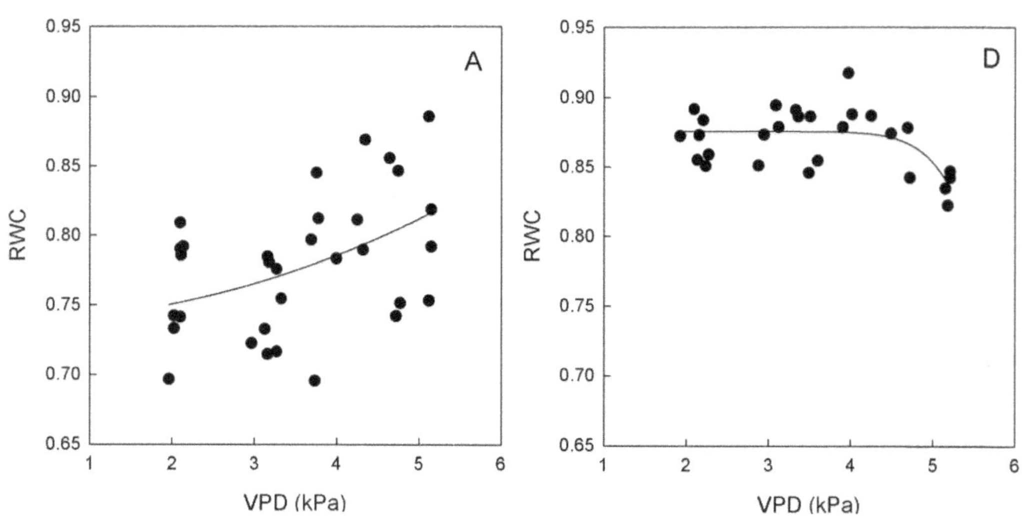

Figure 5. *Cont.*

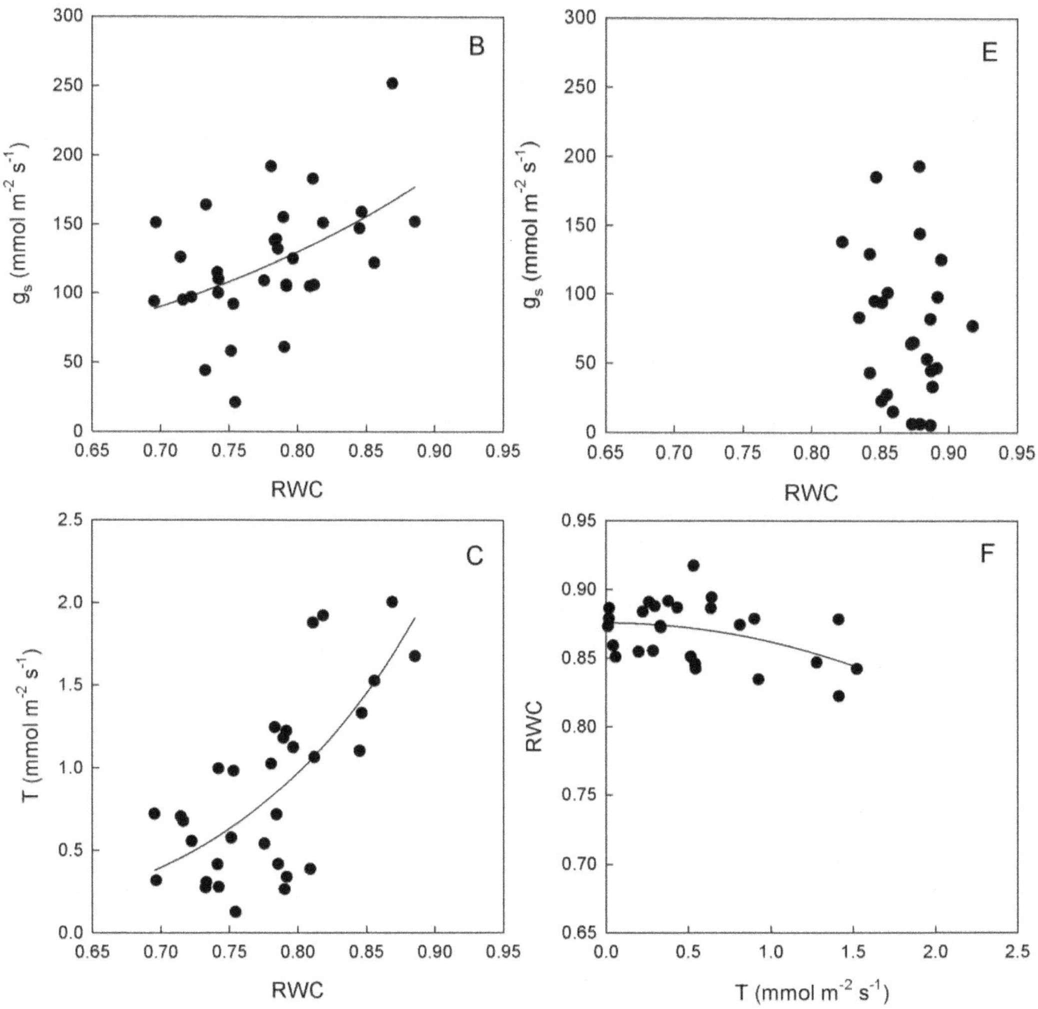

A positive association between RWC and mannitol was observed only in 'Biancolilla' leaves (Figure 6A), while no relationship was detected in 'Cerasuola' leaves (Figure 6C). No relationship was found between RWC and malic acid in either cultivar (Figure 6B,D). In 'Biancolilla', mannitol could play an active osmotic role and regulate leaf hydration level (increase turgor) and, in turn, g_s and T. Besides, it could also explain the unexpected increase of RWC as VPD increases. In particular, the accumulation of mannitol during evening hours seems very interesting and could be particularly important in terms of water and carbon balance during diurnal cycles. The amount of carbon in the form of mannitol accumulated in the evening could in fact be easily and efficiently translocated to sinks for growth and energy needs. At that point, the plant could maximize carbon, water, and energy use efficiency by producing and accumulating mannitol during daylight, when an osmotic function is needed, and translocating and using mannitol during dark hours when ambient conditions are more favorable for sink growth. It has been proposed indeed that mannitol, and polyols in general, can be a direct and efficient means for delivering carbon and energy to plant tissues and organs, as they yield NADH already in the first degradation step [23,24].

Figure 6. Relationships between mannitol, malic acid, and relative water content (RWC) in leaves of (**a**, **b**) 'Biancolilla' and (**c**, **d**) 'Cerasuola' olive trees during a typical summer day in Sicily. In 'Biancolilla', Mannitol = 1066 + 1203 Ln(RWC); $R^2 = 0.511$, $p < 0.001$. Models tested by regression analysis and best equation chosen according to Schwarz's Bayesian Information Criterion.

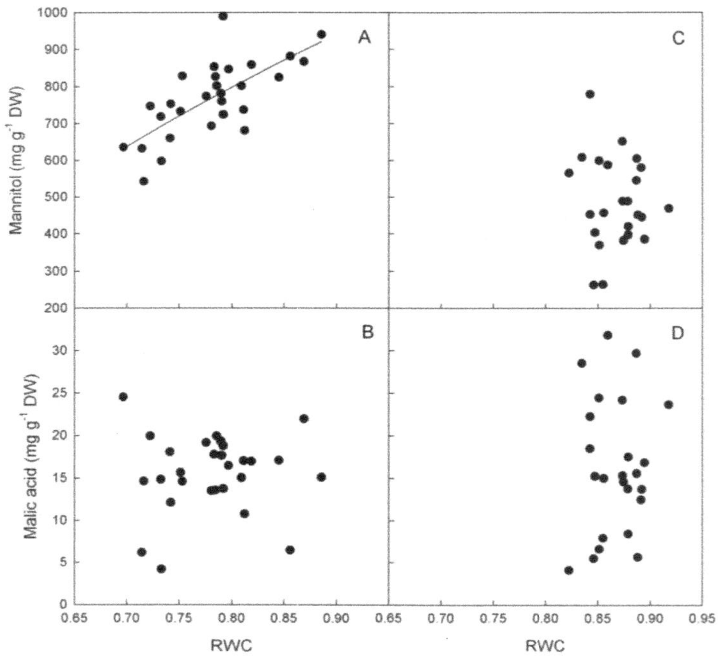

A similar mechanism was not evident in 'Cerasuola' leaves, where, in addition to stomatal control, some factor other than leaf mannitol (e.g., hydraulic conductivity and/or the ability to extract water from the soil) must be responsible for maintaining steadily high hydration levels during day hours.

3. Experimental

For the experiment, eight adult olive trees were selected, four of the cultivar Biancolilla and four of the cultivar Cerasuola. The two cultivars were chosen based on their different leaf mannitol content reported in a previous study [29]. Trees were all of the same age (20 years old), grown in the same grove (near Scillato, Sicily, 37°50'14"N, 13°56'47"E and 400 m a.s.l.), and treated with the same cultural cares and no irrigation.

All measurements and samples were collected on 18 August 2007, during a typical hot and dry summer day and after a period of over a month of no rain and maximum air temperatures above 30 °C. Hourly air temperature, relative humidity and solar radiation were acquired with a local meteorological station and used to calculate vapor pressure deficit (VPD).

Stomatal conductance (g_s) was measured with an AP4 Delta-T porometer (Delta-T Devices, Cambridge, UK) at regular intervals from 8:00 to 20:00 HR (for a total of nine measurements per day) on one sun-exposed, mature, but non-senescent, leaf per tree. Transpiration (T) was estimated by converting VPD into vapor concentration deficit (mol m^{-3}) and dividing by leaf resistance (m s^{-1}). After each g_s measurement, the same leaf was wrapped in parafilm and aluminum foil, detached, stored in an ice-bag, and finally transported to the laboratory for determination of fresh weight (FW) with a four-digit precision scale. Fresh leaves were then rehydrated by immersing their petiole in distilled water for 24 h in the dark and at 8 °C, and weighed again for determination of turgid weight (TW). Finally, leaves were dried in a ventilated oven at 60 °C until constant weight and their dry weight (DW) was recorded. Relative water content (RWC) of each leaf was calculated as (FW − DW)/(TW − DW).

Leaves from the same shoots and opposite to the ones used for g_s and RWC measurements were sampled, transferred to the laboratory and stored at −40 °C for subsequent determination of mannitol and malic acid. Extraction was carried out using about 1 g of leaf blade, cut in small pieces, transferred into 1.5-mL eppendorf tubes and finely ground with a V-shaped pestle in presence of liquid nitrogen. Ground dry tissues were weighed and extracted with 1 mL of 80% (v:v) methanol solution containing 2.2 mg of phenyl-β-glucopiranoside as an internal standard. The homogenate was vortexed for 1 min and centrifuged for 5 min at 3000 g. The supernatant was stored at −40 °C for subsequent quantification of mannitol and malic acid with a Shimadzu 2010 gas chromatograph (GC) mounting an Equity 5 fast column (Sigma Aldrich, St. Louis, MO, USA). Sample preparation and GC method were those described in Lo Bianco *et al.* [29]. Briefly, 50-µL of extract was evaporated to dryness into 2-mL GC vials. Samples were derivatized by adding 75 µL of *N,O*-bis(trimethylsilyl)trifluoroacetamide and 15 µL of piridine, sonicating for 15 min and heating at 75 °C for 120 min. Calibration curves were constructed with separate standards for mannitol and malic acid using reagents from Sigma Aldrich.

Data were analyzed using GLM procedures of Systat software (Systat Software Inc., Richmond, CA, USA) with cultivar and time as main factors and cultivar x time as the sole interaction in the ANOVA model. Associations between response variables and for each cultivar were described by fitting linear or non-linear models with SigmaPlot (Systat Software Inc., Richmond, CA, USA). Schwarz's Bayesian Information Criterion was used to compare regression models and choose best equation.

4. Conclusions

This study highlights two different strategies adopted by the two olive cultivars in order to cope with diurnal changes in atmospheric water demand. Specifically, 'Cerasuola' olive showed a sort of steady, conservative behavior as it tried to maintain favorable hydration levels by consuming less water and doing less in general. Conversely, 'Biancolilla' olive showed a more cyclic, definitely less conservative behavior as it kept stomates open causing some dehydration (day time), but allowing for production and accumulation of mannitol to possibly resume growth as soon as conditions become favorable (night time).

Acknowledgments

Assistance from graduate and undergraduate students was appreciated. This work was in part supported by the FFR 2012 intramural funding from the University of Palermo.

Author Contributions

Dr. Avellone contributed to this study mainly with the analytical determination of mannitol and malic acid in olive leaves. He also developed a new fast method to run samples through gas chromatography, and contributed to the interpretation of analytical outputs. Dr. Lo Bianco designed the experiments, and coordinated sampling and field measurements. He also managed final data statistical analysis, interpretation and presentation in graphs and tables, and carried out most of the manuscript preparation.

Conflicts of Interest

The authors declare no conflict of interest.

References

1. Moriana, A.; Orgaz, F.; Pastor, M.; Fereres, E. Yield responses of a mature olive orchard to water deficits. *J. Am. Soc. Hortic. Sci.* **2003**, *128*, 425–431.
2. Connor, D.J.; Fereres, E. The physiology of adaptation and yield expression in olive. *Hortic. Rev.* **2005**, *31*, 155–229.
3. Xiloyannis, C.; Dichio, B.; Nuzzo, V.; Celano, G. Defence strategies of olive against water stress. *Acta Hortic.* **1999**, *474*, 423–426.
4. Dichio, B.; Xiloyannis, C.; Angelopoulos, K.; Nuzzo, V.; Bufo, S.A.; Celano, G. Drought-induced variations of water relations parameters in *Olea europaea*. *Plant Soil* **2003**, *257*, 381–389.
5. Gucci, R.; Moing, A.; Gravano, E.; Gaudillére, J.P. Partitioning of photosynthetic carbohydrates in leaves of salt-stressed olive plants. *Austr. J. Plant Physiol.* **1998**, *25*, 571–579.
6. Rejšková, A.; Patková, L.; Stodůlková, E.; Lipavská, H. The effect of abiotic stresses on carbohydrate status of olive shoots (*Olea europaea* L.) under *in vitro* conditions. *J. Plant Physiol.* **2005**, *164*, 174–184.
7. Wang, Z.; Stutte, G.W. The role of carbohydrates in active osmotic adjustment in apple under water stress. *J. Am. Soc. Hortic. Sci.* **1992**, *117*, 816–823.
8. Ranney, T.G.; Bassuk, N.L.; Whitlow, T.H. Osmotic adjustment and solute constituents in leaves and roots of water-stressed cherry (*Prunus*) trees. *J. Am. Soc. Hort. Sci.* **1991**, *116*, 684–688.
9. Lo Bianco, R.; Rieger, M.; Sung, S.S. Effect of drought on sorbitol and sucrose metabolism in sinks and sources of peach. *Physiol. Plant* **2000**, *108*, 71–78.
10. Peltier, J.P.; Marigo, D.E; Marigo, G. Involvement of malate and mannitol in the diurnal regulation of the water status in members of Oleaceae. *Trees* **1997**, *12*, 27–34.
11. Bieleski, R.L. Sugar Alcohols. In *Plant Carbohydrates. I. Intracellular Carbohydrates*; Loewus, F.A., Tanner, W., Eds.; Springer-Verlag: Berlin, Germany, 1982; pp. 158–192.
12. Barker, S.A. Acyclic Sugar Alcohols. In *Modern Methods of Plant Analysis*; Paech, K., Tracey, M.V., Eds.; Springer-Verlag: Berlin, Germany, 1955; pp. 55–63.
13. Zimmermann, M.H.; Ziegler, H. List of Sugars and Sugar Alcohols in Sieve-Tube Exudates. In *Transport in Plants. 1. Phloem Transport*; Zimmerman, M.H., Milburn, J.A., Eds.; Springer-Verlag: Berlin, Germany, 1975; pp. 480–503.
14. Conde, C.; Silva, P.; Agasse, A.; Lemoine, R.; Delrot, S.; Tavares, R.; Gerós, H. Utilization and transport of mannitol in *Olea europaea* and implications on salt stress tolerance. *Plant Cell Physiol.* **2007**, *48*, 42–53.

15. Galinski, E.A. Compatible solutes of halophilic eubacteria: Molecular principles, water-solute interaction, stress protection. *Experientia* **1993**, *49*, 487–496.
16. Smirnoff, N.; Cumbes, Q.J. Hydroxyl radical scavenging activity of compatible solutes. *Phytochemistry* **1989**, *28*, 1057–1060.
17. Tarczynski, M.C.; Jensen, R.G.; Bohnert, J.H. Stress protection of transgenic tobacco by production of the osmolyte mannitol. *Science* **1993**, *259*, 508–510.
18. Abebe, T.; Guenzi, A.C.; Martin, B.; Cushman, J.C. Tolerance of mannitol-accumulating transgenic wheat to water stress and salinity. *Plant Physiol.* **2003**, *131*, 1748–1755.
19. Macaluso, L.; Lo Bianco, R.; Rieger, M. Mannitol-producing tobacco exposed to varying levels of water, light, temperature and paraquat. *J. Hortic. Sci. Biotechnol.* **2007**, *82*, 979–985.
20. Karakas, B.; Ozias-Akins, P.; Stushnoff, C.; Suefferheld, M.; Rieger, M. Salinity and drought tolerance of mannitol-accumulating transgenic tobacco. *Plant Cell Environ.* **1997**, *20*, 609–616.
21. Shen, B.; Jensen, R.G.; Bohnert, H.J. Mannitol protects against oxidation by hydroxyl radicals. *Plant Physiol.* **1997**, *115*, 527–532.
22. Lo Bianco, R.; Losciale, P.; Manfrini, L.; Corelli Grappadelli, L. Possible role of mannitol as an oxygen radical scavenger in olive. *Acta Hortic.* **2011**, *924*, 83–88.
23. Pharr, D.M.; Stoop, J.M.H.; Williamson, J.D.; Studer Feusi, M.E.; Massel, M.O.; Conkling, M.A. The dual role of mannitol as osmoprotectant and photoassimilate in celery. *HortScience* **1995**, *30*, 1182–1188.
24. Lo Bianco, R. Carbohydrate Metabolism and Source-Sink Relationships in Peach. In *Tree Growth: Influences, Layers and Types*, Environmental Science, Engineering and Technology Series; Karam, W.P., Ed.; Nova Science Publishers, Inc.: New York, NY, USA, 2009; Chapter 5, pp. 79–91.
25. Lance, C.; Rustin, P. The central role of malate in plant metabolism. *Physiol. Végétale* **1984**, *22*, 625–641.
26. Martinoia, E.; Reutsch, D. Malate compartmentation. Responses to a complex metabolism. *Annu. Rev. Plant Physiol. Plant. Mol. Biol.* **1994**, *45*, 447–467.
27. Marigo, G.; Peltier, J.P. Analysis of the diurnal change in osmotic potential in leaves of *Fraxinus excelsior* L. *J. Exp. Bot.* **1996**, *47*, 763–769.
28. Patonnier, M.P.; Peltier, J.P.; Marigo, G. Drought-induced increase in xylem malate and mannitol concentrations and closure of *Fraxinus excelsior* L. stomata. *J. Exp. Bot.* **1999**, *336*, 1223–1229.
29. Lo Bianco, R.; Panno, G.; Avellone, G. Characterization of Sicilian olive genotypes by multivariate analysis of leaf and fruit chemical and morphological properties. *J. Agric. Sci.* **2013**, *5*, 229–245.
30. Elsheery, N.I.; Cao, K.-F. Gas exchange, chlorophyll fluorescence, and osmotic adjustment in two mango cultivars under drought stress. *Acta Physiol. Plant.* **2008**, *30*, 769–777.
31. Khakwani, A.A.; Dennett, M.D.; Munir, M.; Baloch, M.S. Wheat yield response to physiological limitations under water stress condition. *J. Anim. Plant Sci.* **2012**, *22*, 773–780.
32. Cuevas, M.V.; Torres-Ruiz, J.M.; Álvarez, R.; Jiménez, M.D.; Cuerva, J.; Fernández, J.E. Assessment of trunk diameter variation derived indices as water stress indicators in mature olive trees. *Agric. Water Manag.* **2010**, *97*, 1293–1302.

33. Diaz-Espejo, A.; Nicolas, E.; Fernandez, J.E. Seasonal evolution of diffusional limitations and photosynthetic capacity in olive under drought. *Plant Cell Environ.* **2007**, *30*, 922–933.
34. Dichio, B.; Margiotta, G.; Xiloyannis, C.; Bufo, S.A.; Sofo, A.; Cataldi, T.R.I. Changes in water status and osmolyte contents in leaves and roots of olive plants (*Olea europaea* L.) subjected to water deficit. *Trees* **2009**, *23*, 247–256.
35. Streck, N.A. Stomatal response to water vapor pressure deficit: An unsolved issue. *Rev. Bras. Agrociência* **2003**, *9*, 317–322.

3

Four New Species of *Nepenthes* L. (*Nepenthaceae*) from the Central Mountains of Mindanao, Philippines

Thomas Gronemeyer [1,*], Fulgent Coritico [2], Andreas Wistuba [3], David Marwinski [4], Tobias Gieray [5], Marius Micheler [6], François Sockhom Mey [7] and Victor Amoroso [2]

1 Institute of Molecular Genetics and Cell Biology, Ulm University, 89081 Ulm, Germany
2 Center for Biodiversity Research, Central Mindanao University, University Town, 8710 Musuan, Bukidnon, Philippines
3 Friedhofweg 4, 88437 Maselheim, Germany
4 In der Messe 27, 70327 Stuttgart, Germany
5 Losburgstrasse 39, 73776 Altbach, Germany
6 Frankfurter Ring 3, 80807 Munich, Germany
7 3 rue Frédéric Chopin, 59320 Haubourdin, France

* Author to whom correspondence should be addressed; E-Mail: thomas.gronemeyer@uni-ulm.de

---

Abstract: Together with the islands of Sumatra (Indonesia) and Borneo (Indonesia, Malaysia), the Philippines are the main center of diversity for carnivorous pitcher plants of the genus, *Nepenthes* L. *Nepenthes* are the largest of all carnivorous plants, and the species with the biggest pitchers are capable of trapping and digesting small amphibians and even mammals. The central cordillera of Mindanao Island in the south of the Philippines is mostly covered with old, primary forest and is the largest remaining cohesive, untouched area of wilderness in the Philippines. In a recent field exploration of two areas of the central cordillera, namely Mount Sumagaya and a section of the Pantaron range, four new taxa of *Nepenthes* were discovered. These four remarkable new species, *N. pantaronensis*, *N. cornuta*, *N. talaandig* and *N. amabilis*, are described, illustrated and assessed.

Keywords: carnivorous pitcher plants; *Nepenthes*; biodiversity; Philippines

---

1. Introduction

The Philippines are nowadays considered to host the highest diversity of carnivorous pitcher plants of the genus, *Nepenthes*. Local conflicts have hindered exploration of the islands in the past, and therefore, only 15 distinct species were known by the late 1990s [1]. In the past decade, a series of field expeditions (mainly to Mindanao and Palawan) has led to the discovery of 11 new taxa ([1] and references therein), while two more species, *N. peltata* Sh.Kurata and *N. robcantleyii* Cheek, were described based on cultivated material [2,3]. More recently, *N. viridis* Micheler from Dinagat Island was described [4]. The description of 11 further species based on herbarium material and the resurrection of *N. graciliflora* Elmer [5–11] raises the number of Philippine species to 40, more than on the islands of Borneo and Sumatra. However, Jebb and Cheek note that some of their new taxa (e.g., *N. extincta* Cheek) might be "extinct in the wild, due to habitat destruction" [5].

The islands of Mindanao and Palawan are the Philippines' main centers of diversity for *Nepenthes*. Most species occur in the intermediate tropical highland climate prevalent on the numerous mountains of the islands. Diversity hotspots are mountains with ultramafic soils, such as Mt. Hamiguitan in Davao Oriental province (five species in total, three of these endemic) or the Diwata Mountains in Agusan del Norte.

The most extensive mountain massif on Mindanao is the Pantaron mountain range, the major part of the central cordillera of Mindanao. It runs south from the municipality of Claveria in the north of the island towards the south to the municipality of San Fernando and separates the provinces of Bukidnon in the west and Agusan del Sur in the east. The average altitude of this mountain range is approximately 1,000 m, with summits of around 1,500 m that peak out of the cordillera. Almost the whole area is covered by dipterocarp, lower and upper mountain forest and is designated as a key biodiversity area by the Philippine Department of Environment and Natural Resources. The Pantaron range is flanked by the marshes of the Agusan River in the east and by the Pulangi River in the west, the latter being the longest river in Mindanao. Along the Pulangi River, large areas of lower mountain forest have been cleared for agriculture.

West of the Pulangi River, opposite the Pantaron range, the Tago mountain range marks the western border of the central cordillera. The northern border of the central cordillera consists of a conglomeration of prominent peaks with altitudes significantly above 2,000 m, namely Mt. Sumagaya, Mt. Mangabon and Mt. Kimangkil. While other prominent mountains on Mindanao, like Mt. Apo, Mt. Kitanglad, Mt. Hamiguitan or Mt. Matutum have been explored extensively by local and foreign botanists, the central cordillera has remained unexplored until recently. Large areas of these mountains form part of the ancestral territories of local indigenous tribes, and separatist conflicts between the New People's Army (NPA) and the military long hindered safe access to this remote area.

A minor peak in the Pantaron range in the San Fernando area was explored by one of the authors in 2008 [12], and Mt. Kiamo in the Tago range was explored in 2011. During the exploration of Mt. Kiamo, the common *N. mindanaoensis* Sh.Kurata, *N. merrilliana* Macfarlane and *N. truncata* Macfarlane were found, as well as two remarkable new taxa that were previously unknown, *N. pulchra* Gronem. and *N. ceciliae* Gronem. [13,14]. *N. pulchra* is an unusually large species that is related to *N. petiolata* Danser, whereas the gracile *N. ceciliae* shares features with *N. copelandii* Merr. ex Macfarlane and was subsequently grouped with the *N. alata* complex of species [5,8].

These findings, together with the identification of several species, including *N. surigaoensis* Elmer, on minor peaks in the Pantaron range near San Fernando [12], led to the assumption that the Pantaron range and the adjacent mountains might represent a huge area of diversity for carnivorous pitcher plants.

In order to verify this theory, an area of the Pantaron range approximately 60 km north of San Fernando and Mt. Sumagaya, one of the high peaks forming the northern border of the central cordillera, were chosen for further field research. This expedition led to the discovery of four additional new *Nepenthes* taxa. The taxonomic descriptions of these new species are provided in this article.

2. Results and Discussion

2.1. Taxonomic Description of Nepenthes pantaronensis

Nepenthes pantaronensis Gieray, Gronem., Wistuba, Marwinski, Micheler, Coritico, V.B. Amoroso, spec. nov.

Diagnosis:

Differs from *N. pulchra* Gronem. in having 2 longitudinal nerves (*N. pulchra*: 3–4), 2 non-fringed or barely fringed wings on the lower pitchers and no wings on the upper pitchers (*N. pulchra*: fringed wings on the lower pitchers and wings reduced to ribs on the upper pitchers) and in having basal stem leaves with a canaliculate petiole (*N. pulchra*: broad-winged petiole).

Type:

Philippines, Mindanao Island, Bukidnon Province, Pantaron mountain range, Mt. Gaka (1,390 m) near Sitio Mahayag (Barangay St. Peter, Malaybalay City, Bukidnon Province, Philippines), 15.08.2012, *T. Gronemeyer and F. Coritico*, holotype CMUH00008625, Central Mindanao University Herbarium (CMUH), Musuan, Bukidnon, Philippines.

Etymology:

The specific epithet denotes that *N. pantaronensis* was discovered in the Pantaron mountain range.

Description:

The *stem* is up to 3 m long, cylindrical in the cross-section, 0.9–1 cm in diameter, with internodes up to 13 cm long.

Leaves of the basal stem are linear to oblong, up to 28 cm long and 5 cm wide, with two longitudinal veins on each side of the midrib and numerous pinnate veins running obliquely towards the leaf margins. The apex of the lamina is acute, the base attenuate and forming a canaliculate petiole that is strongly decurrent down the full length of the internode. The tendrils are up to 34 cm long and mostly coiled. All parts of the foliage may be yellowish green.

Leaves of the climbing stem form a broad-winged petiole that is also strongly decurrent. In all other respects, they are consistent with the leaves of the basal stem.

Lower pitchers are up to 35 cm tall and 6 cm wide, though usually smaller. The bottom quarter to third of the pitcher is variably inflated, sometimes to the extent that the basal part is bulbous. Above the broad lower section, the pitcher narrows, becoming cylindrical or very slightly infundibular towards the pitcher opening. Wings are usually reduced to ribs, although occasionally, some fringes are present below the pitcher opening. The pitcher opening is oblique and up to 5 cm wide. The peristome is flattened, up to 2.5 cm wide and particularly broad around the sides and below the lid. The peristome is lined with ribs up to 2 mm high, spaced up to 2 mm apart. The ribs are elongated on the inner edge of the peristome to form narrow teeth up to 3 mm long. The lid is ovate, up to 7 cm long and 6 cm wide, with glands densely and evenly distributed over the lower surface. A prominent, triangular, hooked and keeled appendage up to 5 mm long is present on the underside of the lid. The spur is unbranched and up to 5 mm long. The exterior of the lower pitchers is greenish or slightly orange, with narrow orange-red blotches. The interior of the pitcher is white or light yellowish green, lined with small, angular blotches of purple-red. The peristome is red, variably striped with bands of yellow, orange and purple. The lower and upper surface of the lid is bright red.

Upper pitchers are up to 40 cm long and 5 cm wide, though usually smaller. The bottom fifth to quarter of the trap is inflated and bulbous, narrowing above this part before becoming cylindrical towards the pitcher opening. Some of the upper pitchers are narrower above the bottom fifth to quarter of the trap, while others are almost cylindrical with almost no narrowing. Wings are indiscernible. The peristome is up to 1 cm wide, cylindrical or slightly flattened, lined with ribs up to 1 mm high, spaced up to 1 mm apart. In all other aspects, including coloration and color variability, the upper pitchers are consistent with the lower ones, although some upper pitchers are entirely green with no blotches.

The *inflorescence* is a panicle, up to 29 cm long, composed of a 19.5 cm-long scape and an additional 9.5 cm rachis bearing pedicels evenly scattered along its length. The pedicels are about 2 cm long, branched and one- or two-flowered. Bracts are absent.

Indumentum is present across the foliage, inflorescence and pitchers, consisting of orangey brown hairs up to 1 mm long.

N. pantaronensis (Figure 1; Appendix Figure A1) is closely related to *N. pulchra* [14] and *N. petiolata* [15]. The most important morphological and ecological differences between these three species are summarized in Table 1. Following the interpretation of Jebb and Cheek [5], these three species belong to the *Reginae* group of species of Danser [15] and represent a group of species on their own among the Philippine species. *N. petiolata* is only known from the Diwata Mountains (including the type locality, Mt. Masay) in the northeast of Mindanao and grows mainly epiphytically. *N. pulchra* represents the most western species and is currently known only from Mt. Kiamo in the Tago range, where it grows terrestrially on ultramafic soils. *N. pantaronensis* is located in between these two species (although geographically closer to *N. pulchra*) and represents a true intermediate between these two; it grows both epiphytically and terrestrially (although not on ultramafic soils, but in an "epiphyte like" mossy substrate) and has both canaliculate and winged petioles, the former on the basal stem and the latter on the climbing stem.

Figure 1. *N. pantaronenesis*. (**a**) Upper pitcher with leaf; (**b**) habit with female inflorescence and upper pitcher; (**c**) partial inflorescence with female flowers; (**d**) upper surface of lid and spur.

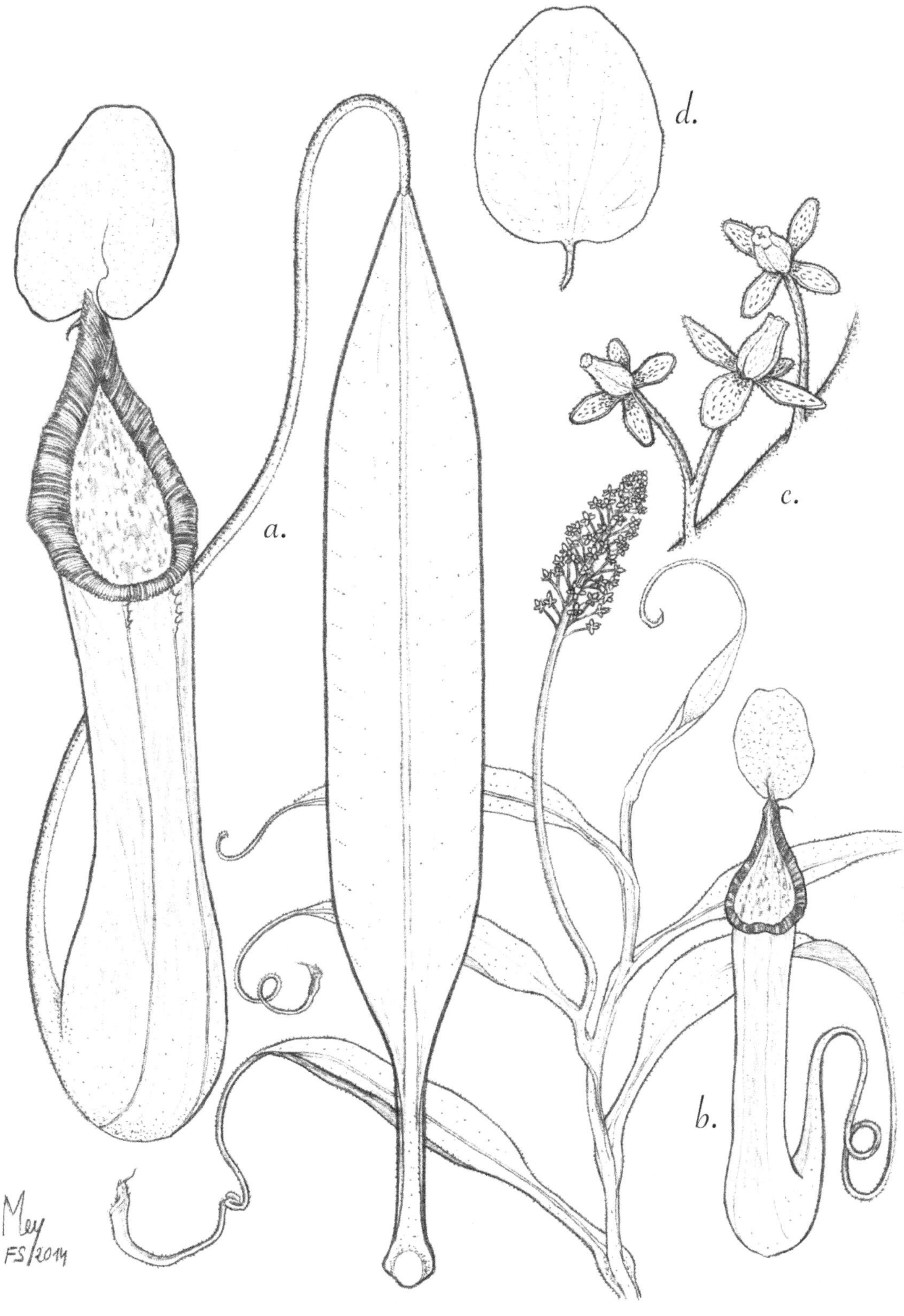

Table 1. The main morphological differences between *N. pantaronensis*, *N. pulchra* and *N. petiolata* (data from [14,15]).

	N. pantaronensis	*N. pulchra*	*N. petiolata*
Lower Pitchers			
Overall form	Cylindrical, bottom quarter to third variably inflated	Cylindrical, bottom quarter to third variably inflated, often bulbous	Cylindrical, bottom third ovate and variably inflated
Dimension	Up to 35 cm tall and 6 cm wide, though usually smaller	Up to 35 cm tall and 6 cm wide, though usually smaller	Usually up to 21 cm tall and 4.5 cm wide
Wings	Seldom fringed, mostly only ribs	Fringed wings	Fringed wings
Peristome	Flattened, up to 2.5 cm wide	Flattened, up to 2.5 cm wide	Loosely cylindrical
Appendage	Very prominent, triangular, hooked appendage	Rounded, triangular appendage; sometimes only keel present	Only keel; no appendage
Upper pitchers			
Overall form	Bottom fifth to quarter inflated and bulbous, narrowing above this part before becoming cylindrical towards the pitcher opening	Bottom fifth to quarter inflated and bulbous or near spherical, and slightly infundibular above towards the pitcher opening	Cylindrical, bottom fifth to quarter ovate and slightly swollen
Dimension	Up to 40 cm long and 5 cm wide, usually smaller	Up to 34 cm long and 7 cm wide, occasionally up to 42 cm long	Up to 18 cm long and 5 cm wide, occasionally up to 25 cm long
Wings	Indiscernible	Hardly discernible; reduced to ribs if present	Hardly discernible; reduced to ribs if present
Peristome	Cylindrical or slightly flattened, 1–2 cm wide	Flattened, 3 cm wide	Loosely cylindrical, 1–2 cm wide
Appendage	Not hooked, sometimes reduced to a keel	Hooked and rounded, up to 5 mm long near base of the lid	No appendage
Leaf			
	Petiole canaliculate on the nonclimbing stem and broad-winged on the climbing stem. Decurrent down full length of internode. Leaves with two longitudinal nerves.	Petiole broadly winged, decurrent down full length of internode, often part way down previous internode as well. Leaves with 3–4 longitudinal nerves.	Petiole canaliculate or winged, clasping the stem. Leaves with four longitudinal nerves.
Habitat and Distribution			
	Epiphytic on high trees in lower montane forest at the type locality. Pantaron range and adjacent mountains.	Strictly terrestrial in clearings and montane scrub. Currently only known from Mt. Kiamo.	Epiphytic or rarely terrestrial in lower and upper montane forest. Diwata mountains.

2.2. Taxonomic Description of Nepenthes cornuta

Nepenthes cornuta Marwinski, Coritico, Wistuba, Micheler, Gronem., Gieray, V.B.Amoroso, spec. nov.

Diagnosis:

Differs from *N. copelandii* Macfarlane in having a narrower lamina, upper pitchers with a distinctive, swollen base and an almost completely cylindrical upper two-thirds (*N. copelandii*: upper pitchers strongly infundibular with a narrow base and proportionally a much wider opening) and noticeably smaller lower pitchers lacking wings (*N. copelandii*: wings always present on lower pitchers).

Type:

Philippines, Mindanao Island, Bukidnon Province, Pantaron mountain range, trail from Sitio Mahayag (Barangay St. Peter, Malaybalay City) to Sitio Balaudo, 15.08.2012, *T. Gronemeyer and F. Coritico*, holotype CMUH00008547, Central Mindanao University Herbarium (CMUH), Musuan, Bukidnon, Philippines.

Etymology:

The specific epithet, *cornuta* (lat. cornu = horn), refers to the plant's horn-shaped upper pitchers.

Description:

The *stem* is up to 3 m long and 7 to 8 mm in diameter; the internodes are 4 to 5 cm long.

Leaves of the climbing stem are petiolate and linear to lanceolate, up to 30 cm long and 4.5 cm wide. The apex of the leaf blade is acute, the leaf base broadly attenuate and petiolate. The petiole is up to 4 cm long, canaliculate and amplexicaul, clasping between one third and half of the stem. Three parallel longitudinal nerves run on each side of the midrib, on the outer half of the lamina. The tendrils can reach up to 24 cm in length. The foliage and the midrib are green or yellowish green, while the stem and the tendrils can take on a slight red hue when growing in full sunlight.

Lower pitchers are up to 15 cm long and 3.4 cm wide. The pitcher opening is ovate and up to 2.8 cm in diameter. The pitcher's bottom third is bulbous; above this part, the pitchers narrow, sometimes forming a faint hip and becoming cylindrical to slightly infundibular towards their opening. The lid is up to 3.2 cm long, 2.8 cm wide and ovate. The appendage is reduced to a rudimentary keel of up to 1 mm. The spur is up to 3 mm long. The gland distribution is unknown. The peristome has a cylindrical cross-section of up to 5 mm in diameter, positioned at an angle and is lined with very fine ribs. The wings of the pitchers are reduced to ridges, usually without fringes. Occasionally, a few fringes can appear on the top end of those ridges, just below the peristome. The pitchers are green in their bottom third, sometimes taking on a yellowish hue towards the pitcher opening, whilst the upper two-thirds of the pitchers are mottled with red or maroon blotches. The peristome is yellowish to orange or reddish with fine red stripes. The lid is red to maroon with yellowish blotches.

Upper pitchers are up to 20 cm long and 4.5 cm wide, the pitcher opening being the widest part. The bottom third is swollen to varying degrees, sometimes displaying a faint hip, and tapers towards the tendril, giving the pitcher its typical, horn-shaped appearance. Above this part, the pitcher narrows faintly and becomes almost completely cylindrical or slightly infundibular towards its opening. The narrowing between the swollen base and the upper part of the pitcher is less pronounced than in the lower pitchers. The lid is up to 5 cm in diameter, ovate to cordate and lacks an appendage. The glands are evenly distributed across the lower surface of the lid. The spur is up to 8 mm long. The peristome is oblique and has a cylindrical cross-section of 5 to 7 mm in diameter. The wings are reduced to inconspicuous vestigial ridges. The pitchers are always green or yellow, except for the interior and the bottom side of the lid, which can have red to maroon speckles. The peristome is green or yellow, often with fine red stripes.

The *inflorescence* is a predominantly two-flowered rachis up to 40 cm long sitting on a scape of up to 21 cm. The individual pedicel can measure up to 15 mm. Bracts are absent.

Indumentum is absent from stems, leaves, pitchers and flowers, but the developing pitcher buds are covered with minute, orangey brown hairs.

N. cornuta (Figure 2, Appendix Figure A2) belongs to the *N. alata* complex of species that was recently installed by Jebb and Cheek [5,8]. Together with *N. copelandii* [16] and *N. ceciliae* [13], it forms a subgroup with the upper pitchers, which are not widest in their basal half [5]. The morphological differences between these species are listed in Table 2.

Figure 2. *N. cornuta.* Habit, showing the stem with upper pitchers (**top**). Top view of the rosette with lower pitchers (**bottom**).

Table 2. The main morphological differences between *N. cornuta*, *N. copelandii* and *N. ceciliae* (data from [13,16]).

	N. cornuta	*N. copelandii*	*N. ceciliae*
Lower Pitchers			
Overall form	Elongate, with a bulbous bottom third, narrowing above this part, sometimes with a faint hip, cylindrical to slightly infundibular towards the pitcher opening	Elongate, almost cylindrical, narrowing only faintly above the bottom third of the pitcher	Squat, with a bulbous bottom half, the upper half widening slightly towards the opening
Dimensions	Up to 15 cm long and 3.4 cm wide in the lower part	Up to 26 cm long and 5.5 cm wide at the peristome	Up to 10 cm long and 3 cm wide
Wings	Reduced to ridges	Present, up to 1.4 cm wide, fringes up to 12 mm long	Present, up to 4 mm wide, fringes up to 3 mm long
Peristome	Cylindrical, up to 5 mm wide	Loosely cylindrical, up to 1.2 cm wide	Flattened, up to 4 mm wide
Appendage	Reduced to a rudimentary keel	Absent or reduced to a rudimentary keel	Present, 2 mm long
Upper pitchers			
Overall form	Elongate, bottom third variably swollen, faint narrowing towards the middle, cylindrical to slightly infundibular in the upper half, the peristome being the widest part	Entirely infundibular, the peristome being the widest part, no narrowing in the middle	Infundibular bottom third, cylindrical or slightly infundibular upper part, the peristome being the widest part, no narrowing in the middle
Dimensions	Up to 20 cm long and 4.5 cm wide at the peristome	Up to 12 cm long and 4.5 cm wide at the peristome	Up to 10 cm long and 4 cm wide at the peristome
Wings	Reduced to faint ridges	Absent	Reduced to faint ridges
Appendage	Absent	Absent or reduced to a rudimentary keel	Present, 2 mm long
Habitat and Distribution			
	Terrestrial on ultramafic soils on clearings or open scrub. Pantaron range.	Terrestrial or epiphytic. Mt. Apo and Mt. Pasian.	Strictly terrestrial on ultramafic soils. Currently only known from Mt. Kiamo.

2.3. Taxonomic Description of Nepenthes talaandig

Nepenthes talaandig Gronem., Coritico, Wistuba, Micheler, Marwinski, Gieray, V.B.Amoroso, spec. nov.

Diagnosis:

Differs from *N. cornuta* Marwinski in having bulbous lower pitchers with a flattened, crenellated peristome (*N. cornuta*: slender lower pitchers; cylindrical towards the pitcher opening; with cylindrical peristome) and having a winged petiole that clasps the stem (*N. cornuta*: canaliculate petiole).

Type:

Philippines, Mindanao Island, Bukidnon Province, Pantaron mountain range, trail from Sitio Mahayag (Barangay St. Peter, Malaybalay City) to Sitio Balaudo, 15.08.2012, *T. Gronemeyer and F. Coritico*, holotype CMUH00008624, Central Mindanao University Herbarium (CMUH), Musuan, Bukidnon, Philippines.

Etymology:

The specific epithet was chosen to acknowledge the indigenous tribe of the Talaandig. *N. talaandig* occurs on the ancestral territory of the Talaandig communities of east Bukidnon.

Description:

The stem is up to 8 m long, cylindrical in the cross-section and 7–8 mm in diameter. The internode length is 3.5–6 cm.

Leaves of the climbing stem are elliptic, up to 35 cm long and 4.5 cm wide, with four equally distributed longitudinal veins on each side of the midrib and numerous pinnate veins running obliquely towards the leaf margins. The apex of the lamina is acute, the base attenuate and forming a narrow-winged petiole 6 cm in length that clasps one third of the stem. Pitchers are on tendrils 16–20 cm long.

Lower pitchers are bulbous or ovate, 10 cm tall and 5 cm wide. The pitcher narrows slightly towards the opening. Wings are up to 7 mm wide, fringed with filaments up to 7 mm long and run down the front of the trap. The pitcher opening is at a prominent angle, up to 3.5 cm wide and acuminates towards the lid. The peristome is flattened, up to 1.5 cm wide and crenellated. It extends into a neck towards the lid and is densely lined with ribs approximately 1 mm high. The lid is ovate to cordate, up to 5.5 cm long and 5 cm wide, with glands evenly distributed across the lower surface. A rounded, triangular appendage up to 3 mm long, sitting on a keel, is present on the underside of the lid. The spur is branched and up to 6 mm long. The exterior of the lower pitchers is rusty red to violet, with narrow blotches of reddish purple. The interior of the pitcher is white and sometimes lined with angular blotches of purple. The peristome is dark red and not striped.

Rosette pitchers, which are squatter and considerably smaller than the lower pitchers, have been observed *in situ* on young plants. In all other respects, they resemble the lower pitchers.

Upper pitchers are slender, around 20 cm long and 7 cm wide. The bottom third of the trap is slightly inflated. The pitcher narrows towards a cylindrical middle section before it widens towards the oblique pitcher opening; the latter being 4.5 cm wide. Wings are reduced to ribs throughout. The peristome is not crenellated, is cylindrical to flattened, up to 1.2 cm wide and extends into a short neck. The color of the peristome is yellow with discrete stripes of red. The lid is ovate to orbicular with a diameter of 4 cm. A prominent triangular appendage of 2.5 mm on a keel is present; the spur is 6 mm long and branched. The upper pitchers are colored yellowish-green, often with red speckles or blotches. The pitcher interior is cream-colored without blotches.

The *inflorescence* is a panicle, composed of a rachis up to 40 cm long sitting on a 20-cm scape. It bears branched two-flowered partial peduncles 5 mm long. Bracts are present.

Indumentum is sparse, if present at all.

N. talaandig (Figure 3, Appendix Figure A3) is not directly related to any other Philippine species, due to its unique ovate, bulbous lower pitchers with a prominently angled mouth that acuminates towards the lid. The lower pitchers bear superficial similarities with those of some Sumatran species, such as *N. longifolia* Nerz or *N. bongso* Korth., whereas the upper traps share similarities with those of species of the *N. alata* group (usually ovoid-cylindrical upper pitchers with an appendage on the basal part of the lower surface) [5].

Figure 3. *N. talaandig*. (**a**) Climbing stem with leaf; (**b**) upper pitcher; (**c**) lower pitcher. (**d**) Detail of the upper pitcher peristome; (**e**) detail of the lower pitcher peristome; (**f**) the underside of the lid; (**g**) appendage and spur.

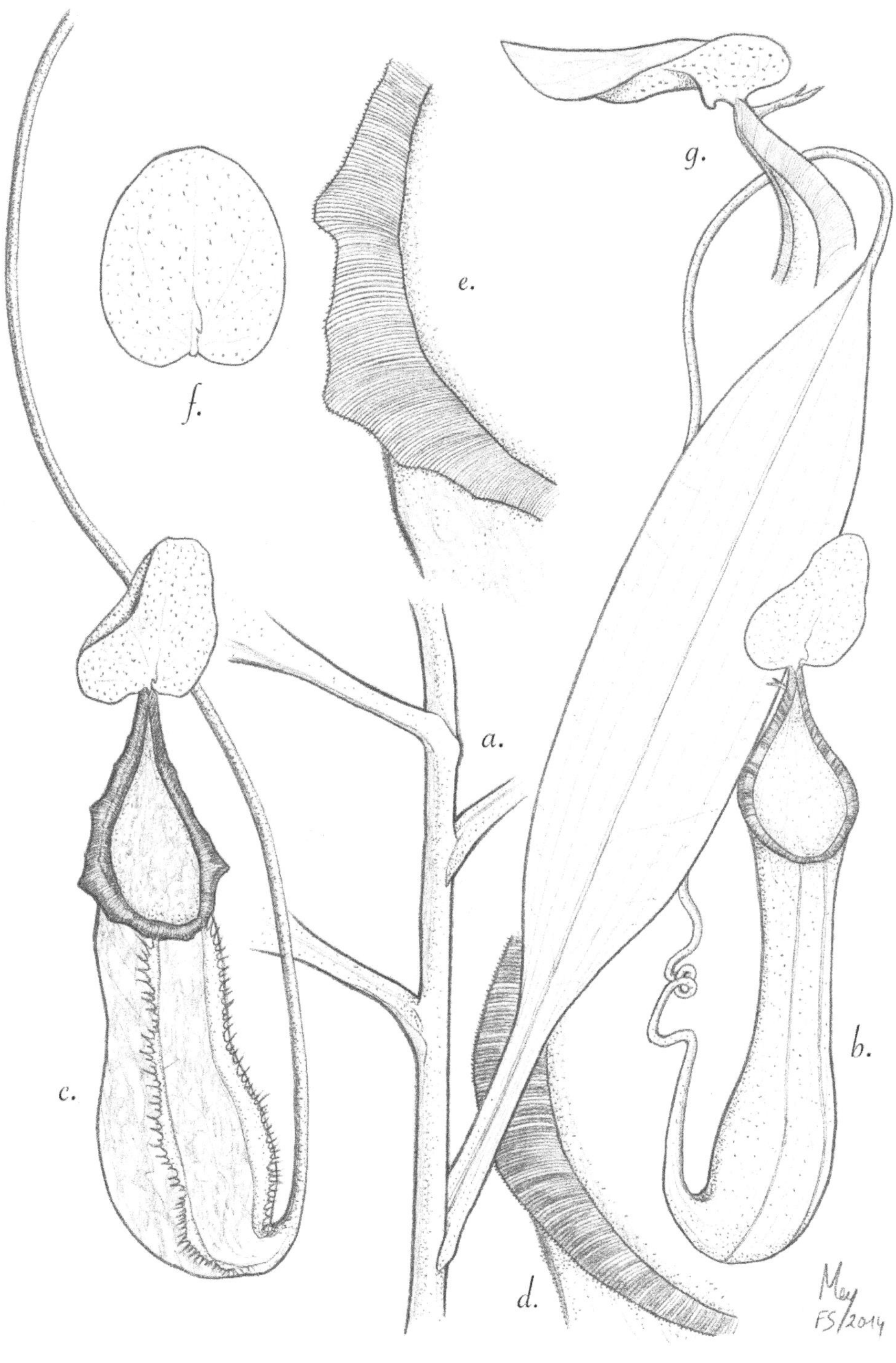

2.4. Taxonomic Description of Nepenthes amabilis

Nepenthes amabilis Wistuba, Gronem., Micheler, Marwinski, Gieray, Coritico, V.B.Amoroso, spec. nov.

Diagnosis:

Differs from *N. pantaronensis* Gieray in having spathulate-ovate, approximately 10 cm long and 3.5 cm wide leaves with an obtuse apex (*N. pantaronensis:* up to 28 cm long and narrowly acute apex) and having mostly cylindrical to slightly infundibular upper pitchers that are only 10–15 cm tall and distinctly contracted in the region below the peristome (*N. pantaronensis:* clearly inflated in the lowest quarter and up to 40 cm tall).

Type:

Philippines, Mindanao, Mt. Sumagaya, trail from Barangay Mat-I (Municipality of Claveria) to the summit 19.08.2012, *T. Gronemeyer and A. Wistuba*, holotype CMUH00008635 (male flower), isotype CMUH00008637 (female flower), Central Mindanao University Herbarium (CMUH), Musuan, Bukidnon, Philippines.

Etymology:

The specific epithet, *amabilis* (lat. amabilis = lovely), refers to the extraordinary beauty of the compact specimens with very colorful pitchers and mostly striped peristomes that were observed *in situ*.

Description:

The *stems* are often very short. Plants start to grow upright at a young age and are supported by surrounding vegetation, mostly low shrubs. The stem of short plants is green and often almost without any significant internodes. Occasionally, in more shady locations and where the stems can scramble along other plants as support, plants develop climbing stems that reach 2–3 m in length. These vines are almost round in the cross-section, 8 mm in diameter, green to dark red and with internodes that are 10–15 cm long.

Leaves of the vines and short stems are broadly spathulate-ovate, 10 cm long and 3.5 cm wide, with a broad, winged and 1 cm-wide petiole. The leaves are green with yellowish to reddish midribs. The leaf attachment is decurrent, running 1–3 cm down the stem. One to two longitudinal nerves run parallel to the midrib in the outer half of the lamina. Pitchers are attached to tendrils up to 13 cm long.

Lower pitchers of rosette plants are unknown; plants with lower pitchers could not be identified with certainty at the type locality.

Upper pitchers of short and climbing stems are 10–15 cm high, approximately 3 cm wide, mostly cylindrical, slightly infundibular in the lower third and distinctly contracted in the region below the peristome. The bottom fifth is often slightly ventricose with a slight waist above. The pitcher wings are reduced to ribs. Only occasionally in some pitchers, approximately 5 mm-wide wings bearing two or three fringes measuring approximately 5 mm in length are visible close to the peristome. The peristome is cylindrical and approximately 4 mm wide. The spur on the pitcher lid is unbranched, filiform and 3 mm long. The pitcher mouth is round, slightly elevated at the front and elongated towards the back with a slightly developed neck. The lids are 3 cm long, 3.5 cm wide, orbicular, sometimes with a small appendage near the base of the lid, which is mostly reduced to a keel. The glands are evenly distributed across the lower surface of the lid. The exterior of the upper pitchers is yellowish green to orange, unevenly suffused with red or with red blotches. The interior of the pitcher is whitish with numerous purple blotches that often shine through the pitcher wall. The peristome is yellowish green with numerous red stripes. The lid is yellowish with numerous red blotches.

The *inflorescence* is a panicle composed of a 15 cm-long scape and an additional 15 cm rachis bearing two-flowered, 20 mm-long pedicels. The anther column is up to 14 mm and the ovary 5 mm long, containing ripe seeds; the tepals are up to 7 mm long and ovate.

A prominent *indumentum* is present across the foliage, the inflorescence, especially the margins of young leaves, and pitchers, consisting of brownish hair up to 1 mm long.

N. amabilis (Figure 4 and Appendix Figure A4) is apparently not related to *N. pantaronensis*, with which it grows sympatrically on Mt. Sumagaya, nor to any other species occurring on Mindanao. Thus, it is unlikely that this species could be confused with another in the field.

2.5. Habitat

N. cornuta and *N. talaandig* occur in the Pantaron mountain range at altitudes around 1,000 m. Most observed populations occur terrestrially on ultramafic soil in clearings and on bright ridges in lower montane forest.

At the type locality, *N. pantaronensis* grows at intermediate altitudes as a strict epiphyte on moss-laden branches of large trees in lower montane forest.

On Mt. Sumagaya, smaller, stunted stands of *N. pantaronensis* occur terrestrially in lower and upper montane forest in mossy and open substrate.

The three species grow in the Pantaron range sympatrically with *N. surigaoensis* and *N. truncata.* Although no natural hybrids with any of these species have been observed *in situ*, hybridization events are likely.

N. amabilis grows exclusively in open areas of grass and shrub vegetation on Mt. Sumagaya from 1,600 m up to the summit, which stands at 2,247 m. It is completely absent from the lower and upper montane forest that covers the slopes of this mountain. Plants of possible hybrid origin with *N. pantaronensis* have been observed *in situ*.

2.6. Infauna

At the type locality of *N. cornuta*, several ant colonies (*Crematogaster* sp.) were found in dead or dying pitchers of this species. The ants obstruct the pitcher opening with a greyish matter, probably made out of vegetal detritus and dirt. This 'lid' has small entry holes along the pitcher walls. The bottom of the pitchers is pierced with one or two round holes, draining them of their digestive fluid. These holes also serve as "emergency exits" for the ants when disturbed. The presence of those insects on some pierced functional pitchers without a nest might suggest that the ants first pierce and drain functional pitchers before colonizing them. A similar colonization has been observed before in other *Nepenthes* species, e.g., *N. macfarlanei* Hemsl. in Malaysia [17] and *N. maxima* Reinw. on the lower slopes of the Maoke Mountains in West Papua (personal observation by the authors).

Figure 4. *N. amabilis*. (**a**) Habit with upper pitchers; (**b**) leaf with upper pitcher; (**c**) detail of peristome; (**d**) male inflorescence; (**e**) partial inflorescence with male flowers; (**f**) infructescence; (**g**) the upper surface of lid with spur; (**h**) the underside of the lid. Rosette plants with lower pitchers could not be identified with certainty at the type locality and are thus not depicted.

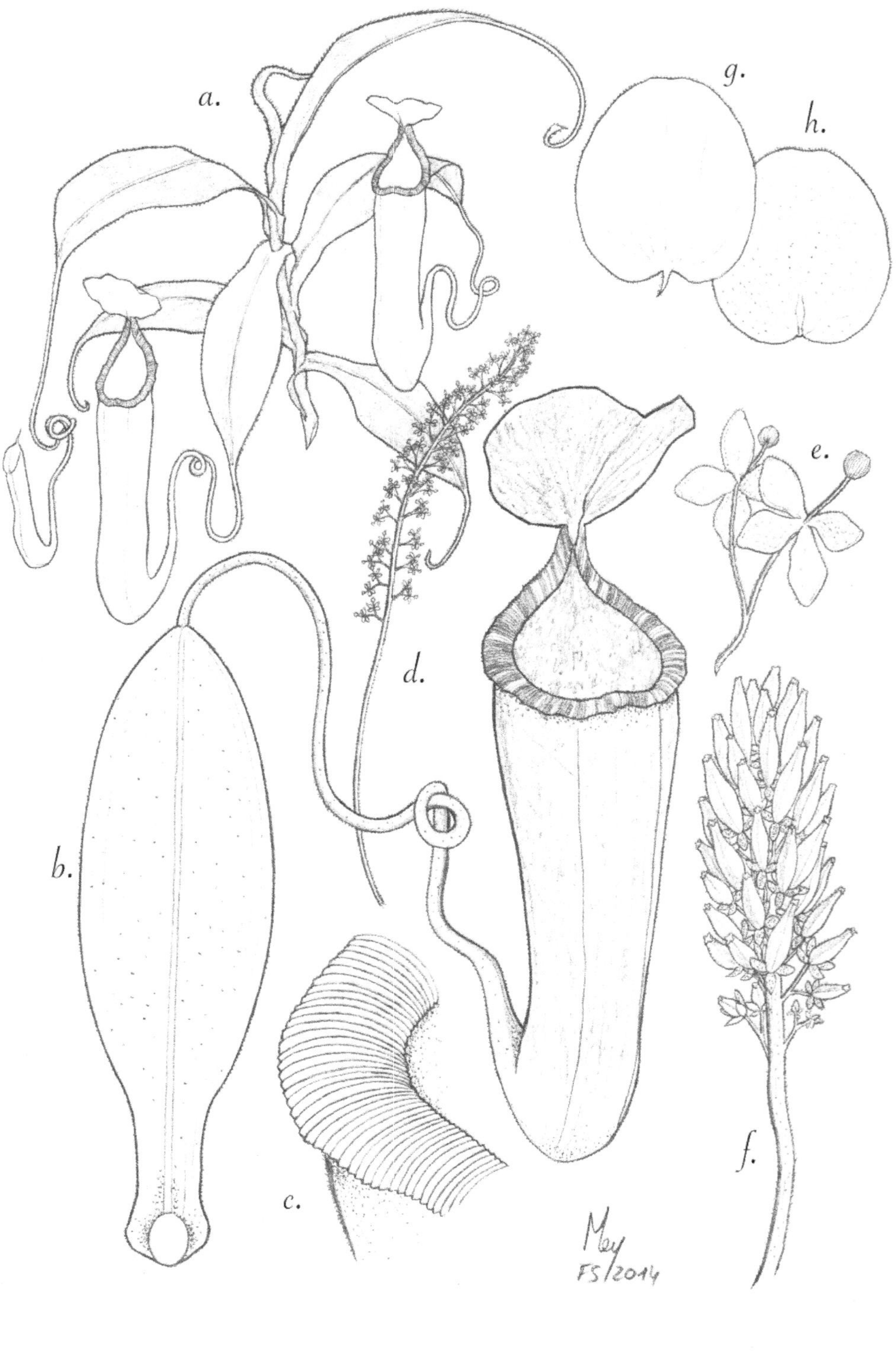

2.7. Conservation Status

The Pantaron range is currently visited only by the indigenous people of the Talaandig tribe, who cross the mountains to reach remote, temporarily inhabited villages or with the purpose of hunting and collecting rattan or resin. The impact of poaching on the population of *Nepenthes* spp. is low.

However, the foothills of the Pantaron range are completely deforested for agriculture, and illegal logging may take place in the forests. Therefore, habitat destruction has the biggest impact on the population of the *Nepenthes* spp. in this region.

It may be anticipated that both *N. talaandig* and *N. cornuta* are widely distributed in the Pantaron range, as they do not occur on isolated peaks. Since only few data have been recorded about the distribution of *N. talaandig* and *N. cornuta*, these two taxa are assessed here as VU (vulnerable) according to the IUCN Red List criteria [18].

N. pantaronensis is known both from the Pantaron range and Mt. Sumagaya, and thus, a wider distribution is documented. It is assessed here as NT (near threatened).

N. amabilis is currently only known from the upper slopes of Mt. Sumagaya. It has not been recorded in the explored areas of the Pantaron range, nor on Mt. Kiamo in the Tago range. Mt. Sumagaya is only visited by the indigenous people of the Higaonon tribe, who cross the mountains to hunt deer, and by dispersed groups of the NPA who hide in the mountains. The impact of poaching is low. However, as no comprehensive data about the distribution of this taxon have been recorded, it is classified here as EN (endangered).

3. Experimental Section

Field research was carried out during August, 2012. All data and dimensions were either collected in the field from live plants or from the respective herbarium types cited above.

The Pantaron range in the area of Barangay St. Peter (Malaybalay City) was explored after an access permit was granted by the Bukidnon Resource Management Foundation and the Biodiversity Management Bureau (BMB) of the Department of Environment and Natural Resources (DENR), and with the consent of the local indigenous people of the Talaandig tribe.

Mount Sumagaya was explored with the consent of the local military authorities in Barangay Mat-I (Municipality of Claveria, Misamis Oriental Province, Philippines) and the local indigenous people of the Higaonon tribe.

Herbarium specimens were collected under a collection permit granted to VB (Permit No. MOA 10-2-5 from 2012).

Cropped plant material was placed between paper sheets *in situ* and processed subsequently using standard methods at the herbarium of the Central Mindanao University.

Photographs were made from suitable, representative plant specimens *in situ*.

Image credits: Botanical drawings in Figures 1, 3 and 4: FSM. Botanical drawing in Figure 2: DM. Photos in Appendix Figures A1–4: AW.

4. Conclusions

Carnivorous pitcher plants are widespread within the Malesian biogeographical region comprising the Malayan peninsula and the islands of Sumatra, Borneo, Sulawesi and the Philippines. Especially in the Philippines, the majority of species are known from centers of diversity and endemism. Several species are restricted to one single mountain top or a mountain ridge.

A field research trip to the formerly unexplored central cordillera of Mindanao led to the discovery of four new taxa of carnivorous pitcher plants, namely *N. amabilis*, *N. cornuta*, *N. pantaronensis* and *N. talaandig*.

While *N. cornuta* and possibly *N. talaandig* belong to the *N. alata* group of species, *N. pantaronensis* is clearly related to *N. pulchra* and *N. petiolata* and, thus, belongs to the *Reginae* group.

N. amabilis stands clearly on its own among the Philippine *Nepenthes* species.

Other species in the central cordillera include *N. ceciliae*, *N. pulchra*, *N. surigaoensis* and *N. truncata*. However, the latter two species are widespread and also occur in other regions of Mindanao.

As with many other species of *Nepenthes* in the Philippines, the future of all species of the central cordillera is highly dependent on habitat conservation. Mt. Kiamo is not currently a protected area and, thus, the conservation status of the two endemic species, *N. ceciliae* and *N. pulchra*, needs to be monitored closely in the future.

If the current status “key biodiversity area” for the Pantaron range and the Kimangkil massif, including Mt. Sumagaya, leads to permanent protection combined with careful monitoring of the relevant protective measures, at least the four endemic *Nepenthes* species we described in this article would be assured preservation in the long term.

Acknowledgments

The authors are deeply grateful to Oda Beltran and Nelson Agpawa from the Bukidnon Resource Management Foundation for making the exploration of the Pantaron range and Mt. Sumagaya possible. The staff of the BMB and the DENR is acknowledged for assistance with the permits. The authors thank all guides from the Talaandig and Higaonon tribes for their hospitality and company during the expedition. Cristopher “Tupi” Salem is acknowledged for his assistance with the organization on site.

Author Contributions

T.G., F.C., A.W., D.M., T.G. and M.M. conducted the field research under the guidance of V.A. T.G., A.W., D.M. and T.G. wrote the manuscript. F.C. prepared and maintained the herbarium specimens. M.M., F.S.M. and V.A. reviewed and improved the manuscript. F.S.M. and D.M. made the botanical illustrations.

Conflicts of Interest

The authors declare no conflict of interest.

References

1. McPherson, S. *New Nepenthes*, 1st ed.; Redfern Natural History: Poole, UK, 2012.
2. Kurata, S. *Nepenthes peltata* (Nepenthaceae), a new species of pitcher plant from the Philippines. *J. Insectivorous Plant Soc.* **2008**, *59*, 12–17.
3. Cheek, M. *Nepenthes robcantleyi* sp. nov. (Nepenthaceae) from Mindanao, Philippines. *Nord. J. Bot.* **2013**, *29*, 677–681.
4. Micheler, M.; Gronemeyer, T.; Wistuba, A.; Marwinski, M.; Suarez, W.; Amoroso, V. *Nepenthes viridis*, eine neue *Nepenthes*-Art von der Insel Dinagat, Philippinen. *Das Taublatt.* **2012**, *76*, 4–21.
5. Cheek, M.; Jebb, M. Recircumscription of the *Nepenthes alata* group (Caryophyllales: Nepenthaceae), in the Philippines, with four new species. *Eur. J. Taxon.* **2013**, *69*, 1–23.
6. Cheek, M.; Jebb, M. *Nepenthes ramos* (Nepenthaceae), a new species from Mindanao, Philippines. *Willdenowia* **2013**, *43*, 107–111.
7. Cheek, M.; Jebb, M. *Nepenthes ultra* (Nepenthaceae), a new species from Luzon, Philippines. *Blumea* **2013**, *58*, 241–244.
8. Cheek, M.; Jebb, M. Typification and redelimitation of *Nepenthes alata* with notes on the *Nepenthes alata* group, and *Nepenthes negros* sp. nov. from the Philippines. *Nord. J. Bot.* **2013**, *31*, 616–622.
9. Cheek, M.; Jebb, M. *Nepenthes samar* (Nepenthaceae), a new species from Samar, Philippines. *Blumea* **2013**, *58*, 82–84.
10. Cheek, M.; Jebb, M. Identification and typification of *Nepenthes blancoi*, with *Nepenthes abalata* sp. nov. from the western Visayas, Philippines. *Nord. J. Bot.* **2013**, *31*, 151–156.
11. Cheek, M.; Jebb, M. The *Nepenthes micramphora* (Nepenthaceae) group, with two new species from Mindanao, Philippines. *Phytotaxa* **2013**, *151*, 25–34.
12. Gronemeyer, T. Hochland *Nepenthes* auf Mindanao. *Das Taublatt.* **2009**, *63*, 18–39.
13. Gronemeyer, T.; Coritico, F.; Micheler, M.; Marwinski, M.; Acil, R.; Amoroso, V. *Nepenthes ceciliae*, a new pitcher plant species from Mount Kiamo. In *New Nepenthes*, 1st ed.; McPherson, S., Ed.; Redfern Natural History: Poole, UK, 2012; pp. 412–423.
14. Gronemeyer, T.; McPherson, S.; Coritico, F.; Micheler, M.; Marwinski, M.; Amoroso, V. *Nepenthes pulchra*, a new pitcher plant species from Mount Kiamo, Mindanao. In *The New Nepenthes*, 1st ed.; McPherson, S., Ed.; Redfern Natural History: Poole, UK, 2012; pp. 424–439.
15. Danser, B.H. The Nepenthaceae of the Netherlands Indies. *Bull. du Jardin. Botanique. de Buitenzorg Série III* **1928**, *9*, 249–438.
16. Macfarlane, J.M. *Nepenthes copelandii*. In *Das Pflanzenreich*, III, IV; Engler, A., Ed.; Verlag Wilhelm Engelmann: Leipzig, Germany, 1912; Heft 36, 1–91.
17. Schmid-Hollinger, R. *Nepenthes macfarlanei*: Ameisennest in alter Kanne. *Das Taublatt.* **2010**, *66*, 39–42.
18. IUCN. The IUCN Red List of Threatened Species. Version 2013.2. Available online: http://www.iucnredlist.org (accessed on 21 November 2013).

Appendix

Figure A1. *N. pantaronensis*: photos of a lower (**left**) and upper pitcher (**right**) taken at the type locality.

Figure A2. *N. cornuta*: photos of a lower (**left**) and upper pitcher (**right**) (with *Crematogaster* sp.) taken at the type locality.

Figure A3. *N. talaandig*: Photos of a lower (**left**) and upper pitcher (**right**) taken at the type locality.

Figure A4. *N. amabilis*: photos of two pitchers taken at the type locality. In the **left** panel, the distinct leaf shape can be seen.

4

Regulation by Light of Chemotaxis to Nitrite during the Sexual Life Cycle in *Chlamydomonas reinhardtii*

Elena Ermilova * and Zhanneta Zalutskaya

Laboratory of Adaptation in Microorganisms, Saint-Petersburg State University, Oranienbaumskoe Schosse 2, Stary Peterhof, Saint-Petersburg 198504, Russia; E-Mail: z.zalutskaya@bio.spbu.ru

* Author to whom correspondence should be addressed; E-Mail: Ermilova@bio.spbu.ru

Abstract: Nitrite plays an important role in the nitrogen metabolism of most cells, including *Chlamydomonas reinhardtii*. We have shown that vegetative cells of *C. reinhardtii* are attracted by nitrite. The *Nia1nit2* mutant with defects in genes encoding the nitrate reductase and regulatory protein NIT2 respectively was found to exhibit normal chemotaxis to nitrite. The data suggest that chemotaxis events appear to be specific and independent of those involved in nitrate assimilation. Unlike vegetative cells and noncompetent pregametes, mature gametes did not show chemotaxis to nitrite. Just like gamete formation, the change in chemotaxis mode is controlled by the sequential action of two environmental cues, removal of nitrogen from the medium and light. Comparative analysis of wild-type and RNAi strains with reduced level of phototropin has indicated that switch-off of chemotaxis towards nitrite is dependent on phototropin. The studies revealed individual elements of the phototropin-dependent signal transduction pathway involved in the blue-light-controlled change in chemotaxis mode of *C. reinhardtii* during gamete formation: three protein kinases, one operating against signal flux and two that promote signal transduction. We have proposed a working model for the signaling pathway by which blue light controls chemotaxis towards attractants, which are nitrogen sources, during pregamete-to-gamete conversion of *C. reinhardtii*.

Keywords: chemotaxis; *Chlamydomonas*; phototropin; protein kinases

1. Introduction

Algae use light not only as a source of energy for photosynthesis but also as a source of environmental information to regulate a variety of processes, including pigment biosynthesis, chloroplast movement, circadian rhythms, transport activity, cell cycle, motile behaviour and development [1–9]. One of these phenomena, the differentiation of vegetative cells into sexually mature cells has been widely studied in *Chlamydomonas* and may proceed in two steps [10]. The depletion of a usable nitrogen source induces sexually incompetent gametes, named pregametes in the dark [11]. In a second step, these pregametes may be converted into mating-competent gametes by light irradiation [12]. The flavin-based blue light receptor phototropin (Phot) in *C. reinhardtii* controls the progression of the sexual life cycle of this green alga [6]. Phot in *C. reinhardtii* also functions in transcriptional regulation [13] and in the control of phototaxis that desensitizes the eyespot when blue light intensities increase [14]. The interacting proteins of Phot involved in downstream signaling are not known in this alga. Analysis of the signaling pathway by which light controls the conversion of pregametes to gametes using pharmacological compounds suggests the participation of protein kinases and cAMP in this signaling cascade [15]. Thus, a protein kinase C (PKC)-like activity operates against signal flux while a protein tyrosine kinase (PTK) stimulates signal flux. The intracellular accumulation of cAMP, either by feeding or by the use of phosphodiesterase inhibitors, inhibited the light-dependent conversion of pregametes to gametes suggesting a role of this cyclic nucleotide as a negative regulator.

In *C. reinhardtii*, vegetative cells, pregametes and mature gametes are motile. Vegetative cells and mating incompetent pregametes exhibit chemotaxis towards ammonium and nitrate [2,16,17]. Chemotaxis to these ions enables vegetative cells to orient themselves towards the nitrogen sources. Gametic differentiation has been shown to be associated with changes not only in the cells' biochemistry and subcellular morphology [10] but also in chemotactic behavior [2]. Mature gametes are not attracted by ammonium and nitrate [2,17]. Like the development of mating competence, the loss of chemotaxis requires the action of two environmental signals: lack of a nitrogen source and blue light. We have shown that the photoreceptor, which mediates the loss of chemotaxis to these attractants during gametogenesis, is Phot [17,18]. Ammonium, nitrate and another nitrogen source, nitrite, also play an important role in sexual differentiation of *C. reinhardtii* [19]. Vegetative cells only in the absence of these nitrogen sources initiate the program of sexual differentiation [12,20]. We suggest that loss of chemotaxis towards nitrogen sources by gametes may be one of prerequisites to their successful mating at locations accessible to light. However, chemotactic behaviour to nitrite has not been characterized. Therefore, the goals of this study were to (1) test chemotaxis to nitrite during sexual life cycle; (2) analyze the involvement of Phot in control of chemotaxis to nitrite; and (3) reveal individual elements of the signal transduction pathway involved in the blue-light-controlled change in chemotaxis to nitrite and nitrate. For the analyses, we used RNAi strains with reduced levels of phototropin, chemotaxis assay and pharmacological compounds known to target defined signaling steps. Our results demonstrate that vegetative cells and pregametes of *C. reinhardtii* are attracted by nitrite and signals generated by activation of phototropin feeds into a linear signal chain that switches off chemotaxis towards ammonium, nitrate and nitrite.

2. Results

2.1. Nitrite as Chemoattractant

Earlier experiments showed that *C. reinhardtii* vegetative cells were attracted to nitrate [17]. We wondered whether *C. reinhardtii* exhibits chemotaxis to nitrite. Strain CC-124 (*nia1*, *nit2*) accumulated in capillaries containing KNO_2 (Figure 1). The peak accumulation of the cells was observed when the capillary contained 5 mM of KNO_2. The threshold concentration (the lowest concentration needed to elicit an observable response) was approximately 0.5 mM. At the same time *C. reinhardtii* did not react chemotactically to KCl and KH_2PO_4. It means that nitrite but not potassium acts as an attractant for chemotactic response.

Figure 1. Reaction of vegetative cells of strain CC-124 to nitrite (KNO_2) (■), nitrate (KNO_3) (■), KCl (■) and KH_2PO_4 (■). Data are means ± SD of 5 independent experiments.

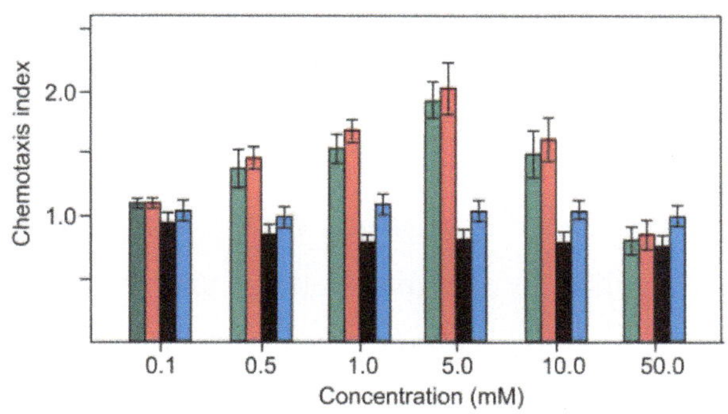

We were surprised that the CC-124 strain, which does not express many of the genes required for nitrate/nitrite assimilation (*Nia1*, *Nrt2.1*, *Nrt2.2*, *Nii1*, *Nar2* and *Nar1.1*) [21], showed normal chemotactic responses both to nitrite and nitrate. These data suggest that nitrate assimilation is unlikely to play a role in mediating the chemotaxis to nitrate/nitrite. These results were supported by the competition tests when both media, in the bath and in the capillary, contained the second attractant of the same concentration (5 mM KNO_2 or 5 mM KNO_3) (Figure 2). Nitrate completely prevented chemotaxis to nitrite. Additionally chemotaxis to nitrate was also affected by nitrite. These data indicate that the same components may be shared in the control of chemotaxis to both anions nitrate and nitrite.

2.2. Role of Light in Change of Chemotaxis to Nitrite during Gametogenesis

The ability of unicellular organisms to differentiate in response to nutrient availability is essential to their survival in a changing environment [22]. The initial step in the sexual life cycle of *C. reinhardtii* is gametogenesis. During gametogenesis, vegetative cells are transformed into mating-competent gametes. Effective formation of mating ability depends on removal of nitrogen from the medium and on the presence of light [10].

We wondered next if pregametes and gametes react chemotactically to nitrite. Figure 3 shows that pregametes are attracted to nitrite. Even though acetate as source of carbon and energy was present in the nitrogen-free medium, cells were not capable to complete the program of differentiation and were not competent for mating. The results demonstrate that nitrogen deprivation is not enough to change the mode of chemotactic behavior to nitrite. Additionally the data indicate that light is not essential for chemotaxis towards nitrite.

Figure 2. Effect of nitrate and nitrite to the chemotactic responses of CC-124 vegetative cells in different media. Vegetative cells in TAP-N (■); TAP-N supplied with nitrate (5 mM) (■), and TAP-N supplemented with nitrite (5 mM) (■) were assayed for chemotaxis to 5 mM nitrite or 5 mM nitrate as indicated. Data are means ± SD of 3 independent experiments.

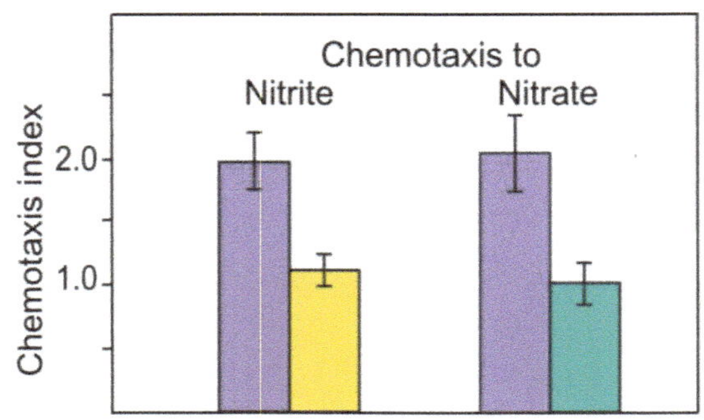

Figure 3. Chemotactic responses of pregametes (●) and gametes obtained from pregametes (○) towards 5 mM nitrite. Vegetative cells of *mt*⁻ were resuspended in TAP-N at time 0. The percentage of mating (Δ) is given for a comparison. At the times indicated, cells were mixed with an excess of sexually competent mating partner, and incubated in the dark. After 1h, the cells were fixed by the addition of glutaraldehyde and relative mating was determined as described in the Experimental Section. Data are means ± SD of triplicate determination from a representative experiment.

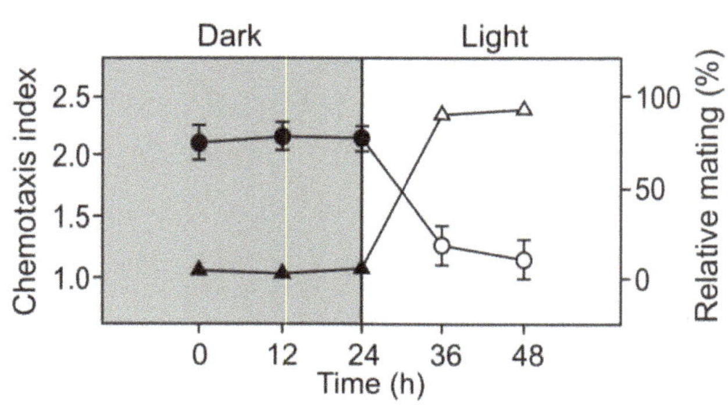

Upon incubation in the light, pregametes are converted into mating-competent gametes [23]. We next tested whether these two environmental signals might also act on change of chemotaxis to nitrite sequentially. For these tests, pregametes, generated from vegetative cells by incubation in the dark without a utilizable nitrogen source, were exposed to light. As shown in Figure 3, light induced the switch-off of to chemotaxis in cells. Therefore, chemotactic behavior to nitrite may be differentiated according to the development of competence ability. We assume that chemotaxis towards nitrite, nitrate and ammonium in gametes can be blocked by using identical light-dependent signaling pathways.

2.3. *Effect of Reduced Phototropin Levels on the Kinetics of Changes in Chemotaxis to Nitrite*

Absorption of blue light by phototropin results in the activation of signaling pathways that control the switch-off of chemotaxis to ammonium and nitrate during gametogenesis [17,18]. RNA-interference strains with reduced levels of the blue-light receptor phototropin showed an attenuated inactivation of chemotaxis to these attractants. To determine if phototropin was involved in changes in chemotactic activity to nitrite, we tested behavior of two strains that exhibited reduced phototropin levels. The RNAi20 strain, having severely reduced phototropin levels [18], exhibited a switching-off of chemotaxis at a fluence rate of 60 μmol m^{-2} s^{-1}, however, with a delay of about 2 h when compared to the parental wild-type strain CC-124 (Figure 4).

Figure 4. Chemotaxis of strains CC-124 (**A**); RNAi20 (**B**) and RNAi30 (**C**) to nitrite. Pregametes, generated from synchronously growing cells by incubation in TAP-N medium in the dark for 18 h, at time 0 were shifted into white light at the fluence rates of 60 μmol m^{-2} s^{-1} (○) and of 30 μmol m^{-2} s^{-1} (●). Relative amounts of phototropin in different transformants harbouring RNAi constructs as compared to the strain CC-124 (**D**).

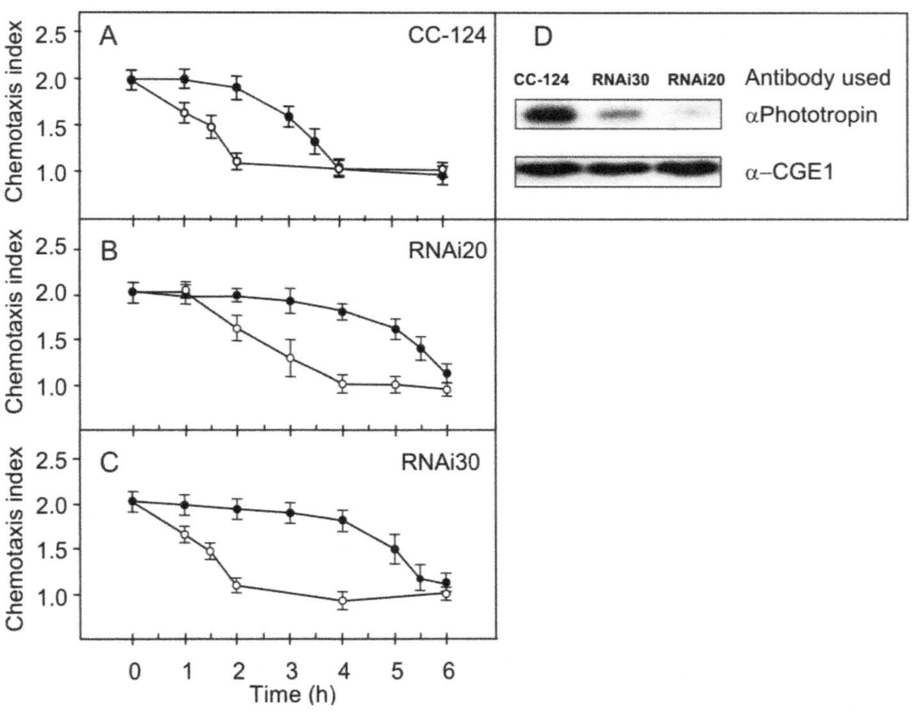

At a fluence rate of 30 μmol m^{-2} s^{-1}, the wild-type strain showed a distinct decrease in the chemotaxis index after 3 h in nitrogen-free medium; with the RNAi20 strain this was observed only after 5 h. At a fluence rate of 60 μmol m^{-2} s^{-1}, the strain RNAi30 exhibited kinetics of loss of chemotaxis to nitrite that was essentially the same as those observed for the wild-type strain (Figure 4C). At a fluence rate of 30 μmol m^{-2} s^{-1}, however, this strain showed a reduced loss of chemotaxis when compared to parental strain CC-124. As in the case of chemotactic behavior to ammonium, the kinetics observed for strain RNAi30 were faster than those for strain RNAi20 and thus were intermediate between those of the wild-type strain and RNAi20. As shown earlier [18], both strains exhibited reduced levels of phototropin, although the Phot levels of the RNAi20 strain were distinctly lower than those of RNAi30. Thus, the kinetics of the loss of chemotaxis to nitrite approximately correlates with the level of Phot protein detected by specific antibodies. We conclude that phototropin is the photoreceptor used by blue light to control the switch-off of chemotaxis in response to nitrite during gametogenesis.

2.4. Effects of Pharmacological Compounds on the Kinetics of Changes in Chemotaxis to Nitrite in the Strain RNAi20

Analysis of the signaling pathway by which light controls the switch-off of chemotaxis to ammonium suggests the participation of protein kinases and cAMP in this signaling cascade [24]. To determine whether the switch-off of chemotaxis to ammonium and nitrite is controlled by identical key phototropin-controlled signaling components, we analyzed chelerythrine, staurosporine, papaverine, IBMX (3-isobutyl-1-methylxanthine) and db-cAMP, compounds shown to activate loss of chemotaxis to ammonium [24], for their ability to compensate for the slowed-down loss of chemotaxis seen in strain RNAi20.

Staurosporine at low concentrations is a specific inhibitor of PKC, as is chelerythrine. After addition of staurosporine or chelerythrine, the kinetics of loss of chemotaxis to nitrite in the light observed for strain RNAi20 were faster than those seen without treatment (Figure 5A).

Figure 5. Effects of various pharmacological compounds on the light-induced loss of chemotaxis towards nitrite in pregametes of phototropin-reduced strain RNAi20. (**A**) Effect of staurosporine (♦), chelerythrine (●) or SC-9 (▲) on the loss of chemotaxis in pregametes of RNAi20 that were shifted into white light at time 0; (**B**) Treatment with papaverine (●), IBMX (♦) or db-cAMP (■) on the loss of chemotaxis in pregametes of RNAi20 that were shifted into white light at time 0. Compounds were added to pregametes 10 min prior to start of illumination. The concentrations used were 1 μM for chelerythrine, 50 μM for papaverine, 50 μM for IBMX, 20 nM for staurosporine and and 200 μM for SC-9. Non-treated cells (●) were used as control. The fluence rate was 60 μmol m^{-2} s^{-1}.

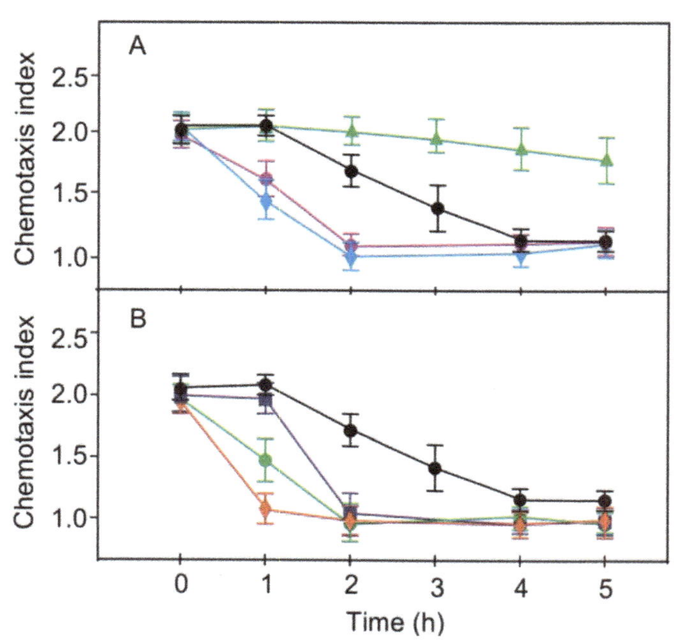

Both inhibitors at the concentrations used did not influence cellular motility (data not shown). Treatment of pregametes with the naphthalene sulfonamide SC-9, a PKC activator, specifically inhibited loss of chemotaxis to nitrite in the light (Figure 5A). The combined data suggest that a PKC-like component may be involved in signaling that results in the loss of chemotaxis to nitrite.

Treatment of pregametes with two phosphodiesterase inhibitors, papaverine and IBMX, assumed to result in elevated levels of cAMP, switched off chemotaxis towards nitrite more rapid than in untreated cultures and similar to those observed for the parental strain (Figure 5B). An increase in cAMP levels as a result of phosphodiesterase inhibition by papaverine and IBMX is known to activate protein kinase A (PKA) [15]. An involvement of cAMP in the control of chemotaxis is supported by the observation that the feeding of db-cAMP caused a faster switching off of chemotaxis than that seen without treatment (Figures 4A and 5B). These results suggest that a PKC and a PKA, targeted by staurosporine and compounds that elevate cAMP levels, respectively, may be components of the same signaling pathway that is involved in the control of the switch-off of chemotaxis to nitrite via phototropin-perceived blue light. The same results were obtained also for effects of these pharmacological compounds on the kinetics of changes in chemotaxis to nitrate (Figure 6).

Figure 6. Effects of various pharmacological compounds on the light-induced loss of chemotaxis towards nitrate in pregametes of phototropin-reduced strain RNAi20. (**A**) Effect of staurosporine (♦), chelerythrine (●) or SC-9 (▲) on the loss of chemotaxis in pregametes of RNAi20 that were shifted into white light at time 0.; (**B**) Treatment with papaverine (●), IBMX (♦) or db-cAMP (■) on the loss of chemotaxis in pregametes of RNAi20 that were shifted into white light at time 0. Compounds were added to pregametes 10 min prior to start of illumination. The concentrations used were 1 μM for chelerythrine, 50 μM for papaverine, 50 μM for IBMX, 20 nM for staurosporine and and 200 μM for SC-9. Non-treated cells (●) were used as control. The fluence rate was 60 μmol m^{-2} s^{-1}.

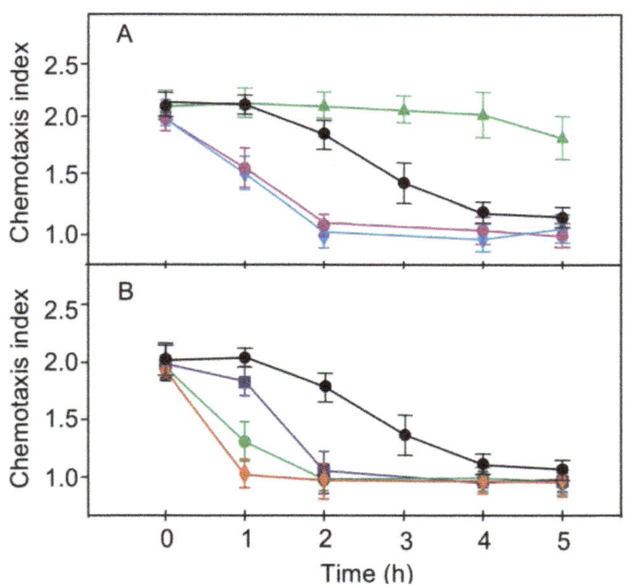

We next tested genistein, a protein Tyr kinase inhibitor that negatively regulated the loss of chemotaxis to ammonium in the light [24]. Addition of genistein to pregametes of RNAi20 completely prevented the light-induced switch-off of chemotaxis to nitrite (Figure 7). We assume that phototropin and protein kinases are components of a linear signaling pathway that is involved in the control of the switch-off of chemotaxis.

Figure 7. Treatment of pregametes of strain RNAi20 in the light with combinations of pharmacological compounds that either activate or inhibit the loss of chemotaxis in pregametes towards nitrite (■) and nitrate (■). Compounds were added to pregametes 10 min prior to start of illumination at the fluence rates of 60 μmol m^{-2} s^{-1} for 2 h. The genistein concentration was 200 μM. The concentrations used for other compounds were the same as those given in the legend of Figure 5. Non-treated cells were used as control.

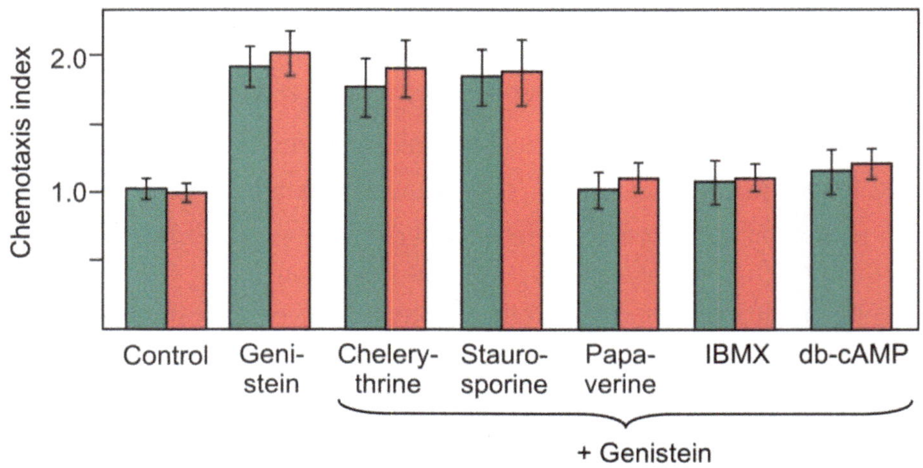

In order to gain information on the sequence of the protein kinases in this signaling pathway, we employed a combined application of the activators and inhibitors defined above. Since chelerythrine and staurosporine activated loss of chemotaxis in RNAi20, we used these compounds in combination with genistein that prevented loss of chemotaxis to nitrite in the light (Figure 7). Genistein completely blocked the activation seen with PKC inhibitors chelerythrine and staurosporine. This implies that the target of genistein in the signaling pathway is located downstream from the PKC. Genistein, however, did not prevent the switch-off of chemotaxis to nitrite promoted by addition of papaverine, IBMX or db-cAMP in pregametes. This indicates that a cAMP-targeted PKA operates downstream from the PTK inactivated by genistein. The effects were registered only when the assays were performed within 6 h after removal of the nitrogen source from the medium. No effect of the inhibitors on chemotaxis to nitrite in vegetative cells was observed. The effects monitored in pregametes of RNAi20 thus appear not to be caused by interference of these compounds with the signaling mechanism of chemotaxis (data not shown).

We were also interested in determining whether this signaling pathway is identical to that for the control of blue-light-mediated loss of chemotaxis to nitrate (Figure 7). Taken together these studies suggest that (i) a PKC-like component with a role as negative regulator is involved in the control of chemotaxis to nitrite/nitrate; (ii) a PTK transduces the signal downstream from the PKC-like component; (iii) a PKA transduces the signal downstream from the PTK.

3. Discussion

Ammonium is the preferred nitrogen source and chemoattractant for the unicellular green alga *Chlamydomonas reinhardtii*. In the absence of ammonium, vegetative cells can utilize nitrate and nitrite. In *C. reinhardtii* nitrate is also a signalling molecule that controls chemotactic behaviour [17]. Here we have shown that vegetative cells of *C. reinhardtii* sense nitrite and are attracted by this anion as well (Figure 1). Potassium did not act as an attractant for motile cells.

In an effort to identify if components of nitrate transport/assimilation are required for chemotaxis to nitrate/nitrite, the strain CC-124 with defects in nitrate reductase and NIT2 protein was tested. It was found to exhibit normal chemotaxis to nitrite and nitrate (Figure 1). In *Chlamydomonas*, the NIT2 transcription factor, which contains GAF and Leu zipper domains, is required for nitrate regulation of gene expression [25]. The data suggest that chemotaxis events appear to be specific and independent of those involved in nitrate assimilation. This conclusion is supported by the facts, that (1) nitrate competitively inhibits the nitrite chemotactic activity and vice versa (Figure 2) and (2) pregametes, cells generated in the dark, also demonstrated normal chemotactic activity (Figure 3). More recently, in *Arabidopsis*, nitrate transporter NRT1.1 (CHL1) was identified as a nitrate sensor at high nitrate concentrations [26]. It was demonstrated that NO_3^- transport activity is not required for the sensing function of CHL1. There exists a single copy of NRT1 in the *C. reinhardtii* genome [27]; whether or not this NRT1 protein has a transceptor function as proposed in plants or any other function remains to be addressed in *C. reinhardtii*. Nitrate transporters NRT2.1 and NRT2.2 were shown to be key elements for sensing at low nitrate concentrations to activate gene expression for nitrate assimilation [28]. These transporters dependent on NIT2 activation are however dispensable for nitrate/nitrite sensing in chemotaxis.

During sexual differentiation, *C. reinhardtii* changes its chemotactic behavior not only to ammonium and nitrate but also to nitrite (Figure 3). Unlike vegetative cells and noncompetent pregametes, mature gametes did not show chemotaxis to nitrite. This change in chemotactic behavior requires the sequential action of two environmental cues: nitrogen deprivation and light. Thus, cells starved for nitrogen in the dark (pregametes) still exhibit chemotaxis (Figure 3). Irradiation of pregametes causes both a loss of chemotaxis to nitrite and a gain of mating competence. We conclude that chemotaxis of *Chlamydomonas* towards inorganic nitrogen sources is a property of vegetative cells and pre-gametes but not of gametic cells.

We show here that a reduction in phototropin levels led to a delay in the light-induced switch-off of chemotaxis not only to ammonium [18] and nitrate [17] but also to nitrite (Figure 4). By using a strain strongly reduced in phototropin (RNAi20), we discovered a compensation of this defect upon application of compounds that appeared to activate signal flux: the PKC inhibitors (staurosporine or chelerythrine) or the PKA activators (phosphodiesterase inhibitors or db-cAMP) (Figures 5 and 6). These protein kinases are thus proposed to be downstream components of the phototropin-activated signaling pathway that controls the switch-off of chemotaxis to nitrite and to nitrate. In this pathway, a PKC-like component is proposed to act as negative regulator that attenuates transduction of the light signal. Since genistein inhibited the switch-off of chemotaxis to nitrite/nitrate induced by staurosporine or chelerythrine, a PTK appears to be positioned downstream from the PKC-like activity (Figure 7). A PKA may be positioned downstream from this PTK since conditions that resulted in a rise of intracellular cAMP levels, and thus in an activation of PKA, could be shown to promote the switch-off of chemotaxis even when PTK was inhibited by genistein (Figure 7). Thus, this signaling pathway is identical to that proposed for the control of blue-light-mediated loss of chemotaxis to ammonium [24].

The combined data have been summarized in a model (Figure 8). We assume that at least three protein kinases are components of a linear phototropin-dependent signalling chain controlling the switch-off of chemotaxis. In some respects this blue light signaling pathway shares features with the one required for the formation of sexual competence [29,30]. Both processes employ the same photoreceptor and at least two components, PKC and PKT, of the downstream signalling pathway. Why does *Chlamydomonas* exhibit this intriguing coupling of phototropin-dependent light control for disappearance of chemotaxis towards the nitrogen sources and for gamete formation? It was shown that ammonium and nitrate/nitrite interfered with gametic differentiation [2,10]. Therefore, in gametes, loss of chemotaxis towards nitrogen sources, which are known to play a key role in the repression of the gametic differentiation, may be viewed as one of the cellular adaptations to changing environmental conditions. Seen from this point of view, the coordinated but opposite regulation of chemotaxis towards ammonium, nitrate and nitrite and gamete formation may be of advantage for the alga.

Figure 8. A model for the signaling pathway by which blue light controls chemotaxis towards nitrogen sources in *C. reinhardtii* during pregamete-to-gamete conversion. Black arrows indicate a stimulating effect of pharmacological compounds, T-lines indicate an inhibiting effect. Green arrows indicate stimulation of switch-off of chemotaxis, red T-line indicates exhibiting of switch-off of chemotaxis.

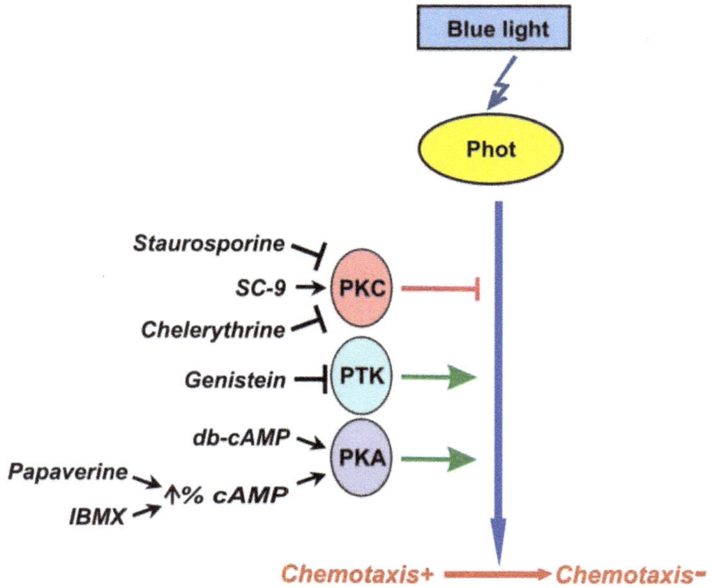

4. Experimental

4.1. Strains and Culture Conditions

Chlamydomonas reinhardtii strain CC-124 (*nia1*, *nit2*, *mt*$^-$) have been described elsewhere [31]. Another wild-type strain tested was CC-620 (*mt*$^+$), obtained from Dr. S. Purton, University College London, GB. Strains RNAi20 and RNAi30 with reduced levels of phototropin [30] were obtained from

Dr. C. Beck. Cells were grown at 22 °C under a 12-h light/12-h dark regime in Tris-acetate-phosphate (TAP) medium [32] or in acetate-free TAP (TMP) medium.

4.2. Reagents

All pharmacological compounds were prepared as stocks and stored at −70 °C in the dark. Papaverine was dissolved in water. Other compounds were dissolved in DMSO when an organic solvent was required, or in TAP-N. Papaverine, 3-isobutyl-1-methylxanthine, chelerythrine, genistein, staurosporine, SC-9, and db-cAMP were purchased from Sigma. The components were tested at the concentrations that previously have been shown to be effective in mammalian or plant cell systems (Table 1) and affected pregamete-to-gamete conversion in *C. reinhardtii* [15]. All reagents used did not influence the cellular swimming speed (data not shown).

Table 1. Pharmacological compounds that affect light control of chemotaxis.

Compound	Mode of action	Effective concentration	Reference
Papaverine	Phosphodiesterase inhibitor	50 μM	[24,33]
3-isobutyl-1-methylxanthine	Phosphodiesterase inhibitor	50 μM	[15,24]
db-cAMP	PKA activator	15 mM	[34]
Genistein	PTK inhibitor	200 μM	[15,24,35]
Staurosporine	PKA inhibitor	20 nM	[15,36,37]
Chelerythrine	PKC inhibitor	1 μM	[24,38]
SC-9	PKC activator	200 μM	[15,39]

4.3. Generation of Pregametes and Gametes

For the generation of pregametes, liquid cultures of synchronously growing cells at the beginning of the light period were washed twice with nitrogen-free medium (TAP-N or TMP-N), resuspended in the same medium at a density of 0.1–1.0 × 10^6 cells/mL and incubated in the dark for 24 h. Gametes were obtained from pregametes by exposing them to light (fluence rate 60 μmol m^{-2} s^{-1}) for 2 h. The light source used was a 500 W lamp (LOMO, Saint-Petersburg, Russia) mounted in a slide projector.

4.4. Determination of Mating Competence

The percentage of gametes was assayed by mixing the cells to be tested with a threefold excess of gametes of opposite mating type. Mating was allowed to proceed in the dark for one hour and stopped by adding glutaraldehyde (final concentration 0.5%). The number of biflagellate cells and quadriflagellate cells in the mating mixture was recorded microscopically. The percentage of mating competent gametes was calculated as described [11].

4.5. Chemotaxis Assay

Chemotactic responses were tested by counting the number of cells that in darkness swam into rectangular capillaries (260 μm × 450 μm) filled with 3 μL of medium containing either KNO_2 or the tested source. This number was compared to the number of cells entering capillaries filled with nitrogen-free medium [40]. Capillaries were closed at one end with Parafilm. The other end was submerged in the cell suspension for 10 min at 22 °C. In the competition tests KNO_2 or KNO_3 were added to the cell suspension so that the gradient of nitrate or nitrite was generated in the presence of second attractant.

The chemotaxis index (CI) was calculated using the following ratio Equation (1):

$$\frac{\text{number of cells entering capillaries filled with medium containing the tested source}}{\text{number of cells entering capillaries filled with nitrogen-free medium}} \quad (1)$$

Switch-off of chemotaxis was recorded as 1.0. Data are means ± SD of triplicate determinations from a representative experiment from at least three different ones.

4.6. Protein Isolation, SDS-PAGE and Immunoblot Analysis

Chlamydomonas cells (2–4 × 10^6 cells mL^{-1} in 100 mL) were collected by centrifugation (3000 g, 5 min) and resuspended in 0.1 M DTT, 0.1 M Na_2CO_3. Then, 0.66 vol of 5% SDS, 30% sucrose were added. Homogenization of the suspensions was achieved by rapid shaking at room temperature for 20 min. The protein concentration was determined by staining with amido black, using BSA as a standard [41]. After separation of the proteins by SDS-PAGE on a 12% polyacrylamide gel [42], they were transferred to nitrocellulose membranes (Carl Roth, Karlsruhe, Germany) by a semidry blotting (Trans-blot SD BioRad). Blots were blocked in 5% milk in Tris-buffered saline solution with 0.1% Tween 20 prior to 1 h of incubation in the presence of primary antibodies. For this assay, an antibody directed against Chlamydomonas phototropin [6] was used. For a loading control, an antibody that reacts with the co-chaperone CGE1 [43] was employed. The dilutions of the primary antibodies used were as follows: 1:5,000 anti-Phot and 1:6,000 anti-α-CGE1. A 1:10,000 dilution of horseradish peroxidase-conjugated anti-rabbit serum (Sigma) was used as a secondary antibody. The peroxidase activity was detected by an enhanced chemiluminescence assay (Roche).

5. Conclusions

We have shown that vegetative cells of *Chlamydomonas reinhardtii* are attracted by nitrite. The data suggest that chemotaxis events appear to be specific and independent of those involved in nitrate/nitrite assimilation. Chemotaxis towards inorganic nitrogen sources is a property of vegetative cells and pregametes but not of gametic cells. Absorption of blue light by phototropin results in the activation of a signaling pathway that controls the switch-off of chemotaxis. These studies also revealed individual elements of the signal transduction pathway involved in the blue-light-controlled change in the chemotaxis mode of *C. reinhardtii* during gamete formation: three protein kinases, one operating against signal flux and two that promote signal transduction.

Acknowledgments

We thank Christoph Beck for providing antibodies against phototropin and Michael Schroda for providing antibodies against CGE1. This work was supported by grants from the Russian Foundation for Basic Research (No. 13-04-00087-a), and the Ministry of Education and Science of the Russian Federation (No. 8093).

Author Contributions

Zhanneta Zalutskaya carried out chemotaxis assays and immunoblot analysis. Elena Ermilova supervised all work, carried out data analysis and wrote the paper.

Conflicts of Interest

The authors declare no conflict of interest.

References

1. Byrne, T.E.; Wells, M.R.; Johnson, C.H. Circadian rhythms of chemotaxis and of methylammonium uptake in by *Chlamydomonas reinhardtii*. *Plant Physiol*. **1992**, *98*, 879–886.
2. Ermilova, E.V.; Zalutskaya, Z.M.; Lapina, T.V.; Nikitin, M.M. Chemotactic behavior of *Chlamydomonas reinhardtii* is altered during gametogenesis. *Curr. Microbiol*. **2003**, *46*, 261–264.
3. Ermilova, E.V.; Zalutskaya, Z.M.; Nikitin, M.M.; Lapina, T.V.; Fernández, E. Regulation by light of ammonium transport systems in *Chlamydomonas reinhardtii*. *Plant Cell Environ*. **2010**, *33*, 1049–1056.
4. Furukawa, T.; Watanabe, M.; Shihira-Ishikawa, I. Blue and green light chloroplast migration in the centric diatom *Pleurosira laevis*. *Protoplasma* **1998**, *203*, 214–220.
5. Herman, C.A.; Im, C.-S. Light-regulated expression of the *gsa* gene encoding the chlorophyll biosynthetic enzyme glutamate 1-semialdehyde aminotransferase in carotenoid-deficient *Chlamydomonas reinhardtii* cells. *Plant Mol. Biol*. **1999**, *39*, 289–297
6. Huang, K.; Merkle, T.; Beck, C.F. Isolation and characterization of a *Chlamydomonas* gene that encodes a putative blue-light photoreceptor of the phototropin family. *Physiol Plant*. **2002**, *115*, 613–622.
7. Johnson, C.H.; Suzuki, L. Algae know the time of day: Circadian and photoperiodic programs. *J. Phycol*. **2001**, *37*, 933–942.
8. Oldenhof, H.; Bisová, K.; van den Ende, H.; Zachleder, V. Effect of red and blue light on the timing of cyclin-dependent kinase activity and the timing of cell division in *Chlamydomonas reinhardtii*. *Plant Physiol. Biochem*. **2004**, *42*, 341–348.
9. Hegemann, P. Algal sensory photoreceptors. *Annu. Rev. Plant Biol*. **2008**, *59*, 167–189.
10. Beck, C.; Haring, M. Gametic differentiation of *Chlamydomonas*. *Int. Rev. Cytol*. **1996**, *168*, 259–302.
11. Beck, C.F.; Acker, A. Gametic differentiation of *Chlamydomonas reinhardtii*. *Plant Physiol*. **1992**, *98*, 822–826.
12. Treier, U.; Fuchs, S.; Weber, M.; Wakarchuk, W.W.; Beck, C. Gametic differentiation in *Chlamydomonas reinhardtii*: Light dependence and gene expression patterns. *Arch. Microbiol*. **1989**, *152*, 572–577.
13. Im, C.-S.; Eberhard, S.; Huang, K.; Beck, C.F.; Grossman, A.R. Phototropin involvement in the expression of genes encoding chlorophyll and carotenoid biosynthesis enzymes and LHC apoproteins in *Chlamydomonas reinhardtii*. *Plant J*. **2006**, *48*, 1–16.

14. Trippens, J.; Greiner, A.; Schellwat, J.; Neukam, M.; Rottmann, T.; Lu, Y.; Kateriya, S.; Hegemann, P.; Kreimer, G. Phototropin influence on eyespot development and regulation of phototactic behavior in *Chlamydomonas reinhardtii*. *Plant Cell* **2012**, *24*, 4687–4702.
15. Pan, J.M.; Haring, M.A.; Beck, C.F. Dissection of the blue-light dependent signal transduction pathway involved in gametic differentiation of *Chlamydomonas reinhardtii*. *Plant Physiol*. **1996**, *115*, 1241–1249.
16. Ermilova, E.; Nikitin, M.; Fernandez, E. Chemotaxis to ammonium/methylammonium in *Chlamydomonas reinhardtii*: The role of transport systems for ammonium/methylammonium. *Planta* **2007**, *226*, 1323–1332.
17. Ermilova, E.V.; Zalutskaya, Z.M.; Lapina, T.V. Chemotaxis of *Chlamydomonas reinhardtii* to nitrate is changed during gametogenesis. *Protistology* **2009**, *6*, 79–84.
18. Ermilova, E.V.; Zalutskaya, Z.M.; Huang, K.; Beck, C.F. Ptototropin plays a crucial role in controlling changes in chemotaxis during the initial phase of the sexual life cycle in *Chlamydomonas*. *Planta* **2004**, *219*, 420–427.
19. Matsuda, Y.; Shimada, T.; Sakamoto, Y. Ammonium ions control gametic differentiation and dedifferentiation in *Chlamydomonas reinhardtii*. *Plant Cell Physiol*. **1992**, *33*, 909–914.
20. Pozuelo, M.; Merchán, F.; Macías, M.I.; Beck, C.F.; Galván, A.; Fernández, E. The negative effect of nitrate on gametogenesis is independent of nitrate assimilation in *Chlamydomonas reinhardtii*. *Planta* **2000**, *211*, 287–292.
21. Galvan, A.; Fernandez, E. Eukaryotic nitrate and nitrite transporters. *Cell Mol. Life Sci*. **2001**, *58*, 225–233.
22. Lengeler, K.B.; Davidson, R.C.; D'souza, C.; Toshiaki, H.; Shen, W.-C.; Wang, P.; Pan, X.; Waugh, M.; Heitman, J. Signal transduction cascades regulating fungal development and virulence. *Microbiol. Mol. Biol. Rev*. **2000**, *64*, 746–785.
23. Treier, U.; Beck, C.F. Changes in gene expression patterns during the sexual life cycle of *Chlamydomonas reinhardtii*. *Physiol. Plant*. **1991**, *83*, 633–639.
24. Ermilova, E.V.; Zalutskaya, Z.M.; Baibus, D.M.; Beck, C.F. Characterization of phototropin-controlled signaling components that regulate chemotaxis towards ammonium in *Chlamydomonas*. *Protistology* **2006**, *4*, 301–310.
25. Camargo, A.; Llamas, A.; Schnell, R.A.; Higuera, J.J.; Gonzáles-Ballester, D.; Lefevbre, P.A.; Fernández, E.; Galván, A. Nitrate signaling by the regulatory gene NIT2 in *Chlamydomonas*. *Plant Cell* **2007**, *19*, 3491–3503.
26. Ho, C.H.; Lin, C.H.; Hu, H.C.; Tsay, Y.F. CHL1 functions as a nitrate sensor in plants. *Cell* **2009**, *138*, 1184–1194.
27. Fernandez, E.; Llamas, A.; Galvan, A. Nitrogen Assimilation and Its Regulation. In *The Chlamydomonas Sourcebook*, 2nd ed.; Harris, S., Ed.; Elsevier Press: Oxford, UK, 2009; Volume II, pp. 69–113.
28. Llamas, A.; Igeño, M.I.; Galván, A.; Fernández, E. Nitrate signaling on the nitrate reductase gene promoter depends directly on the activity of the nitrate transport systems in *Chlamydomonas*. *Plant J*. **2002**, *30*, 261–271.

29. Pan, J.M.; Haring, M.A.; Beck, C.F. Characterization of blue light signal transduction chains that control development and maintenance of sexual competence in *Chlamydomonas reinhardtii*. *Plant Physiol.* **1997**, *115*, 1241–1249.

30. Huang, K.; Beck, C.F. Phototropin is the blue-light receptor that controls multiple steps in the sexual life cycle of the green alga *Chlamydomonas reinhardtii*. *Proc. Natl. Acad. Sci. USA* **2003**, *100*, 6269–6274.

31. Harris, E.H. *The Chlamydomonas Sourcebook*; Academic Press: San Diego, CA, USA, 1989; pp. 1–24.

32. Gorman, D.S.; Levine, R.P. Cytochrome and plastocyanin: Their sequences in the photosynthetic electron transport chain of *Chlamydomonas reinhardtii*. *Proc. Natl. Acad. Sci. USA* **1965**, *54*, 1665–1669.

33. Pasquale, S.M.; Goodenough, U.W. Cyclic AMP functions as a primary sexual signal in gametes of *Chlamydomonas reinhardtii*. *J. Cell Biol.* **1987**, *105*, 2279–2292.

34. Lando, M.; Abemayor, E.; Verity, M.A.; Sidell, N. Modulation of intracellular cyclic adenosine monophosphate levels and the differentiation response of human neuroblastoma cells. *Cancer Res.* **1990**, *50*, 722–777.

35. Bowler, C.; Yamagata, H.; Neuhaus, G.; Chua, N.H. Phytochrome mediated signal transduction pathways are regulated by reciprocal control mechanisms. *Genes Dev.* **1994**, *8*, 2188–2202.

36. Zhang, Y.; Snell, W.J. Flagellar adhesion-dependent regulation of *Chlamydomonas* adenylyl cyclase *in vitro*: A possible role for protein kinases in sexual signaling. *J. Cell Biol.* **1994**, *125*, 617–624.

37. Meggio, F.; Donella Deana, A.; Ruzzene, M.; Brunati, A.M.; Cesaro, L.; Guerra, B.; Meyer, T.; Mett, H.; Fabro, D.; Furet, P.; *et al.* Different susceptibility of protein kinases to staurosporine inhibition. *Eur. J. Biochem.* **1995**, *234*, 317–322.

38. Herbert, J.M.; Augereau, J.M.; Gleye, J.; Maffrand, J.P. Chelerythrine is a potent and specific inhibitor of protein kinase C. *Biochem. Biophys. Res. Commun.* **1990**, *172*, 993–999.

39. Ito, M.; Tanaka, T.; Inagaki, M.; Nakanishi, K.; Hidaka, H.N. (6-Phenylhexyl)-5-chloro-1-naphthalenesulfonamide, a novel activator of protein kinase C. *Biochemistry* **1986**, *25*, 4179–4184.

40. Ermilova, E.V.; Zalutskaya, Z.M.; Munnik, T.; van den Ende, H.; Gromov, B.V. Calcium in the control of chemotaxis in *Chlamydomonas*. *Biol. Bratislava* **1998**, *53*, 577–581.

41. Popov, N.; Schmitt, S.; Matthices, H. Eine störungsfreie Mikromethode zur Bestimmung des Proteingahalts in Gewebshomogenaten. *Acta Biol. Germ.* **1975**, *34*, 1441–1446.

42. Laemmli, U.K. Cleavage of structural proteins during the assembly of the head of bacteriophage T4. *Nature* **1970**, *227*, 680–685.

43. Schroda, M.; Vallon, O.; Whitelegge, J.P.; Beck, C.F.; Wollman, F.-A. The chloroplastic GrpE homolog of *Chlamydomonas*. Two isoforms generated by differential splicing. *Plant Cell* **2001**, *13*, 2823–2839.

5

Structural Diversity and Function of Xyloglucan Sidechain Substituents

Alex Schultink [1], Lifeng Liu [2], Lei Zhu [2] and Markus Pauly [1,2,*]

[1] Department of Plant and Microbial Biology, University of California, Berkeley, CA 94720, USA; E-Mail: aschult@berkeley.edu

[2] Energy Biosciences Institute, University of California, Berkeley, CA 94720, USA; E-Mails: lifeng.liu@berkeley.edu (L.L.); lei.zhu@berkeley.edu (L.Z.)

* Author to whom correspondence should be addressed; E-Mail: mpauly69@berkeley.edu

External Editor: Philip Harris

Abstract: Xyloglucan (XyG) is a hemicellulose found in the cell walls of all land plants including early-divergent groups such as liverworts, hornworts and mosses. The basic structure of XyG, a xylosylated glucan, is similar in all of these plants but additional substituents can vary depending on plant family, tissue, and developmental stage. A comprehensive list of known XyG sidechain substituents is assembled including their occurrence within plant families, thereby providing insight into the evolutionary origin of the various sidechains. Recent advances in DNA sequencing have enabled comparative genomics approaches for the identification of XyG biosynthetic enzymes in *Arabidopsis thaliana* as well as in non-model plant species. Characterization of these biosynthetic genes not only allows the determination of their substrate specificity but also provides insights into the function of the various substituents in plant growth and development.

Keywords: xyloglucan; hemicellulose; plant cell walls; polysaccharide; glycosyltransferase

1. Introduction

The cells of plants are encased in a wall, a sophisticated composite material that is composed of the polysaccharides cellulose, various non-cellulosic polysaccharides including hemicelluloses and pectic polysaccharides, structural glycoproteins and, in mature secondary walls, the polyphenol lignin [1,2]. One of the hemicelluloses present in the walls of all land plants (embryophytes) is xyloglucan (XyG). XyG is the most abundant hemicellulose in the primary walls, *i.e.*, the walls of elongating cells [1], of many plants and can also occur as a storage amyloid in seeds [3].

In primary walls, XyG cross-links cellulose microfibrils [4], forming a strong but extensible XyG-cellulose network [2,5–8]. XyG may act as a spacer-molecule preventing aggregation of cellulose microfibrils [9,10] or a mediator-molecule enabling interactions between cellulose microfibrils and other wall matrix polymers [11,12]. Once deposited into the wall, XyG can undergo metabolism and turnover, which is thought to play a role in cell elongation [13]. Despite its prominence in models of the primary cell wall, an *Arabidopsis* mutant lacking detectable XyG has surprisingly only minor growth phenotypes suggesting that XyG is not as critical for wall architecture and function as once thought [14]. Indeed, recent data from mechanical testing experiments on primary walls using XyG modifying enzymes point towards a more limited but nevertheless significant structural role of XyG in wall mechanics [15].

In some plant species including tamarind (*Tamarindus indica*), nasturtium (*Tropaeolum majus*) and *Hymenaea courbaril*, XyG functions as a seed storage polymer [3,16,17]. In these species, XyG is deposited in large quantities during the development of the seed and upon germination is degraded to provide energy for the emerging seedling [18].

XyG is used in various food, industrial and pharmaceutical applications [19]. Seed storage XyG is often used for these purposes as it is easier to extract than XyG embedded in the wall matrix of cells in vegetative organs and is highly abundant in certain species. XyG isolated from tamarind seed has been tested successfully as a carrier for specific drug delivery systems [20,21] and the administration of antibiotics and treatment of ulcers [22]. Chemical derivatives of XyG oligosaccharides can also be used as biodegradable surfactants [23] and to activate cellulosic surfaces with a variety of functional groups [24].

2. XyG Structural Nomenclature and Presently Known Sidechain Diversity

XyG consists of a β-1,4 glucan chain that is decorated with a certain pattern of glycosyl groups [25,26]. The identity and position of the glycosyl units attached to a particular glucosyl backbone residue can be diverse and a one letter code for the various sidechains has been established, which allows for a concise depiction of specific XyG structures [27]. In this system, a backbone glucosyl residue without any further substitution is depicted by the letter G and a backbone glucosyl residue with an additional xylosyl unit at the O-6 position is described as X. X groups can carry additional glycosyl-moieties and so far 17 different sidechain structures have been identified, each with a unique one-letter abbreviation (Figure 1) [28]. The glycosyl units found in these various sidechains include D-galactose, D-xylose, L-arabinopyranose, L-arabinofuranose, D-galacturonic acid, L-fucose, and L-galactose.

Figure 1. XyG sidechain composition and diversity. A one letter nomenclature has been developed for all XyG sidechains identified to date [27,28]. The backbone β-(1,4)-glucosyl residues are identified in black. Acetylated sidechains are indicated by an underline, as in L or S. The directionality of the glycosidic bonds is indicated by an arrow, with the number referring to the hydroxyl-group of the sugar (arrow-head) that is linked to the anomeric carbon (C1) of the other sugar. Similar oligosaccharide structures are boxed.

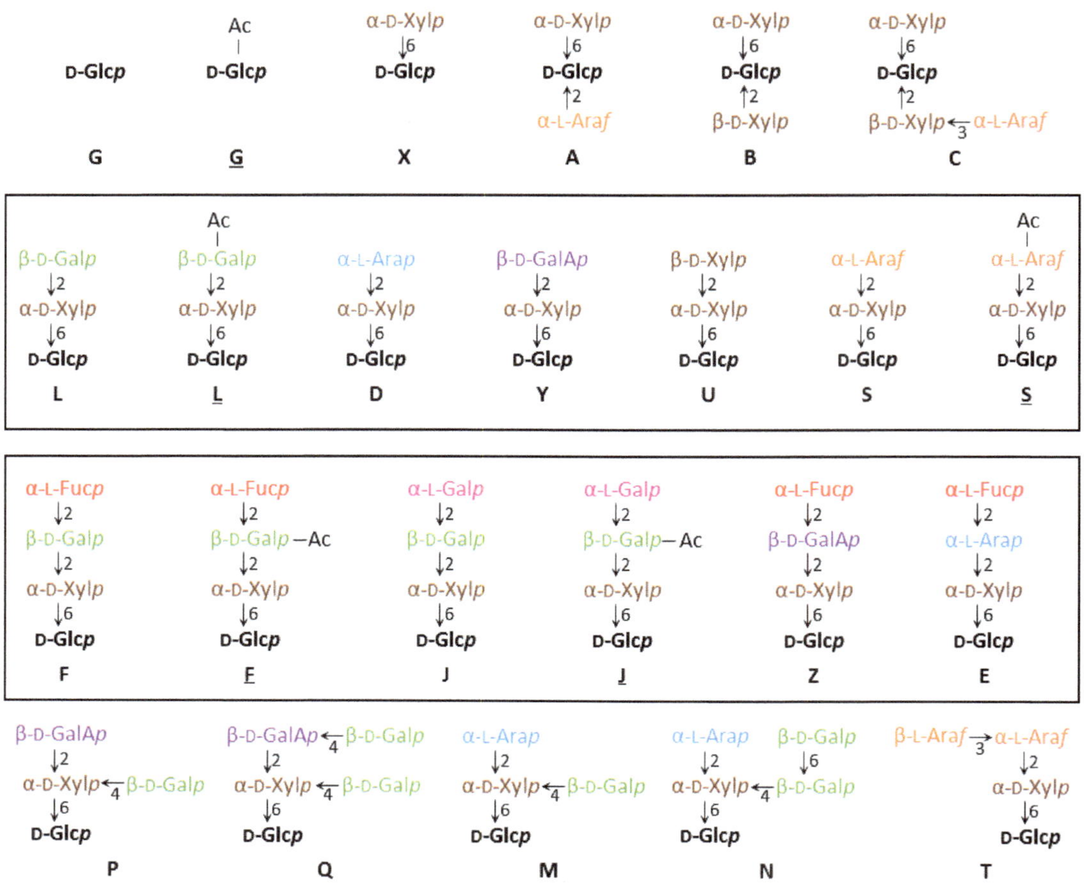

XyG sidechains can be placed into groups based on similarities in structure. For example, the L, S, D, Y and U units differ only in the identity of the terminal glycosyl group linked to the O-2 position of the xylosyl-residue (D-galactose, L-arabinofuranose, L-arabinopyranose, D-galacturonic acid, and D-xylose, respectively). A similar grouping can be composed of the F, E, J and Z sidechains, which contain terminal α-(1,2)-L-fucosyl-residues or the conformationally analogous L-galactosyl-residue on the L, D, and Y units.

XyG sidechains can be further modified by the addition of one or multiple acetyl groups on specific glycosyl groups [29,30]. Acetylated sidechains are depicted with an underline, such as in S and L. To date only three distinct XyG glycosyl units have been found to be acetylated. These include unsubstituted glucosyl backbone residues (G) [31,32], the arabinofuranosyl-residue of S groups [33], and the D-galactosyl-residue of L, F and J groups [29,30,34]. Acetyl groups have been shown to migrate between hydroxyl-groups within a particular glycosyl unit to reach a distribution equilibrium [35–37]. Some plant species including pea (*Pisum sativum*) have been shown to have two or more acetyl-substituents on a single galactosyl-residue [38]. A distinction between singly and doubly acetylated sidechains is not made under the current nomenclature system. NMR evidence has suggested the existence of methylated XyG sidechains in certain species, however specific structures of these groups have not been elucidated [39].

3. XyG Biodiversity

Although antibodies have indicated the presence of epitopes that are recognized by XyG antibodies in the walls of aquatic green algae from the Charophyta [40] and these algae possess homologs of genes known to be involved in XyG biosynthesis [41], specific XyG structures have hitherto only been characterized in land plants (embryophytes) (Figures 2 and 3). Besides the defining X units, further sidechain extension by a galactosyl-moiety is a very common feature and seems to occur in all plant species suggesting an important functional role. The fucosyl-residue seems to first appear in the hornworts and is a common feature in nearly all vascular plants investigated. Interestingly, a small group of eudicots including the Lamiales, which include basil (*Ocimum basilicum*), olive (*Olea europaea*), and plantain (*Plantago major*), the Solanales, which include tomato (*Solanum lycopersicum*), potato (*Solanum tuberosum*), and tobacco (*Nicotiana tabacum*), do not contain fucosylated XyG in their leaf tissue [42]. However, fucosylated XyG has been found in the pollen tube of *Nicotiana alata* [43] indicating that this XyG structure has been retained in some reproductive tissues and hinting at a functional significance in tip-growing tissues.

Figure 2. XyG structures of non-eudicot land plants. In each order or clade, XyG xylosylation motifs and sidechains are summarized from that reported for individual species within the group. See Table S1 for a detailed listing with plant families and species names and Figure 3 for the XyG structures of eudicots. The phylogenetic tree is adapted and modified from Hsieh and Harris [44] and that proposed by the Angiosperm Phylogeny Group III [45].

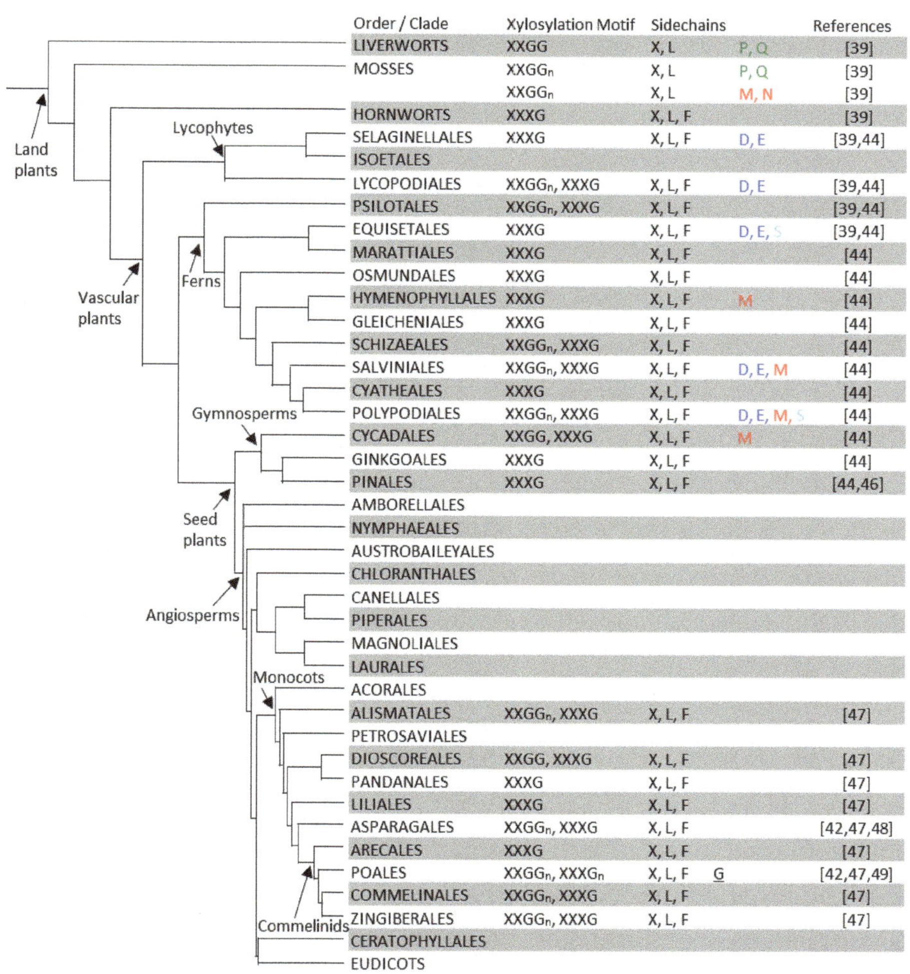

Figure 3. XyG structures of eudicot plants. In each order or family, XyG xylosation motifs structures and sidechains are summarized from individual species or tissue types. Two rows for a plant order indicates that different XyG structures were found in different tissues of a species of this order. For details see Table S1. The phylogenetic tree is adapted and modified from that proposed by the Angiosperm Phylogeny Group III [45].

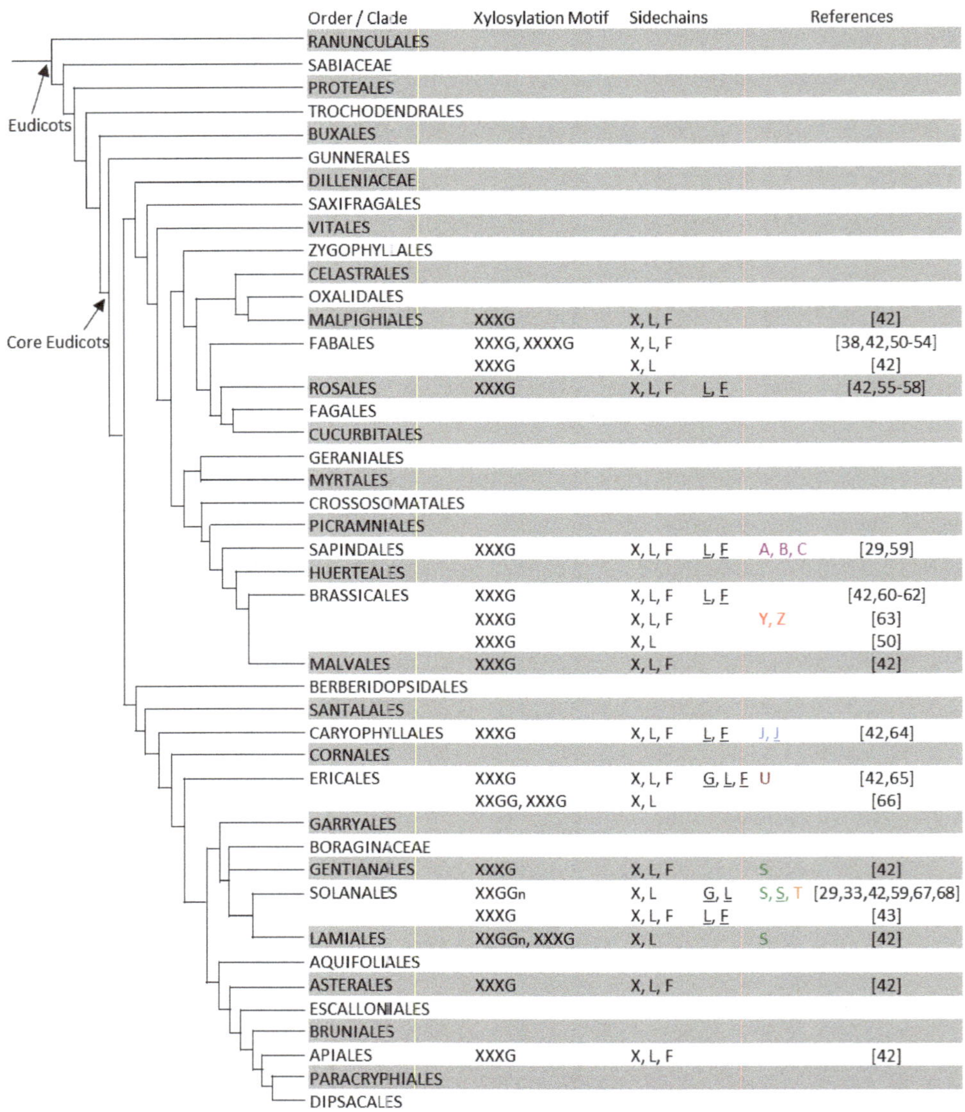

Unlike X, L and F, which are found in many or all clades throughout the land plant phylogeny, the occurrence of some other sidechains appears restricted to one or a few clades. For example, the P and Q sidechains have so far only been found in the liverwort and moss clades [39]. The P and Q sidechains are also unique as they contain branched sidechains with galactosyl-units bound to the xylosyl-residues at the O-4 position. This branching moiety is also present in the M and N sidechains of XyG found in the moss *Physcomitrella patens* [39]. Several species of ferns and a cycad also contain the M sidechain, a moiety that has not been reported to be present in any angiosperm species to date [39,44]. Sidechains containing arabinopyranosyl-residues (D and E) have been found in seedless vascular plants including lycophytes and ferns [39,44], but not in seed plants. The S sidechain containing arabinofuranosyl-residues has been found in the ferns (Equisetales and Polypodiales) [39], but is lacking in higher plant orders including the gymnosperms, monocots, and most of the eudicots with the exception of the subgroup of the Gentianales,

Lamiales, and Solanales. The phylogenetic distance between this subgroup and the seedless plants suggests that the glycosyltransferases that lead to XyG arabinofuranosylation have evolved independently at least twice [69].

The J sidechain contains an L-galactosyl-residue instead of the conformationally analogous fucosyl-residue found in F sidechains. XyG containing the J sidechain has been identified in jojoba seeds (*Simmondsia chinensis*) [70]. In suspension cultures of sycamore (*Acer pseudoplatanus*) the sidechains A, B, and C were found [29,59], each of which contain an additional sugar residue on the glucosyl backbone at the O-2 position.

Another feature of XyG is the *O*-acetyl-moiety. *O*-acetyl-substituents have only rarely been reported as this substituent is lost when XyG is extracted with alkali, a common method for XyG extraction from wall materials [71]. The function of the acetyl-substituent on XyG is not well understood but may influence polymer solubility as well as enzymatic accessibility. Removal of *O*-acetyl-substituents on XyG from the galactosyl-sidechains does not lead to any change in the binding capacity to cellulose [37].

XyG sidechains are not randomly positioned but are organized into XyG oligosaccharide motifs. These oligosaccharide motifs can be observed when XyG is digested with endoglucanases [72], which cleave XyG at unsubstituted glucosyl-residues. The oligosaccharides released by digestion of the XyG of Arabidopsis and many other dicot species share the common xylosylation motif of "XXXG", where the backbone largely consists of three xylosylated (and possibly further glycosylated) glucosyl units followed by one unsubstituted glucosyl residue [73]. A reduction in XyG backbone xylosylation is observed in the Poaceae, such as barley (*Hordeum vulgare*) and rice (*Oryza sativa*), which harbor a xylosylation motif of "$XXGG_n$" with n typically ranging from one to three [42]. A decreased degree of xylosylation is also observed in the Solanaceae, such as tobacco and tomato [42,67], which have "XXGG" as the dominant XyG xylosylation motif (Figures 2 and 3, Table S1). In plants with the "$XXGG_n$" motif *O*-acetylation of the glucan backbone appears to occur as an alternative to xylosylation at some positions. Additional xylosylation motifs have also been identified including XXXXG and XXXXXG as part of seed storage XyG [50]. While the majority of XyG oligosaccharides released from a particular species and tissue follow the dominant xylosylation pattern, there are exceptions. For example, XyG digestions of both monocot and dicot species including Arabidopsis were found to contain minor amounts of the XXG oligosaccharide [47,74]. These structures may result from partial degradation and reincorporation of XyG oligosaccharides within the wall [75].

The structure of XyG can vary within a single plant species depending on the tissue, cell type, and even position within the wall. One of the most prominent examples is represented by seed XyG. The storage XyG of several species including nasturtium and tamarind does not harbor fucosyl- or acetyl-groups, whereas these structures are present in the XyG found in the primary wall of these species [50]. Seed cotyledon parenchyma cells from these plants produce both types of XyG, which are deposited in different layers of the wall [76].

Other specialized tissues have also been reported to have XyG structures differing from the most prevalent form of XyG in a particular species. The Y sidechain, containing a charged galacturonic acid, was identified from Arabidopsis root tissue and appears to be present only in root hairs [63]. In the Solanaceae, where the dominant XyG xylosylation motif is XXGG and fucosylation is typically not observed, fucosylated XXXG-type XyG has been observed specifically in pollen tubes [43]. Cell-specific differences in XyG fucosylation have also been observed via antibody labeling in certain species within the Poaceae, in which phloem cells appear to have fucosylated XyG whereas neighboring ones do not [77].

While the above examples represent qualitative differences in XyG structure within a plant, quantitative differences in the relative abundance of specific XyG sidechains and oligosaccharide motifs are also common. For example, in pea XLFG oligosaccharide motifs make up approximately 25% of the XyG in leaves whereas they only account for 5% of the XyG in etiolated stems [38].

4. XyG Biosynthesis

Numerous proteins involved in XyG biosynthesis have been identified (Figure 4) [26]. All these proteins have been shown to be localized in the Golgi, where XyG is synthesized [78,79]. The XyG glucan backbone is believed to be synthesized by proteins of the cellulose synthase-like C (CSLC) family. The *TmCSLC4* gene is highly expressed in nasturtium (*Tropaeolum majus*) seeds, which contain a large abundance of galactosylated XyG. Heterologous expression of *TmCSLC4* and its ortholog *AtCSLC4* in *Pichia pastoris* resulted in the production of β-(1,4) glucan *in vitro* [80]. Proteins from glycosyltransferase (GT) family 34 have been shown to act as XyG xylosyltransferases (XXT) [81–83]. Of the five XXTs in Arabidopsis, XXT1, XXT2 and XXT5 have the most significant effect on XyG biosynthesis when knocked out [14,84–86]. A double *xxt1 xxt2* knockout mutant lacked detectable XyG and exhibited abnormal root hairs and slow growth compared to wild type plants [14,86]. Overexpression of *AtXXT3* could partially restore XyG abundance in *xxt2*, *xxt5* and *xxt2 xxt5* mutants indicating it is likely to represent a XyG xylosyltransferase [85]. Heterologously expressed AtXXT4 showed the ability to transfer UDP-xylose onto cellohexaose, as previously shown for AtXXT1 and AtXXT2, indicating it is also likely to be xylosyltransferase involved in XyG biosynthesis [85]. Recently, a putative rice XXT (OsXXT1) restored XyG structures and eliminated root hair defects when expressed in the Arabidopsis *xxt1 xxt2* mutant indicating that OsXXT1 has the same activity as its Arabidopsis XXT orthologs [87]. In the Solanaceae and Poaceae, the glucan backbone of XyG is *O*-acetylated (G̲) and has a lower degree of xylosylation ($XXGG_n$ type XyG; Figure 4). A XyG backbone *O*-acetyltransferase has not been identified to date and the mechanism of the reduction in XyG backbone xylosylation remains elusive.

The xylosyl substituents are often additionally glycosylated, most commonly with a galactosyl-residue. MURUS 3 (MUR3) from the GT family 47 exhibits XyG β-(1,2)-galactosyltransferase activity and is responsible for galactosylation of the third xylosyl residue of the XXXG XyG oligosaccharide motif (*i.e.*, the product is XXLG) [88] while Xyloglucan L-sidechain Galactosyltransferase Position 2 (XLT2), another GT47 galactosyltransferase, seems to be specific for galactosylating the second xylosyl residue (*i.e.*, XXXG will be converted to XLXG) based on genetic experiments in Arabidopsis [89]. These enzymes thus exhibit regio-selectivity in that they add the same glycosyl-residue to the same glycosyl-acceptor, but the acceptor-moiety has to be in a specific structural environment on the XyG. The determinant of the GT protein structure for this regio-selectivity is not known and it will be interesting to crystalize these proteins to discover the mechanism and specific recognition motifs. Regio-selectivity cannot be excluded for other XyG GTs such as the xylosyltransferases. The double mutant *xlt2 mur3* lacks substitution of the xylosyl-residues thus harboring a XyG made nearly entirely of XXXG units and resulting in a dwarfed phenotype in Arabidopsis [89]. MUR3 orthologs from other species such as *Eucalyptus grandis* and *Solanum lycopersicum* restore the galactosylated XyG when expressed in the *mur3.1* or *xlt2 mur3* Arabidopsis mutants, suggesting that they exhibit the same activity as AtMUR3 [69,90]. Two GT47 homologs of AtXLT2 in tomato were found to result in the production of arabinosylated XyG when expressed in Arabidopsis, suggesting that both proteins, named Xyloglucan S-sidechain Transferase (XST) 1 and 2, represent arabinofuranosyltransferases and lead to the synthesis of the S sidechain found in tomato [69]. Overexpressing XST1 or XST2 in the *xlt2 mur3* double mutant rescued the dwarfed mutant phenotype indicating an equivalent function in plant growth for the L and S sidechains in this case [69]. A root hair specific gene, *Xyloglucan-specific Galacturonosyltransferase1* (*XUT1*), was found to be required for the presence of galacturonic acid (Y sidechains) on XyG in Arabidopsis. Knock-out of the *AtXUT1* gene resulted in shorter root hairs while the aerial part of wild-type and mutant were indistinguishable indicating that this type of acidic XyG structure plays a role specifically in root hair development [63]. MUR3, XLT2, XUT1 and the XSTs all act to transfer glycosyl-moieties to the same O-2 position of the XyG xylosyl-residue and are part of the same subclade of GT47. Hence, it is likely that the proteins responsible for the addition of other glycosyl groups to this position, including the arabinopyranosyl- and xylosyl- of D and U sidechains, respectively, will also be identified from this GT47 subclade.

Figure 4. Schematic representation of different types of XyG and known proteins involved in its synthesis and modification. Detailed linkage information of XyG structures can be found in Figure 1. *At*, *Arabidopsis*; *Os*, *Oryza sativa*; *Tm*, *Tropaeolum majus*; *Pr*, *Pinus radiata*; *Eg*, *Eucalyptus grandis*; *Sl*, *Solanum lycopersicum*. GT, glycosyltransferase; GH, glycoside hydrolase. Arrows and scissors indicate the site of glycosyltransferase or glycosyl hydrolase activity, respectively.

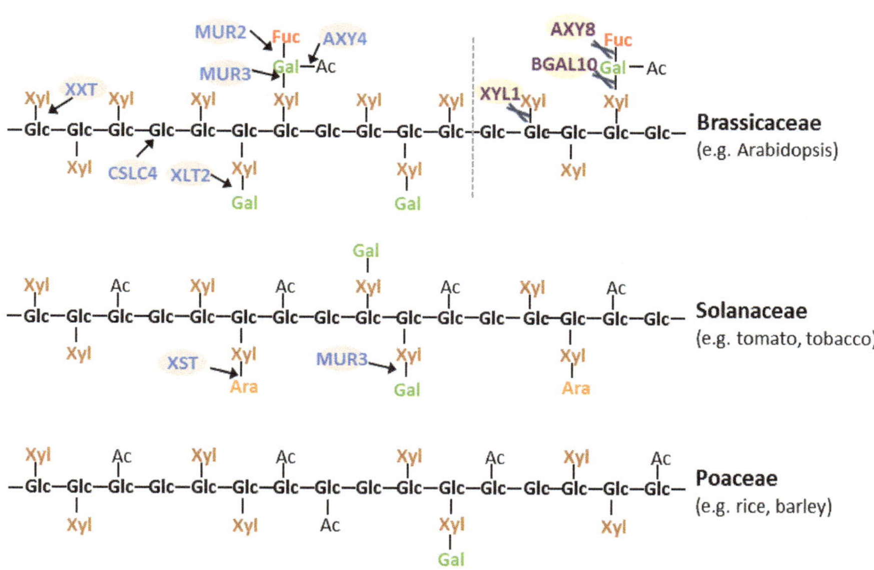

Activity	Gene Name	Species	CAZy Family	Evidence	References
Glucan synthase	CSLC4	*At, Tm*	GT2	*in vitro* activity	[80]
Xylosyltransferase	XXT1-5	*At, Os, Pr*	GT34	loss of function, *in vitro* activity	[14,81-87]
Galactosyltransferase	MUR3	*At, Eg, Sl, Tm*	GT47	loss of function, *in vitro* activity	[69,88-90]
Galactosyltransferase	XLT2	*At, Tm*	GT47	loss of function	[69,89]
Arabinosyltransferase	XST1, XST2	*Sl*	GT47	gain of function	[69]
Galacturonsyltransferase	XUT1	*At*	GT47	loss of function	[63]
Fucosyltransferase	MUR2/FUT1	*At*	GT37	loss of function, *in vitro* activity	[91,92]
Acetyltransferase	AXY4, AXY4L	*At*	-	loss of function	[93]
Xylosidase	AXY3/XYL1	*At*	GH31	loss of function, *in vitro* activity	[94,95]
Fucosidase	AXY8	*At*	GH95	loss of function, *in vitro* activity	[75]
Galactosidase	BGAL10	*At*	GH35	loss of function, *in vitro* activity	[96]

In Arabidopsis, one of the galactosyl-residues can be fucosylated by the α-(1,2)-fucosyltransferase Fucosyltransferase1 (FUT1), a member of GT family 37 [91,92]. Genetic disruption of the gene encoding FUT1 in the *mur2* mutant results in non-fucosylated XyG [92]. Despite this alteration in structure, *mur2* plants display a normal growth habit and wall strength, although ultrastructural analysis revealed an alteration in trichome structure [92]. When the abundance of the donor substrate of FUT1, GDP-fucose, is reduced by genetic alteration of the nucleotide sugar conversion pathway in the *mur1* mutant, the fucosylation of XyG is nearly abolished and a L-galactosyl residue is added to the galactosyl-moiety instead of the fucosyl-substituent to form a J sidechain [92,97,98]. In this case, the availability of donor substrate rather than the identity of the glycosyltransferase determines the sidechain present. FUT1 also exhibits promiscuity for the acceptor substrate as it can utilize a galacturonosyl group to produce the Z sidechain [63]. Based on the similarity in structure, it appears likely that FUT1 or another GT37 enzyme is involved in producing E sidechains found in some ferns and lycophytes. Despite the tolerance for alternative donor and acceptor residues, XyG fucosylation is regio-specific and typically occurs only on the last xylosyl-residue of the XyG oligosaccharide as in the XyG oligosaccharide motif XLFG, but not on the middle galactosyl-position. An XFFG unit has been found in the XyG of suspension cultured cells of sycamore (*Acer pseudoplatanus*) [99], but this structure has not been observed in a plant tissue and it is unclear if MUR2 or another fucosyltransferase is responsible for this transfer.

The galactosyl-residue can also be *O*-acetylated in Arabidopsis. A putative XyG *O*-acetyltransferase has been identified as Altered Xyloglucan 4 (AXY4), a protein which belongs to the Trichome Birefrigence-like (TBL) family [93]. It is not clear onto which position of the galactosyl-residue the acetyl group is initially transferred. AXY4 is required for XyG acetylation in leaves whereas a related TBL gene, *Altered Xyloglucan 4-like* (*AXY4L*), is responsible for XyG acetylation in seeds. Absence of the *O*-acetyl substituent on XyG in the various tissues was not reported to affect tissue growth or morphology. In fact, a naturally occurring ecotype of Arabidopsis, Ty-0, found in the Scottish highlands, has a defect in AXY4 and a decrease in XyG acetylation, suggesting that XyG acetylation is at least partially dispensable for Arabidopsis in this particular environment [93].

XyG is synthesized in the Golgi and exported to the apoplast via exocytosis where it can undergo further metabolism, turn-over and/or maturation [13,100]. Known XyG modifying enzymes and hydrolases include XyG endotransglucosylase/endohydrolases (XTHs), and glycosidases such as the xylosidase Altered Xyloglucan 3 (AXY3)/Xylosidase 1 (XYL1), the fucosidase Altered Xyloglucan 8 (AXY8), and the galactosidase Beta-galactosidase 10 (BGAL10) [75,94–96,101–105]. XTHs are capable of cleaving and reforming the backbone glycosidic bonds of XyG polymers in rapidly growing plant tissues [104,105]. It is thought that this class of enzymes might be responsible for incorporating nascent XyG polymers into the existing cellulose/XyG network. While these enzyme activities have been found in the apoplast [106,107], it is possible that they are also present within the endomembrane system during polysaccharide biosynthesis and transport. XyG isolated from Arabidopsis *axy8* microsomes was found to have a higher degree of fucosylation than that of wild type plants, suggesting AXY8 is active on XyG both in the apoplast and in the endomembrane system [75]. Knocking out the xylosidase XYL1/AXY3 increased the overall abundance of XyG xylosylation in its corresponding Arabidopsis mutant, while the abundance of galactosylated XyG structures was increased in the *bgal10* mutant. Both glycosidases seem to impact plant development as siliques are dwarfed in these mutants [94–96]. The Arabidopsis *axy8* mutant displays an increased abundance of fucosylated XyG oligosaccharides, but unexpectedly results in XyG that contains unusual XyG oligosaccharides such as XFG, GFGXXXG, and GFGXXFG [75]. When both the xylosidase and fucosidase were knocked-out in the *axy8 axy3* double mutant, the XyG structure consists of a 1:1 ratio of the XyG oligosaccharide motifs XXXG and XXFG in contrast to XLFG, XXFG, XXLG, XLXG and XXXG motifs usually found in Arabidopsis tissues. This data provides evidence that apoplastic glycosidases also play an important role in structurally diversifying XyG oligosaccharide motifs found in the wall [75].

5. Outlook

Initially, XyG research was focused on the description of XyG structure, thereby leading to the discovery of sidechain diversity and the distinct ordering of the sidechains that make up various XyG oligosaccharide motifs. While more than 20 different sidechain structures have been identified to date, there remain many plant families and entire orders for which the XyG structure has not been reported (Figures 2 and 3, Table S1) suggesting that further diversity may yet be discovered. Additional challenges remain to understand the order in which the oligosaccharide motifs are linked together and their functional significance. There is evidence that XyG consists of covalently-linked polysaccharide domains that bind to cellulose microfibrils interlaced with domains that span between adjacent microfibrils [15,108]. Different quantities of various XyG oligosaccharide motifs have been observed between these domains [7], however drawing functional conclusions is difficult and the overall patterning of the oligosaccharide motifs is still unclear.

With the advent of genetics and genomics many of the genes involved in XyG biosynthesis and metabolism have now been identified and characterized. This knowledge, together with the corresponding plant mutants, allows XyG structure to be altered *in planta* giving crucial insights into the function of XyG structural features in plant growth and development. While still in its infancy, a closer phylogenetic and protein structural investigation of the GTs involved in XyG biosynthesis will allow us to better understand the molecular basis for GT donor and acceptor specificity as well as the evolutionary history of the corresponding genes. The availability of powerful methods for analyzing XyG structure as well as the ability of plants to tolerate aberrant XyG structures make it uniquely suited as a model plant polysaccharide, which can be used to gain insight into the biosynthesis, deposition, metabolism and evolution of non-cellulosic polysaccharides in the plant cell wall.

Acknowledgments

We would like to mention the funding sources that supported the authors. Lifeng Liu and Lei Zhu were supported by a grant from the Energy Biosciences Institute; Alex Schultink by the Dickinson Chair for wood science and technology for Markus Pauly.

Author Contributions

Alex Schultink, Lifeng Liu, Lei Zhu and Markus Pauly wrote the paper.

Conflicts of Interest

The authors declare no conflict of interest.

References

1. Carpita, N.C.; Gibeaut, D.M. Structural models of primary-cell walls in flowering plants—Consistency of molecular-structure with the physical-properties of the walls during growth. *Plant J.* **1993**, *3*, 1–30.
2. Somerville, C.; Bauer, S.; Brininstool, G.; Facette, M.; Hamann, T.; Milne, J.; Osborne, E.; Paredez, A.; Persson, S.; Raab, T.; *et al.* Toward a systems approach to understanding plant-cell walls. *Science* **2004**, *306*, 2206–2211.
3. Kooiman, P. The constitution of tamarindus-amyloid. *Recl. Trav. Chim. Pay. B* **1961**, *80*, 849–865.
4. Keegstra, K.; Talmadge, K.W.; Bauer, W.; Albersheim, P. The structure of plant cell walls III. A model of the walls of suspension-cultured sycamore cells based on the interconnections of the macromolecular components. *Plant Physiol.* **1973**, *51*, 188–197.
5. Albersheim, P.; Darvill, A.; Roberts, K.; Sederoff, R.; Staehelin, A. *Plant Cell Walls: From Chemistry to Biology*; Garland Science, Taylor and Francis Group, LLC: New York, NY, USA, 2011; pp. 1–430.
6. Hayashi, T. Xyloglucans in the primary-cell wall. *Annu. Rev. Plant Phys.* **1989**, *40*, 139–168.
7. Pauly, M.; Albersheim, P.; Darvill, A.; York, W.S. Molecular domains of the cellulose/xyloglucan network in the cell walls of higher plants. *Plant J.* **1999**, *20*, 629–639.
8. McCann, M.; Roberts, K. Architecture of the primary cell wall. In *The Cytoskeletal Basis of Plant Growth and Form*; Academic Press: Waltham, MA, USA, 1991; pp. 109–129.
9. Thompson, D.S. How do cell walls regulate plant growth? *J. Exp. Bot.* **2005**, *56*, 2275–2285.
10. Anderson, C.T.; Carroll, A.; Akhmetova, L.; Somerville, C. Real-time imaging of cellulose reorientation during cell wall expansion in Arabidopsis roots. *Plant Physiol.* **2010**, *152*, 787–796.
11. Talbott, L.D.; Ray, P.M. Molecular size and separability features of pea cell wall polysaccharides Implications for models of primary wall structure. *Plant Physiol.* **1992**, *98*, 357–368.
12. Ha, M.-A.; Apperley, D.C.; Jarvis, M.C. Molecular rigidity in dry and hydrated onion cell walls. *Plant Physiol.* **1997**, *115*, 593–598.
13. Takeda, T.; Furuta, Y.; Awano, T.; Mizuno, K.; Mitsuishi, Y.; Hayashi, T. Suppression and acceleration of cell elongation by integration of xyloglucans in pea stem segments. *Proc. Natl. Acad. Sci. USA* **2002**, *99*, 9055–9060.
14. Cavalier, D.M.; Lerouxel, O.; Neumetzler, L.; Yamauchi, K.; Reinecke, A.; Freshour, G.; Zabotina, O.A.; Hahn, M.G.; Burgert, I.; Pauly, M.; *et al.* Disrupting two *Arabidopsis thaliana* xylosyltransferase genes results in plants deficient in xyloglucan, a major primary cell wall component. *Plant Cell* **2008**, *20*, 1519–1537.
15. Park, Y.B.; Cosgrove, D.J. A revised architecture of primary cell walls based on biomechanical changes induced by substrate-specific endoglucanases. *Plant Physiol.* **2012**, *158*, 1933–1943.
16. Meier, H.; Reid, J. Reserve polysaccharides other than starch in higher plants. In *Plant Carbohydrates I*; Springer: Heidelberg, Germany, 1982; pp. 418–471.
17. Rao, P.; Srivastava, H. Tamarind. In *Industrial Gums*, 2nd ed.; Academic Press: New York, NY, USA, 1973; pp. 369–411.

18. Edwards, M.; Bowman, Y.; Dea, I.; Reid, J. A β-D-Galactosidase from nasturtium (*Tropaeolum majus* L.) cotyledons. Purification, properties, and demonstration that xyloglucan is the natural substrate. *J. Biol. Chem.* **1988**, *263*, 4333–4337.
19. Mishra, A.; Malhotra, A.V. Tamarind xyloglucan: A polysaccharide with versatile application potential. *J. Mater. Chem.* **2009**, *19*, 8528–8536.
20. Mishra, M.; Khandare, J. Evaluation of tamarind seed polysaccharide as a biodegradable carrier for colon specific drug delivery. *Int. J. Pharm. Pharm. Sci.* **2011**, *3*, 139–142.
21. Rolando, M.; Valente, C. Establishing the tolerability and performance of tamarind seed polysaccharide (TSP) in treating dry eye syndrome: Results of a clinical study. *BMC Ophthalmol.* **2007**, *7*, 5.
22. Ghelardi, E.; Tavanti, A.; Celandroni, F.; Lupetti, A.; Blandizzi, C.; Boldrini, E.; Campa, M.; Senesi, S. Effect of a novel mucoadhesive polysaccharide obtained from tamarind seeds on the intraocular penetration of gentamicin and ofloxacin in rabbits. *J. Antimicrob. Chemother.* **2000**, *46*, 831–834.
23. Greffe, L.; Bessueille, L.; Bulone, V.; Brumer, H. Synthesis, preliminary characterization, and application of novel surfactants from highly branched xyloglucan oligosaccharides. *Glycobiology* **2005**, *15*, 437–445.
24. Brumer, H., 3rd; Zhou, Q.; Baumann, M.J.; Carlsson, K.; Teeri, T.T. Activation of crystalline cellulose surfaces through the chemoenzymatic modification of xyloglucan. *J. Am. Chem. Soc.* **2004**, *126*, 5715–5721.
25. Scheller, H.V.; Ulvskov, P. Hemicelluloses. *Annu. Rev. Plant Biol.* **2010**, *61*, 263–289.
26. Pauly, M.; Gille, S.; Liu, L.; Mansoori, N.; de Souza, A.; Schultink, A.; Xiong, G. Hemicellulose biosynthesis. *Planta* **2013**, *238*, 627–642.
27. Fry, S.C.; York, W.S.; Albersheim, P.; Darvill, A.; Hayashi, T.; Joseleau, J.P.; Kato, Y.; Lorences, E.P.; Maclachlan, G.A.; Mcneil, M.; *et al.* An unambiguous nomenclature for xyloglucan-derived oligosaccharides. *Physiol. Plantarum* **1993**, *89*, 1–3.
28. Tuomivaara, S.T.; Yaoi, K.; O'Neill, M.A.; York, W.S. Generation and structural validation of a library of diverse xyloglucan-derived oligosaccharides, including an update on xyloglucan nomenclature. *Carbohydr. Res.* **2014**, doi:10.1016/j.carres.2014.06.031.
29. Kiefer, L.L.; York, W.S.; Darvill, A.G.; Albersheim, P. Xyloglucan isolated from suspension-cultured sycamore cell-walls is *O*-acetylated. *Phytochemistry* **1989**, *28*, 2105–2107.
30. York, W.S.; Oates, J.E.; van Halbeek, H.; Darvill, A.G.; Albersheim, P.; Tiller, P.R.; Dell, A. Location of the *O*-acetyl substituents on a nonasaccharide repeating unit of sycamore extracellular xyloglucan. *Carbohydr. Res.* **1988**, *173*, 113–132.
31. Sims, I.M.; Munro, S.L.; Currie, G.; Craik, D.; Bacic, A. Structural characterisation of xyloglucan secreted by suspension-cultured cells of *Nicotiana plumbaginifolia*. *Carbohydr. Res.* **1996**, *293*, 147–172.
32. York, W.S.; Kolli, V.S.K.; Orlando, R.; Albersheim, P.; Darvill, A.G. The structures of arabinoxyloglucans produced by solanaceous plants. *Carbohydr. Res.* **1996**, *285*, 99–128.
33. Jia, Z.; Cash, M.; Darvill, A.G.; York, W.S. NMR characterization of endogenously *O*-acetylated oligosaccharides isolated from tomato (*Lycopersicon esculentum*) xyloglucan. *Carbohydr. Res.* **2005**, *340*, 1818–1825.

34. Maruyama, K.; Goto, C.; Numata, M.; Suzuki, T.; Nakagawa, Y.; Hoshino, T.; Uchiyama, T. *O*-acetylated xyloglucan in extracellular polysaccharides from cell-suspension cultures of *Mentha*. *Phytochemistry* **1996**, *41*, 1309–1314.
35. Kamerling, J.P.; Schauer, R.; Shukla, A.K.; Stoll, S.; Halbeek, H.; Vliegenthart, J. Migration of *O*-acetyl groups in *N,O*-acetylneuraminic acids. *Eur. J. Biochem.* **1987**, *162*, 601–607.
36. Kabel, M.A.; de Waard, P.; Schols, H.A.; Voragen, A.G. Location of *O*-acetyl substituents in xylo-oligosaccharides obtained from hydrothermally treated *Eucalyptus* wood. *Carbohydr. Res.* **2003**, *338*, 69–77.
37. Pauly, M. *Development of Analytical Tools to Study Plant Cell Wall Xyloglucan*; Shaker Verlag GmbH: Herzogenrath, Germany, 1999; pp. 1–252.
38. Pauly, M.; Qin, Q.; Greene, H.; Albersheim, P.; Darvill, A.; York, W.S. Changes in the structure of xyloglucan during cell elongation. *Planta* **2001**, *212*, 842–850.
39. Peña, M.J.; Darvill, A.G.; Eberhard, S.; York, W.S.; O'Neill, M.A. Moss and liverwort xyloglucans contain galacturonic acid and are structurally distinct from the xyloglucans synthesized by hornworts and vascular plants. *Glycobiology* **2008**, *18*, 891–904.
40. Popper, Z.A.; Fry, S.C. Primary cell wall composition of bryophytes and charophytes. *Ann. Bot.* **2003**, *91*, 1–12.
41. Mikkelsen, M.D.; Harholt, J.; Ulvskov, P.; Johansen, I.E.; Fangel, J.U.; Doblin, M.S.; Bacic, A.; Willats, W.G. Evidence for land plant cell wall biosynthetic mechanisms in charophyte green algae. *Ann. Bot.* **2014**, *114*, 1217–1236.
42. Hoffman, M.; Jia, Z.; Pena, M.J.; Cash, M.; Harper, A.; Blackburn, A.R., 2nd; Darvill, A.; York, W.S. Structural analysis of xyloglucans in the primary cell walls of plants in the subclass Asteridae. *Carbohydr. Res.* **2005**, *340*, 1826–1840.
43. Lampugnani, E.R.; Moller, I.E.; Cassin, A.; Jones, D.F.; Koh, P.L.; Ratnayake, S.; Beahan, C.T.; Wilson, S.M.; Bacic, A.; Newbigin, E. *In vitro* grown pollen tubes of *Nicotiana alata* actively synthesise a fucosylated xyloglucan. *PLoS One* **2013**, *8*, e77140.
44. Hsieh, Y.S.; Harris, P.J. Structures of xyloglucans in primary cell walls of gymnosperms, monilophytes (ferns *sensu lato*) and lycophytes. *Phytochemistry* **2012**, *79*, 87–101.
45. Bremer, B.; Bremer, K.; Chase, M.; Fay, M.; Reveal, J.; Soltis, D.; Soltis, P.; Stevens, P. An update of the Angiosperm Phylogeny Group classification for the orders and families of flowering plants: APG III. *Bot. J. Linn. Soc.* **2009**, *161*, 105–121.
46. Kakegawa, K.; Edashige, Y.; Ishii, T. Xyloglucan from xylem-differentiating zones of *Cryptomeria japonica*. *Phytochemistry* **1998**, *47*, 767–771.
47. Hsieh, Y.S.; Harris, P.J. Xyloglucans of monocotyledons have diverse structures. *Mol. Plant* **2009**, *2*, 943–965.
48. O'Neill, M.A.; York, W.S. The composition and structure of plant primary cell walls. In *The Plant Cell Wall*, Blackwell Publishing: Oxford, UK, 2003; pp. 1–54.
49. Gibeaut, D.M.; Pauly, M.; Bacic, A.; Fincher, G.B. Changes in cell wall polysaccharides in developing barley (*Hordeum vulgare*) coleoptiles. *Planta* **2005**, *221*, 729–738.
50. Buckeridge, M.S. Seed cell wall storage polysaccharides: Models to understand cell wall biosynthesis and degradation. *Plant Physiol.* **2010**, *154*, 1017–1023.

51. Vinueza, N.R.; Gallardo, V.A.; Klimek, J.F.; Carpita, N.C.; Kenttämaa, H.I. Analysis of xyloglucans by ambient chloride attachment ionization tandem mass spectrometry. *Carbohydr. Polym.* **2013**, *98*, 1203–1213.
52. Huisman, M.; Weel, K.; Schols, H.; Voragen, A. Xyloglucan from soybean (*Glycine max*) meal is composed of XXXG-type building units. *Carbohydr. Polym.* **2000**, *42*, 185–191.
53. Alonso-Simón, A.; Neumetzler, L.; García-Angulo, P.; Encina, A.; Acebes, J.; Álvarez, J.; Hayashi, T. Plasticity of xyloglucan composition in bean (*Phaseolus vulgaris*)-cultured cells during habituation and dehabituation to lethal concentrations of dichlobenil. *Mol. Plant* **2010**, *3*, 603–609.
54. Ren, Y.; Picout, D.R.; Ellis, P.R.; Ross-Murphy, S.B.; Reid, J. A novel xyloglucan from seeds of *Afzelia africana* Se. Pers.—Extraction, characterization, and conformational properties. *Carbohydr. Res.* **2005**, *340*, 997–1005.
55. Watt, D.; Brasch, D.; Larsen, D.; Melton, L. Isolation, characterisation, and NMR study of xyloglucan from enzymatically depectinised and non-depectinised apple pomace. *Carbohydr. Polym.* **1999**, *39*, 165–180.
56. Galvez-Lopez, D.; Laurens, F.; Quemener, B.; Lahaye, M. Variability of cell wall polysaccharides composition and hemicellulose enzymatic profile in an apple progeny. *Int. J. Biol. Macromol.* **2011**, *49*, 1104–1109.
57. Lahaye, M.; Falourd, X.; Quemener, B.; Ralet, M.C.; Howad, W.; Dirlewanger, E.; Arús, P. Cell wall polysaccharide chemistry of peach genotypes with contrasted textures and other fruit traits. *J. Agric. Food Chem.* **2012**, *60*, 6594–6605.
58. Ray, S.; Vigouroux, J.; Quémener, B.; Bonnin, E.; Lahaye, M. Novel and diverse fine structures in LiCl-DMSO extracted apple hemicelluloses. *Carbohydr. Polym.* **2014**, *108*, 46–57.
59. York, W.S.; Impallomeni, G.; Hisamatsu, M.; Albersheim, P.; Darvill, A.G. Eleven newly characterized xyloglucan oligoglycosyl alditols: The specific effects of sidechain structure and location on ^{1}H-NMR chemical shifts. *Carbohydr. Res.* **1995**, *267*, 79–104.
60. Louvet, R.; Rayon, C.; Domon, J.-M.; Rusterucci, C.; Fournet, F.; Leaustic, A.; Crépeau, M.-J.; Ralet, M.-C.; Rihouey, C.; Bardor, M. Major changes in the cell wall during silique development in *Arabidopsis thaliana*. *Phytochemistry* **2011**, *72*, 59–67.
61. Lerouxel, O.; Choo, T.S.; Seveno, M.; Usadel, B.; Faye, L.; Lerouge, P.; Pauly, M. Rapid structural phenotyping of plant cell wall mutants by enzymatic oligosaccharide fingerprinting. *Plant Physiol.* **2002**, *130*, 1754–1763.
62. Crombie, H.J.; Chengappa, S.; Hellyer, A.; Reid, J.S. A xyloglucan oligosaccharide-active, transglycosylating β-D-glucosidase from the cotyledons of nasturtium (*Tropaeolum majus* L.) seedlings—Purification, properties and characterization of a cDNA clone. *Plant J.* **1998**, *15*, 27–38.
63. Peña, M.J.; Kong, Y.; York, W.S.; O'Neill, M.A. A galacturonic acid-containing xyloglucan is involved in Arabidopsis root hair tip growth. *Plant Cell* **2012**, *24*, 4511–4524.
64. Mishra, A.; Joshi, M.; Jha, B. Oligosaccharide mass profiling of nutritionally important *Salicornia brachiata*, an extreme halophyte. *Carbohydr. Polym.* **2013**, *92*, 1942–1945.
65. Hilz, H.; de Jong, L.E.; Kabel, M.A.; Verhoef, R.; Schols, H.A.; Voragen, A.G. Bilberry xyloglucan—Novel building blocks containing β-xylose within a complex structure. *Carbohydr. Res.* **2007**, *342*, 170–181.

66. Aboughe-Angone, S.; Nguema-Ona, E.; Ghosh, P.; Lerouge, P.; Ishii, T.; Ray, B.; Driouich, A. Cell wall carbohydrates from fruit pulp of *Argania spinosa*: Structural analysis of pectin and xyloglucan polysaccharides. *Carbohydr. Res.* **2008**, *343*, 67–72.
67. Assor, C.; Quemener, B.; Vigouroux, J.; Lahaye, M. Fractionation and structural characterization of LiCl-DMSO soluble hemicelluloses from tomato. *Carbohydr. Polym.* **2013**, *94*, 46–55.
68. Jia, Z.; Qin, Q.; Darvill, A.G.; York, W.S. Structure of the xyloglucan produced by suspension-cultured tomato cells. *Carbohydr. Res.* **2003**, *338*, 1197–1208.
69. Schultink, A.; Cheng, K.; Park, Y.B.; Cosgrove, D.J.; Pauly, M. The identification of two arabinosyltransferases from tomato reveals functional equivalency of xyloglucan side chain substituents. *Plant Physiol.* **2013**, *163*, 86–94.
70. Hantus, S.; Pauly, M.; Darvill, A.G.; Albersheim, P.; York, W.S. Structural characterization of novel L-galactose-containing oligosaccharide subunits of jojoba seed xyloglucans. *Carbohydr. Res.* **1997**, *304*, 11–20.
71. Selvendran, R.R.; Ryden, P. Isolation and analysis of plant cell walls. In *Methods in Plant Biochemistry*; Dey, P.M., Ed.; Academic Press: New York, NY, USA, 1990; pp. 549–579.
72. Pauly, M.; Andersen, L.N.; Kauppinen, S.; Kofod, L.V.; York, W.S.; Albersheim, P.; Darvill, A. A xyloglucan-specific endo-β-1,4-glucanase from *Aspergillus aculeatus*: Expression cloning in yeast, purification and characterization of the recombinant enzyme. *Glycobiology* **1999**, *9*, 93–100.
73. Obel, N.; Erben, V.; Schwarz, T.; Kühnel, S.; Fodor, A.; Pauly, M. Microanalysis of plant cell wall polysaccharides. *Mol. Plant* **2009**, *2*, 922–932.
74. Watanabe, T.; Shida, M.; Murayama, T.; Furuyama, Y.; Nakajima, T.; Matsuda, K.; Kainuma, K. Xyloglucan in cell walls of rice hull. *Carbohydr. Res.* **1984**, *129*, 229–242.
75. Günl, M.; Neumetzler, L.; Kraemer, F.; de Souza, A.; Schultink, A.; Pena, M.; York, W.S.; Pauly, M. AXY8 encodes an α-fucosidase, underscoring the importance of apoplastic metabolism on the fine structure of *Arabidopsis* cell wall polysaccharides. *Plant Cell* **2011**, *23*, 4025–4040.
76. Marcus, S.E.; Verhertbruggen, Y.; Herve, C.; Ordaz-Ortiz, J.J.; Farkas, V.; Pedersen, H.L.; Willats, W.G.T.; Knox, J.P. Pectic homogalacturonan masks abundant sets of xyloglucan epitopes in plant cell walls. *BMC Plant Biol.* **2008**, *8*, e60.
77. Brennan, M.; Harris, P.J. Distribution of fucosylated xyloglucans among the walls of different cell types in monocotyledons determined by immunofluorescence microscopy. *Mol. Plant* **2011**, *4*, 144–156.
78. Lerouxel, O.; Cavalier, D.M.; Liepman, A.H.; Keegstra, K. Biosynthesis of plant cell wall polysaccharides—A complex process. *Curr. Opin. Plant Biol.* **2006**, *9*, 621–630.
79. Scheible, W.-R.; Pauly, M. Glycosyltransferases and cell wall biosynthesis: Novel players and insights. *Curr. Opin. Plant Biol.* **2004**, *7*, 285–295.
80. Cocuron, J.C.; Lerouxel, O.; Drakakaki, G.; Alonso, A.P.; Liepman, A.H.; Keegstra, K.; Raikhel, N.; Wilkerson, C.G. A gene from the cellulose synthase-like C family encodes a β-1,4 glucan synthase. *Proc. Natl. Acad. Sci. USA* **2007**, *104*, 8550–8555.
81. Faik, A.; Price, N.J.; Raikhel, N.V.; Keegstra, K. An *Arabidopsis* gene encoding an α-xylosyltransferase involved in xyloglucan biosynthesis. *Proc. Natl. Acad. Sci. USA* **2002**, *99*, 7797–7802.
82. Cavalier, D.M.; Keegstra, K. Two xyloglucan xylosyltransferases catalyze the addition of multiple xylosyl residues to cellohexaose. *J. Biol. Chem.* **2006**, *281*, 34197–34207.

83. Ade, C.P.; Bemm, F.; Dickson, J.M.; Walter, C.; Harris, P.J. Family 34 glycosyltransferase (GT34) genes and proteins in *Pinus radiata* (radiata pine) and *Pinus taeda* (loblolly pine). *Plant J.* **2014**, *78*, 305–318.
84. Zabotina, O.A.; van de Ven, W.T.G.; Freshour, G.; Drakakaki, G.; Cavalier, D.; Mouille, G.; Hahn, M.G.; Keegstra, K.; Raikhel, N.V. Arabidopsis XXT5 gene encodes a putative α-1,6-xylosyltransferase that is involved in xyloglucan biosynthesis. *Plant J.* **2008**, *56*, 101–115.
85. Vuttipongchaikij, S.; Brocklehurst, D.; Steele-King, C.; Ashford, D.A.; Gomez, L.D.; McQueen-Mason, S.J. *Arabidopsis* GT34 family contains five xyloglucan α-1,6-xylosyltransferases. *New Phytol.* **2012**, *195*, 585–595.
86. Zabotina, O.A.; Avci, U.; Cavalier, D.; Pattathil, S.; Chou, Y.H.; Eberhard, S.; Danhof, L.; Keegstra, K.; Hahn, M.G. Mutations in multiple *XXT* genes of Arabidopsis reveal the complexity of xyloglucan biosynthesis. *Plant Physiol.* **2012**, *159*, 1367–1384.
87. Wang, C.; Li, S.; Ng, S.; Zhang, B.; Zhou, Y.; Whelan, J.; Wu, P.; Shou, H. Mutation in xyloglucan 6-xylosytransferase results in abnormal root hair development in *Oryza sativa*. *J. Exp. Bot.* **2014**, PMID: 24834920.
88. Madson, M.; Dunand, C.; Li, X.; Verma, R.; Vanzin, G.F.; Caplan, J.; Shoue, D.A.; Carpita, N.C.; Reiter, W.D. The MUR3 gene of Arabidopsis encodes a xyloglucan galactosyltransferase that is evolutionarily related to animal exostosins. *Plant Cell* **2003**, *15*, 1662–1670.
89. Jensen, J.K.; Schultink, A.; Keegstra, K.; Wilkerson, C.G.; Pauly, M. RNA-Seq Analysis of developing nasturtium seeds (*Tropaeolum majus*): Identification and characterization of an additional galactosyltransferase involved in xyloglucan biosynthesis. *Mol. Plant* **2012**, *5*, 984–992.
90. Lopes, F.J.F.; Pauly, M.; Brommonshenkel, S.H.; Lau, E.Y.; Diola, V.; Passos, J.L.; Loureiro, M.E. The *EgMUR3* xyloglucan galactosyltransferase from *Eucalyptus grandis* complements the *mur3* cell wall phenotype in *Arabidopsis thaliana*. *Tree Genet. Genomes* **2010**, *6*, 745–756.
91. Perrin, R.M.; DeRocher, A.E.; Bar-Peled, M.; Zeng, W.Q.; Norambuena, L.; Orellana, A.; Raikhel, N.V.; Keegstra, K. Xyloglucan fucosyltransferase, an enzyme involved in plant cell wall biosynthesis. *Science* **1999**, *284*, 1976–1979.
92. Vanzin, G.F.; Madson, M.; Carpita, N.C.; Raikhel, N.V.; Keegstra, K.; Reiter, W.D. The *mur2* mutant of *Arabidopsis thaliana* lacks fucosylated xyloglucan because of a lesion in fucosyltransferase AtFUT1. *Proc. Natl. Acad. Sci. USA* **2002**, *99*, 3340–3345.
93. Gille, S.; de Souza, A.; Xiong, G.; Benz, M.; Cheng, K.; Schultink, A.; Reca, I.B.; Pauly, M. *O*-acetylation of Arabidopsis hemicellulose xyloglucan requires AXY4 or AXY4L, proteins with a TBL and DUF231 domain. *Plant Cell* **2011**, *23*, 4041–4053.
94. Günl, M.; Pauly, M. *AXY3* encodes a α-xylosidase that impacts the structure and accessibility of the hemicellulose xyloglucan in Arabidopsis plant cell walls. *Planta* **2011**, *233*, 707–719.
95. Sampedro, J.; Pardo, B.; Gianzo, C.; Guitian, E.; Revilla, G.; Zarra, I. Lack of α-xylosidase activity in Arabidopsis alters xyloglucan composition and results in growth defects. *Plant Physiol.* **2010**, *154*, 1105–1115.
96. Sampedro, J.; Gianzo, C.; Iglesias, N.; Guitian, E.; Revilla, G.; Zarra, I. AtBGAL10 is the main xyloglucan beta-galactosidase in Arabidopsis, and its absence results in unusual xyloglucan subunits and growth defects. *Plant Physiol.* **2012**, *158*, 1146–1157.

97. Zablackis, E.; York, W.S.; Pauly, M.; Hantus, S.; Reiter, W.D.; Chapple, C.C.; Albersheim, P.; Darvill, A. Substitution of L-fucose by L-galactose in cell walls of *Arabidopsis mur1*. *Science* **1996**, *272*, 1808–1810.
98. Pauly, M.; Eberhard, S.; Albersheim, P.; Darvill, A.; York, W.S. Effects of the *mur1* mutation on xyloglucans produced by suspension-cultured *Arabidopsis thaliana* cells. *Planta* **2001**, *214*, 67–74.
99. Hisamatsu, M.; Impallomeni, G.; York, W.S.; Albersheim, P.; Darvill, A.G. A new undecasaccharide subunit of xyloglucans with two α-L-fucosyl residues. *Carbohydr. Res.* **1991**, *211*, 117–129.
100. Nishitani, K. Construction and restructuring of the cellulose-xyloglucan framework in the apoplast as mediated by the xyloglucan-related protein family—A hypothetical scheme. *J. Plant Res.* **1998**, *111*, 159–166.
101. Tabuchi, A.; Mori, H.; Kamisaka, S.; Hoson, T. A new type of endo-xyloglucan transferase devoted to xyloglucan hydrolysis in the cell wall of azuki bean epicotyls. *Plant Cell Physiol.* **2001**, *42*, 154–161.
102. Nishitani, K.; Tominaga, R. Endoxyloglucan transferase, a novel class of glycosyltransferase that catalyzes transfer of a segment of xyloglucan molecule to another xyloglucan molecule. *J. Biol. Chem.* **1992**, *267*, 21058–21064.
103. Fry, S.C.; Smith, R.C.; Renwick, K.F.; Martin, D.J.; Hodge, S.K.; Matthews, K.J. Xyloglucan endotransglycosylase, a new wall-loosening enzyme-activity from plants. *Biochem. J.* **1992**, *282*, 821–828.
104. Rose, J.K.C.; Braam, J.; Fry, S.C.; Nishitani, K. The XTH family of enzymes involved in xyloglucan endotransglucosylation and endohydrolysis: Current perspectives and a new unifying nomenclature. *Plant Cell Physiol.* **2002**, *43*, 1421–1435.
105. Eklof, J.M.; Brumer, H. The XTH gene family: An update on enzyme structure, function, and phylogeny in xyloglucan remodeling. *Plant Physiol.* **2010**, *153*, 456–466.
106. Vissenberg, K.; Martinez-Vilchez, I.M.; Verbelen, J.-P.; Miller, J.G.; Fry, S.C. *In vivo* colocalization of xyloglucan endotransglycosylase activity and its donor substrate in the elongation zone of Arabidopsis roots. *Plant Cell* **2000**, *12*, 1229–1237.
107. Iglesias, N.; Abelenda, J.A.; Rodiño, M.; Sampedro, J.; Revilla, G.; Zarra, I. Apoplastic glycosidases active against xyloglucan oligosaccharides of *Arabidopsis thaliana*. *Plant Cell Physiol.* **2006**, *47*, 55–63.
108. Dick-Pérez, M.; Zhang, Y.; Hayes, J.; Salazar, A.; Zabotina, O.A.; Hong, M. Structure and interactions of plant cell-wall polysaccharides by two-and three-dimensional magic-angle-spinning solid-state NMR. *Biochemistry* **2011**, *50*, 989–1000.

6

Quantifying the Effects of Photoperiod, Temperature and Daily Irradiance on Flowering Time of Soybean Isolines

Elroy R. Cober *, Daniel F. Curtis, Douglas W. Stewart and Malcolm J. Morrison

Agriculture and Agri-Food Canada, Eastern Cereal and Oilseed Research Centre (ECORC), 960 Carling Ave, Bldg 110, Ottawa, Ontario K1A 0C6, Canada; E-Mails: curtisseeds@bioflora.ca (D.F.C.); stewartdmej@gmail.com (D.W.S.); malcolm.morrison@agr.gc.ca (M.J.M.)

* Author to whom correspondence should be addressed; E-Mail: elroy.cober@agr.gc.ca

---

Abstract: Soybean isolines with different combinations of photoperiod sensitivity alleles were planted in a greenhouse at different times during the year resulting in natural variation in daily incident irradiance and duration. The time from planting to first flower were observed. Mathematical models, using additive and multiplicative modes, were developed to quantify the effect of photoperiod, temperature, photoperiod-temperature interactions, rate of photoperiod change, and daily solar irradiance on flowering time. Observed flowering times correlated with predicted times ($R^2 = 0.92$, Standard Error of the Estimate (SSE) = 2.84 d, multiplicative mode; $R^2 = 0.91$, SSE = 2.88 d, additive mode). The addition of a rate of photoperiod change function and an irradiance function to the temperature and photoperiod functions improved the accuracy of flowering time prediction. The addition of a modified photoperiod function, which allowed for photoperiod sensitivity at shorter photoperiods, improved prediction of flowering time. Both increasing and decreasing rate of photoperiod change, as well as low levels of daily irradiance delayed flowering in soybean. The complete model, which included terms for the rate of photoperiod change, photoperiod, temperature and irradiance, predicted time to first flower in soybean across a range of environmental conditions with an SEE of 3.6 days when tested with independent data.

Keywords: *Glycine max*; soybean; flowering time; photoperiod; temperature; irradiance; phenology; modeling

1. Introduction

Eight genes, which influence time from planting to first flower, have been identified in soybean [*Glycine max* (L.) Merr.] to date: E1 and E2 [1], E3 [2], E4 [3,4], E5 [5], E6 [6], E7 [7] and E8 [8]. Two of these, E3 [9] and E4 [10] have been identified as phytochrome A genes while E2 has been identified as a GIGANTEA homolog [11] and E1 as an inhibitor of FT [12]. Alleles at these loci, in conjunction with photoperiod [13] and temperature [14], regulate the timing of flowering and maturity of soybean lines. Soybean is a facultative short day plant, so under non-inductive long days (>14 h) and high temperatures (25 to 30 °C) dominant E alleles delay flowering, with the exception of the E6 locus [14]. Lower temperatures reduced the delaying effect of E alleles under non-inductive photoperiods [14]. Isolines with dominant E alleles at two or three loci flowered earlier under low temperature (18 C) and non-inductive photoperiods compared to higher temperature (28 °C) with non-inductive photoperiods, but the time to flowering in both instances was still greater than the time to flower under inductive short photoperiods [14]. This result underscored the importance of the E genes in the adaptation of soybean lines to specific climates and highlighted the requirement for a thorough understanding of their function in controlling flowering.

The majority of work quantifying the effects of E genes on flowering has been done in growth cabinets under constant daylength and temperature or in the field during summer months when much of the phenological development to flowering is close to the summer solstice with relatively long and constant daylengths. If a soybean line is grown at a different latitude or time of the year, different photoperiods and rates of photoperiod change will be encountered. This is a consideration for countries like India, China, and Australia where soybeans can be grown during the winter months, which have short but increasing photoperiods [15–17]. For this reason, it is important to study the flowering time allele responses under increasing and decreasing day lengths.

Two approaches have been used to model temperature and photoperiod effects on soybean phenology. One approach is to sum over separate temperature and photoperiod functions [18–22] with a convenient coefficient for quantifying genetic differences for photoperiod sensitivity. Unfortunately, simple addition doesn't account for temperature effects on photoperiod sensitivity as shown by Cober *et al.* [14] who added a function to account for temperature-photoperiod interactions. A temperature-photoperiod interactive term was also used by Yan and Wallace [23]. The other approach is to multiply temperature and photoperiod functions which would introduce temperature-photoperiod interactions automatically [24–28]. Accounting for the rate of photoperiod change [19] and low solar irradiance values remains un-quantified. It is not known if daily solar irradiance can be low enough to reduce photosynthate production to a level where phenological development is inhibited. In this study, an experiment was devised to expose soybean plants to large changes in day lengths and varying amounts of daily irradiance while measuring the time from planting to first flower. The objective of this study was to identify and quantify the effects of: photoperiod, rate of photoperiod change, and daily irradiance on the time from planting to first flower in soybeans isolines varying in alleles governing photoperiod sensitivity. Modeling time to first flower started with an existing model incorporating temperature and photoperiod and the new model described in this manuscript added a non-linear photoperiod response function, a rate of photoperiod change function, and a daily irradiance function.

2. Results and Discussion

Our original soybean flowering time model had functions for temperature, photoperiod response and genotype specific parameters [29]. The soybean genotypes, maturity gene near-isogenic lines, used in this experiment shared one of two common genetic backgrounds "Harosoy" or "Clark" where combinations of early or later flowering alleles at various flowering loci were backcrossed into the recurrent parent (Table 1). Since we grew plants at a northern latitude with planting dates throughout the year, we produced a flowering time data set which resulted from growth of soybean isolines in a wide range of photoperiods as well as a wide range of rates of photoperiod change (Figure 1) and a range of incident irradiance (Figure 2b,c). Air temperature within the greenhouse was relatively constant except for some days in the late summer and early fall when supplemental heat at night was unavailable (Figure 2a).

Table 1. Harosoy (H) and Clark (C) soybean isolines used in the phenology study with their known maturity genotype and derived photoperiod coefficients with standard errors for the additive (c, Equation (14)) and multiplicative models (c*, Equation (15)). Dependent (Dep.) and independent (Ind.) photoperiod coefficients were derived from two data sets derived from alternate planting dates throughout the year.

Isoline		Maturity genotype †	Photoperiod coefficient (standard error)			
			Additive ($°C^{-1}$ d^{-1} h^{-1})		Multiplicative (h^{-1})	
			Dep.	Ind.	Dep.	Ind.
OT 94-47	H	e1e2e3e4e5e7	1.32 (0.12)	1.53 (0.18)	0.0206 (0.0057)	0.0193 (0.0098)
OT 89-5	H	e1e2e3e4e5E7	1.66 (0.13)	1.94 (0.23)	0.0416 (0.0068)	0.0478 (0.0116)
OT 93-26 ‡	H	E1e2e3e4e5E7	2.32 (0.09)	2.52 (0.15)	0.0806 (0.0048)	0.0822 (0.0069)
OT 93-28 ‡	H	E1e2e3e4e5E7	2.38 (0.09)	2.57 (0.15)	0.0839 (0.0045)	0.0859 (0.0070)
OT 94-41	H	e1e2E3e4e5E7	2.10 (0.09)	2.28 (0.14)	0.0675 (0.0050)	0.0686 (0.0079)
L92-21	C	e1e2E3e4e5E7	1.90 (0.12)	2.17 (0.18)	0.0543 (0.0063)	0.0609 (0.0092)
L62-667	H	e1e2e3E4e5E7	2.00 (0.12)	2.25 (0.18)	0.0624 (0.0067)	0.0663 (0.0104)
L71-920	C	e1e2e3E4e5E7	2.06 (0.10)	2.23 (0.16)	0.0657 (0.0054)	0.0650 (0.0084)
L71-802	H	E1e2e3E4e5E7	3.17 (0.05)	3.28 (0.07)	0.1304 (0.0029)	0.1290 (0.0046)
L80-5914	C	E1e2e3E4e5E7	3.13 (0.06)	3.18 (0.07)	0.1283 (0.0036)	0.1234 (0.0040)
L84-307	H	e1E2e3E4e5E7	2.71 (0.10)	2.81 (0.11)	0.1034 (0.0052)	0.1011 (0.0064)
L63-2404	C	e1E2e3E4e5E7	2.65 (0.07)	2.76 (0.09)	0.0996 (0.0038)	0.0972 (0.0052)
Harosoy	H	e1e2E3E4e5E7	2.25 (0.12)	2.44 (0.18)	0.0772 (0.0068)	0.0780 (0.0090)
L84-307	H	e1e2e3E4E5E7	2.10 (0.12)	2.28 (0.19)	0.0671 (0.0067)	0.0686 (0.0103)
L74-441	H	E1E2e3E4e5E7	3.63 (0.05)	3.73 (0.07)	0.1567 (0.0027)	0.1561 (0.0043)
L67-2324	H	E1e2E3E4e5E7	3.79 (0.06)	3.82 (0.04)	0.1662 (0.0029)	0.1615 (0.0031)
L66-432	C	E1e2E3E4e5E7	3.66 (0.09)	3.75 (0.09)	0.1588 (0.0050)	0.1569 (0.0054)

Table 1. *Cont.*

Isoline	Maturity genotype †		Photoperiod coefficient (standard error)			
			Additive (°C^{-1}d^{-1}h^{-1})		Multiplicative (h^{-1})	
			Dep.	Ind.	Dep.	Ind.
L64-4584	H	e1E2E3E4e5E7	3.17 (0.06)	3.19 (0.08)	0.1311 (0.0034)	0.1243 (0.0043)
Clark	C	e1E2E3E4e5E7	3.14 (0.07)	3.13 (0.09)	0.1294 (0.0041)	0.1203 (0.0054)
L64-4830	H	e1e2E3E4E5E7	3.22 (0.10)	3.12 (0.12)	0.1345 (0.0051)	0.1194 (0.0068)
L94-1110	C	e1e2E3E4E5E7	2.72 (0.15)	2.70 (0.16)	0.1042 (0.0082)	0.0930 (0.0092)
L71L-3004	H	E1E2E3E4e5E7	4.44 (0.05)	4.47 (0.07)	0.2034 (0.0030)	0.2001 (0.0042)
L65-3366	C	E1E2E3E4e5E7	4.23 (0.05)	4.31 (0.09)	0.1916 (0.0031)	0.1910 (0.0050)
L71L-3015	H	E1e2E3E4E5E7	4.21 (0.05)	4.28 (0.09)	0.1907 (0.0026)	0.1889 (0.0050)
L74-66	H	e1E2E3E4E5E7	3.12 (0.12)	3.36 (0.16)	0.1276 (0.0069)	0.1361 (0.0087)
L92-1195	C	e1E2E3E4E5E7	3.21 (0.12)	2.99 (0.13)	0.1310 (0.0069)	0.1114 (0.0080)

† Dominant E alleles are late flowering under non-inductive long photoperiods while recessive e alleles are early flowering under non-inductive long photoperiods; ‡ These isolines have the same maturity genotype but differ for pubescence color.

Figure 1. The rate of photoperiod change and the length of the photoperiod as it varied over a year at Ottawa, Canada.

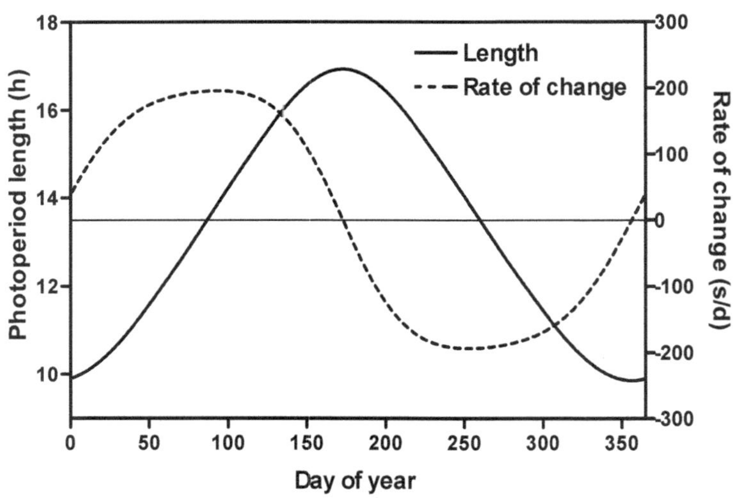

2.1. Irradiance

As a result of a northerly location and winter conditions, we had some days with very low amounts of solar radiation (Figure 2b,c). As solar irradiance increased, the size of the irradiance function decreased exponentially in both the multiplicative and additive models (Figure 3) and as a result the value of the function quickly became small. For the conditions encountered in this experiment, the irradiance functions were important to the accuracy of the models because they reduced the SEEs by 2.24 and 3.77 d for the multiplicative and additive models, respectively (Table 2). An estimate, using Equation (1) (see Section 3, Experimental), of the percent incident irradiance transmitted through the greenhouse was made using daily solar irradiance measured at a weather station approximately 500 m from the greenhouse. In estimating irradiation, Equation (1) accounted for 58 percent of the variation between measured and calculated percent transmission into the greenhouse (P_T) (Figure 4) with a relatively small SEE of 3.65%. Using Equation (1) was far superior to using an average percent transmission (data not shown [30]). Calculated irradiance values inside the greenhouse were about 50% of those outside the greenhouse. Within the greenhouse, irradiance was reduced during the winter months to about 20% of maximum irradiance in the summer (Figure 2b,c).

Figure 2. The (**a**) mean daily temperature in the greenhouse, (**b**) daily amounts of solar irradiance as measured at a weather station and (**c**) daily amounts of solar irradiance as calculated in the greenhouse during the study at Ottawa, Canada. The horizontal rule indicates 10 MJ m^{-2} d^{-1}.

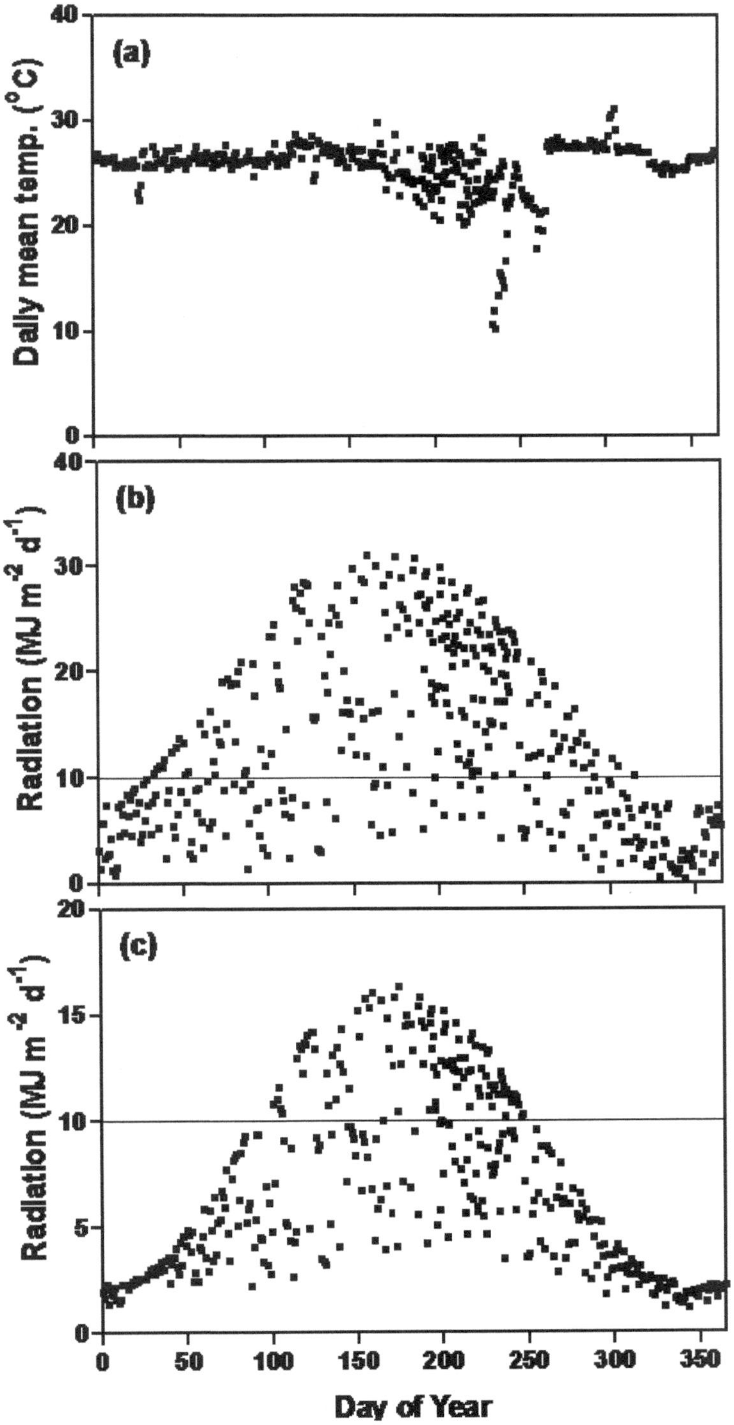

Figure 3. The solar irradiance functions for the multiplicative and additive models which predict flowering time in soybean.

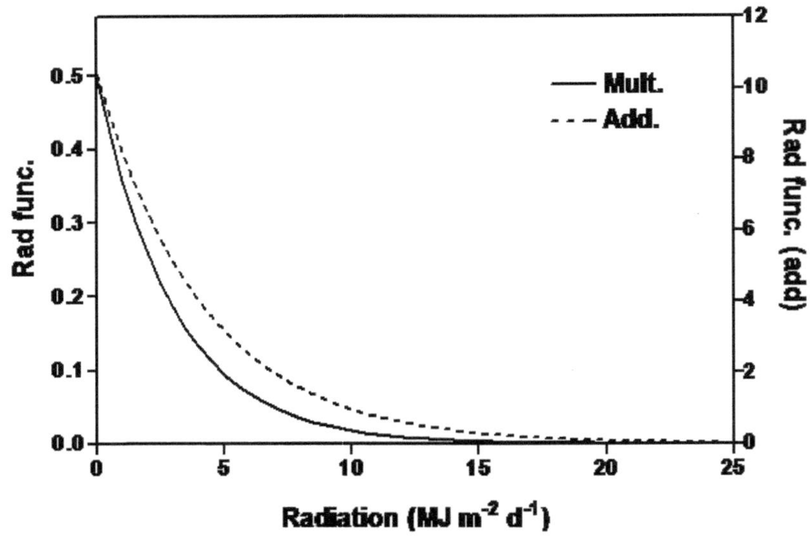

Figure 4. A comparison of measured and calculated percent transmission (PT; Equation (1)) of irradiance in the greenhouse used in the study of soybean flowering time at Ottawa, Canada.

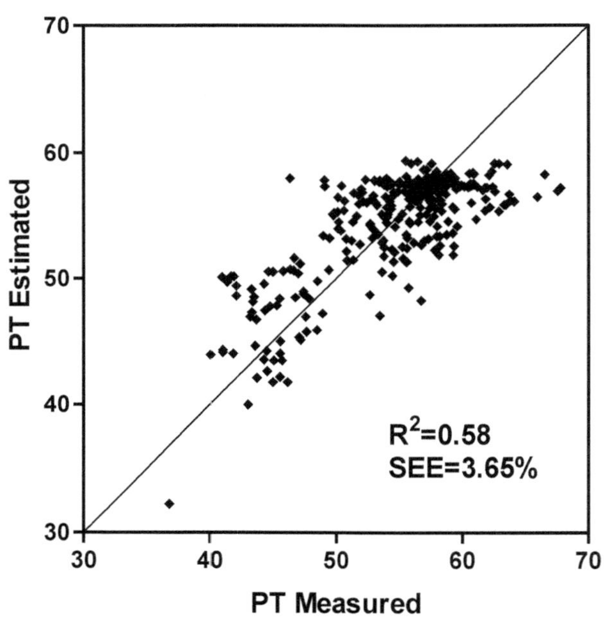

We found the addition of an irradiance parameter improved the model's prediction of first flower. When the daily irradiance was less than 10 MJ m^{-2}, flowering was delayed in our model. Birch *et al.* [31] studied phyllochron length (the time between successive leaf emergence) in maize (*Zea mays* L.) under shading in the field. They found a phyllochron increase of 2 to 4 °C d^{-1} MJ^{-1} of PAR irradiance, as it decreased from 9.6 to 1.1 MJ PAR m^{-2} d^{-1}. Similarly increasing irradiance, to 20 h of 650 μmol m^{-2} s^{-1} radiation from 10 h of 650 μmol m^{-2} s^{-1} plus 10 h of 5–10 μmol m^{-2} s^{-1} radiation, increased the rate of leaf appearance in maize [32]. In a study of quinoa (*Chenopodium quinoa* Willd.) [33], seven cultivars which were sensitive to rate of change of photoperiod were also sensitive to irradiance, where lower irradiance increased the phyllochron. Horticultural researchers have been active in researching irradiance effects on development as it affects commercial greenhouse-produced ornamentals. In a study with yarrow (*Achillea millefolium* L. "Summer Pastels"), Zhang *et al.* [34] used controlled environments and found that irradiance levels of 100 μmol m^{-2} s^{-1} delayed anthesis 20 d compared to irradiance levels of 300 μmol m^{-2} s^{-1}. Mattson and Erwin [35] examined time to flower, as a function of leaf number at first flower, for 41 species of herbaceous ornamentals grown from seed. In this work, short 8 h treatments of ambient irradiance ranging from 222 to 458 μmol m^{-2} s^{-1} were enriched with 50, 100, or 150 μmol m^{-2} s^{-1} additional irradiance, while long photoperiods were provided with the same short day irradiance plus an additional 10 h of 50, 100, or 150 μmol m^{-2} s^{-1} additional irradiance. Of the 41 species, one did not flower, and 28 were insensitive to irradiance, that is increased irradiance under inductive conditions did not reduce the number of leaves at first flower. Ten species showed a facultative irradiance response, that is, there were reduced leaf numbers at first flower as irradiance increased, while two species had increased leaf numbers at flowering as irradiance increased.

In our study, low irradiance delayed soybean flowering. To the best of our knowledge, irradiance functions have not been incorporated into soybean phenological models before this study. Part of the importance of the irradiance function, in our work, was due to the experimental conditions within the greenhouse, where a relatively large number of days were encountered where daily irradiance totals in the greenhouse fell below 10 MJ m^{-2} d^{-1} (Figure 2c) compared to amounts falling on the greenhouse (Figure 2b). During the summer months (days 130 to 270) of this experiment, there were relatively few days with less than 10 MJ m^{-2} d^{-1} of incident irradiance measured at the weather station. Therefore, for most field environments the irradiance effect would be negligible. In maritime climates or years with rainy, cloudy conditions in continental climates, and for some growth room and greenhouse experiments, lower incident solar irradiance could be a factor in determining phenology in soybean and other crops.

2.2. Rate of Photoperiod Change

In this experiment, the addition of the rate of photoperiod change function to the model had a relatively large effect in reducing SEEs by 1.31 and 2.66 d for the multiplicative and additive models respectively (Table 2). The rate of photoperiod change (RPC) coefficients (g and g*) were similar in value and the absolute values of their positive and negative components were not significantly different (Table 3). They have opposite signs to compensate for the change in sign of the rate of photoperiod change. We found that both increasing and decreasing rates of change of photoperiod delayed flowering.

The addition of the rate of photoperiod change (RPC) parameter improved the precision of our soybean flowering time model. Constable and Rose [19], working with a range of soybean cultivars, found the addition of a daylength rate of change parameter also improved their flowering time model. More studies of RPC effects on crop plant development have been done in cereals than in soybean. Kirby *et al.* [36], using nine cultivars of barley (*Hordeum vulgare* L.) found a negative RPC lengthened the phyllocron interval while a positive RPC shortened the phyllocron interval. In contrast, Kernich *et al.* [37] using two barley cultivars found that phyllochron and time to awn initiation was not affected by RPC. In wheat (*Triticum aestivum* L.), Baker *et al.* [38] worked with one cultivar and found the phyllchron was more correlated with RPC than the mean photoperiod and that a negative RPC delayed the phyllocron while a positive RPC increase the rate of leaf emergence. In contrast, Slafer *et al.* [39] found that the phyllochron and final leaf number, as well as time from emergence to terminal spikelet emergence [40] was not affected by RPC. In maize, a short day cereal, Bonhomme *et al.* [41] studied six cultivars and found only the tropical cultivars had significant response to RPC where leaf number was reduced (early flowering) with a positive RPC. Clerget *et al.* [42] studied three tropical sorghum (*Sorghum bicolour* (L.) Moench) cultivars, also short day cereals, and found conflicting results in controlled environments with different RPC. In field studies using different planting dates, they found that the addition of a RPC factor to their photoperiod model improved the prediction of panicle initiation and a negative RPC speeded flowering. We found that both a positive and negative RPC delayed soybean flowering. This is similar to results found in a study of RPC effects on phyllochron of quinoa where seven of nine cultivars studied were sensitive to RPC and both negative and positive RPC lengthened the phyllochron compared to nearly constant photoperiods [33]. There are many conflicting results when considering effects of RPC on either the phyllochron or phenology. An understanding of the effect of RPC is important when modeling plant growth across a range of environmental conditions.

Table 2. Model standard errors of estimate (SEE) and R^2 for complete and reduced models for predicting time to first flower for soybean.

Model		Multiplicative		Additive	
	N	**R^2**	**SEE (d)**	**R^2**	**SEE (d)**
Complete model † (full dataset)	389	0.92	2.84	0.91	2.88
Without non-linear photoperiod function	389	0.90	3.18	0.89	3.34
Without rate of photoperiod change function	389	0.90	4.15	0.87	5.54
Without irradiance function	389	0.87	5.08	0.84	6.64
Without all of the above three functions	389	0.80	7.46	0.75	10.10
Complete model (dependent data)	207	0.92	2.53	0.93	2.89
Complete model (independent data)	187	0.89	3.61	0.90	3.55

† Model with temperature, non-linear photoperiod, rate of photoperiod change, and daily irradiance functions, and a photoperiod coefficient for each isoline.

Table 3. Non-photoperiod coefficients for modeling soybean flowering time (with standard errors (SE) and units) obtained by least squares fitting to observations of times from planting to first flower from a full (N = 389) and reduced (N = 207) dataset. Coefficients are found in Equations (7), (11), (13) and (14) (additive model) and Equations (9), (12) and (15) (multiplicative model). Multiplicative model coefficients are denoted with an * while additive model coefficients do not have an *.

N	Multiplicative Model				Additive Model			
	Coefficient	**Value**	**SE**	**Units**	**Coefficient**	**Value**	**SE**	**Units**
389	b *	0.0021	3.4×10^{-5}	$°C^{-1}$ d^{-1}	b	0.0024	3.8×10^{-5}	$°C^{-1}$ d^{-1}
389	g * (plus)	9.464	1.080	dh^{-1}	g (plus)	11.40	0.983	dh^{-1}
389	g * (neg)	−9.822	1.078	dh^{-1}	g (neg)	−11.02	0.921	dh^{-1}
389	h *	0.499	0.060	h^{-1}	h	10.20	0.438	$°C$ h^{-1}
389	k *	0.333	0.066	d J^{-1}	k	0.228	0.023	d J^{-1}
207	b *	0.0021	4.2×10^{-5}	$°C^{-1}$ d^{-1}	b	0.0025	3.7×10^{-5}	$°C^{-1}$ d^{-1}
207	g * (plus)	7.975	1.461	dh^{-1}	g (plus)	12.92	0.804	dh^{-1}
207	g * (neg)	−11.62	1.353	dh^{-1}	g (neg)	−14.64	0.795	dh^{-1}
207	h *	0.460	0.050	h^{-1}	h	11.94	0.484	$°C$ h^{-1}
207	k *	0.297	0.068	d J^{-1}	k	0.302	0.018	d J^{-1}

2.3. Photoperiod

The small effect of the non-linear photoperiod response (Figure 5) was unexpected as it first appeared that we were obtaining large photoperiod responses (flowering delays) at photoperiods less than 13.5 h. These delays occurred during the short, cloudy days of late fall and the introduction of the irradiance function accounted for these flowering delays much better than the two-line photoperiod response. Our initial photoperiod function had the photoperiod effect decline to nil at 14.3 h.

Figure 5. The original one-line photoperiod function (broken line) and the modified two-line function (solid line) which was used in this study for modeling soybean flowering time.

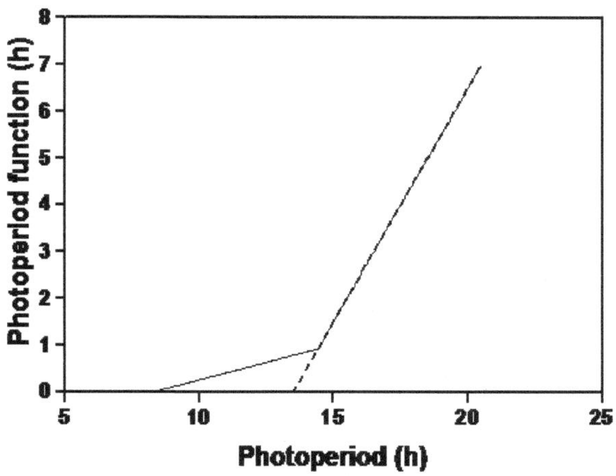

Our current work has showed that a two line function which allowed a photoperiod effect at shorter photoperiods, as low as 8.2 h, somewhat improved our ability to predict flowering time. Sinclair *et al.* [43] working with long juvenile and "normal" lines found the normal line showed a response to photoperiod as it lengthened from 11.5 to 14.4 h. Grimm *et al.* [25] modeled optimum night length, at which rate of development to flowering was fastest, using cultivars ranging from maturity group 000 to VII. When night length was converted to photoperiod, optimum photoperiods were found to be 13.6 to 14 h for the earliest three maturity groups and 12 to 12.8 h for the latest three maturity groups. Grimm *et al.* [25] found that there were significant differences for optimum night length among cultivars. While most soybean crops will not be grown in photoperiods much shorter than 12 h, very short photoperiods could be experienced by subtropical, winter-grown crops.

2.4. Model Development and Extension

When developing these models, we used the whole data set. To verify the independence of the model we divided the data set nearly in half with alternating planting dates assigned to serve as approximations of dependent and independent data sets. The dependent half of the data (n = 207) was used to re-solve for the coefficients in the model. The coefficients generated from the dependent half data set were approximately the same as in the initial set when all the data (n = 389) was used (Tables 4 and 3, Figure 6). When these dependent half-data-set coefficients were used in the model and the model applied to the independent half of the dataset (n = 187), the SEE increased by about a day from the original dependent calculations (Table 2). While there are small differences in coefficients between dependent and independent solutions, the conclusions drawn from the complete dataset appear valid.

Figure 6. Comparisons of the photoperiod coefficients for the (**a**) additive model (c (°C^{-1} h^{-1})) and for the (**b**) multiplicative model (c* (h^{-1})). These coefficients were calculated by least squares using the whole data set (N = 389) and using the subset (N = 207).

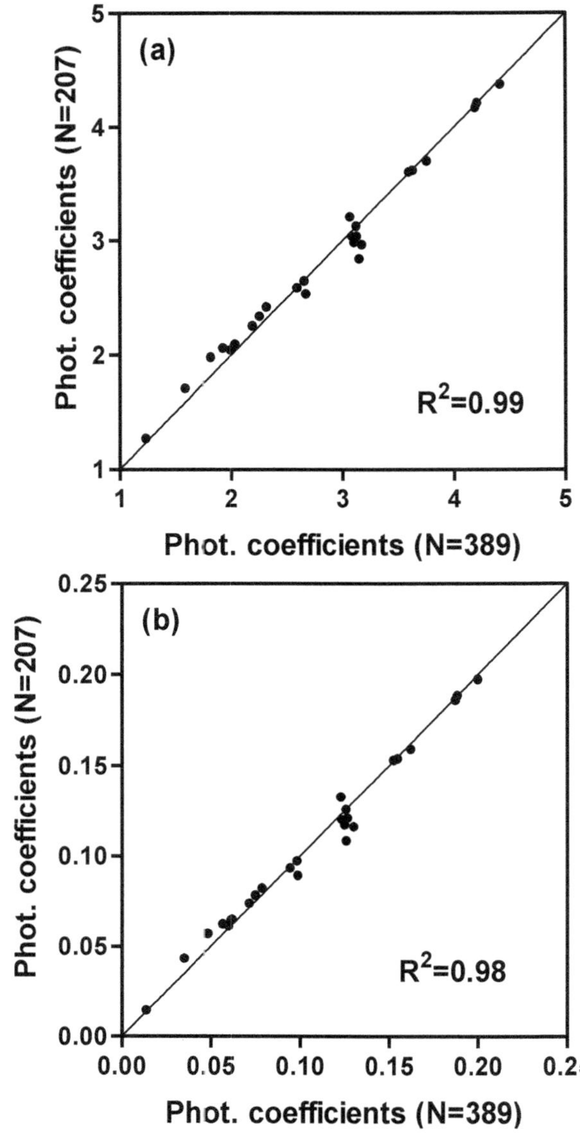

Table 4. Values and standard errors (SE) of the empirical "a" coefficients of Equation (1) which were used to calculate percent transmission of incident radiation into the greenhouse used in the study of soybean flowering time.

Coefficient	Value	SE	Units
a_1	61.17	1.81	%
a_2	−25.47	2.20	%
a_3	−0.061	0.039	% °C^{-1}
a_4	0.439	0.050	% °C^{-1}

When plotting the difference between observed and predicted (using the multiplicative model) days to first flower, we found patterns with day of the year (Figure 7). The differences increased with the photosensitivity of the lines (*i.e.*, from a to d). The same pattern occurred with the additive model (data not shown [30]). This indicated that not all the planting date to flowering time variation was accounted for in these models. This may not be surprising because we did not measure the actual irradiance each plant received. Thus, there could have been azimuthal effects caused by the geometry of the greenhouse and sun angles as they varied throughout the year. In any event the effects are not large, since SEEs of these models were less than three days.

Figure 7. Differences between multiplicative model predictions and observations for soybean isoline flowering time as they varied with planting day of the year for all 26 isolines. Isolines increase in photosensitivity from (**a**) to (**d**). The first seven isolines from Table 1 are grouped in (**a**) (OT 94-47, OT 89-5, OT 93-26, OT 93-28, OT 94-41, 192-21, 162-667), the next seven in (**b**) (L71-802, L80-5914, L84-307, L63-2404, Harosoy, L84-307), the next seven in (**c**) and the last five in (**d**).

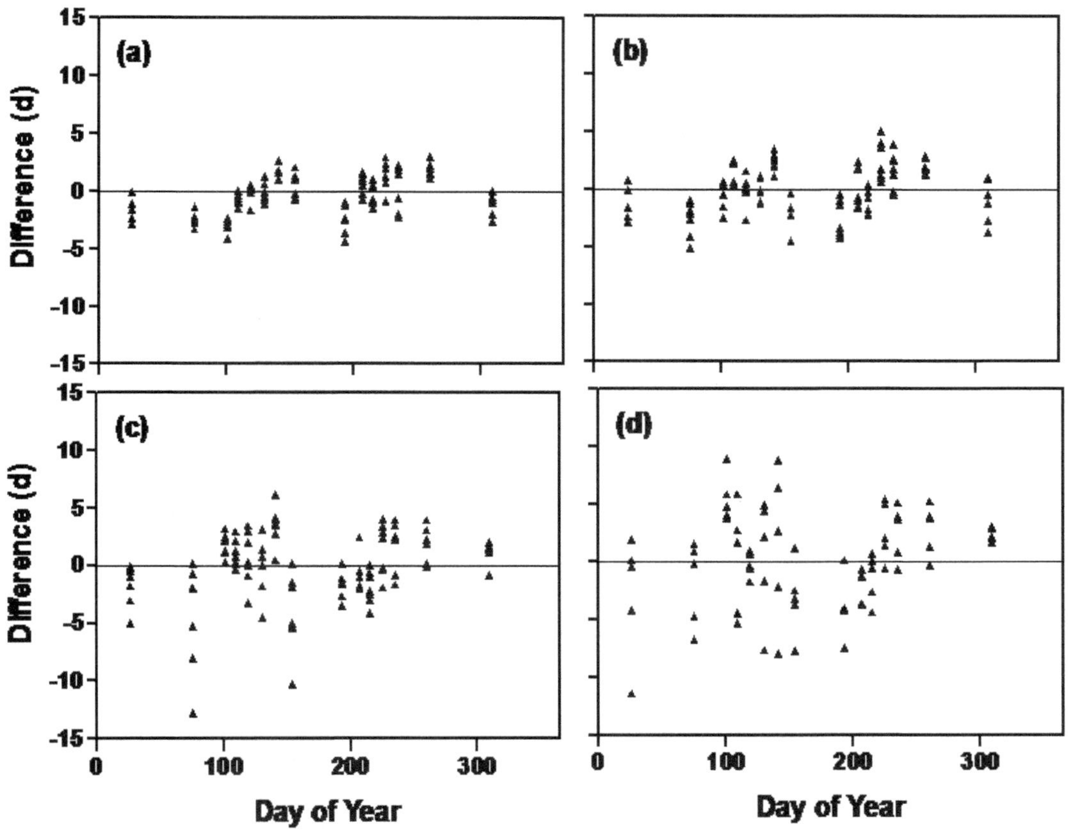

Since the near-isogenic lines share two common backgrounds, with alternative early or late flowering alleles across a range of loci, we were also interested in flowering-time gene effects. To investigate on a gene, rather than isoline basis, multiplicative and additive photoperiod coefficients were plotted for each isoline based on the number of loci with late flowering alleles, *i.e.*, the number of late flowering genes (Figure 8). Based on this classification, the isolines ranged from having zero late flowering alleles (early flowering genotypes) to having late flowering alleles at five loci (late flowering genotypes). The effect of late flowering genes appears additive although some non-additivity was seen especially with larger numbers of late flowering genes.

Figure 8. Photoperiod coefficients for soybean isolines (Y) for the multiplicative and additive models as related to the number of loci (X) in each isoline with dominant (late flowering) genes. For the multiplicative model, the equation is Y = 0.013 + 0.031X, R^2 = 0.79, SEE = 0.023 h^{-1}. For the additive model, the equation is Y = 1.15 + 0.531X, R^2 = 0.79, SEE = 0.389 °C h^{-1}.

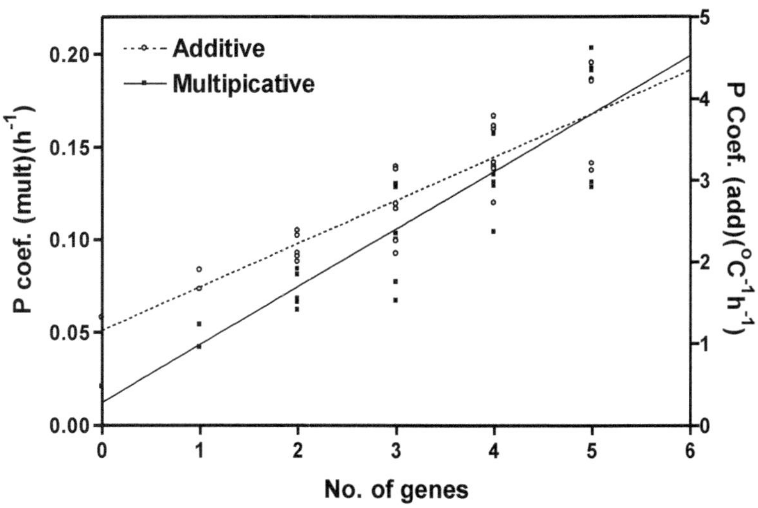

3. Experimental

3.1. Plant Experiments

"Harosoy" and "Clark" near-isogenic soybean lines with different combinations of alleles at six E loci influencing photoperiod sensitivity were grown under greenhouse conditions. There were a total of 17 maturity genotypes in the Harosoy background with nine of these maturity genotypes duplicated in the Clark background (Table 1). E6 isolines were not available. Seeds were planted in vermiculite and seedlings were transplanted to 13 cm diameter pots following unhooking of the hypocotyl. The soil mix consisted of a sterilized mix in a ratio of 3:2:1:0.5:0.5 of loam soil, peat, sand, vermiculite, and crushed brick, respectively. Air temperature in the greenhouse was recorded once each hour during the study. Hourly temperatures were averaged to obtain a mean daily temperature. The date of the first open flower was recorded for each plant.

The plants were grown in a greenhouse under natural daylengths (Figure 1) in Ottawa, Ontario, Canada (45°23'N, 75°63'W). High pressure sodium lamps provided supplemental light when irradiance fell below 100 W m^{-2} outside the greenhouse and were limited to 1 h after sunrise and turned off 1 h before twilight so as not to affect the photoperiod. The output from the high pressure sodium lamps at the plant canopy level was 32.28 W m^{-2}. The instantaneous readings were multiplied by the number of seconds the lamps were on during the day and the values added to the total irradiance calculated for the greenhouse.

The experiment consisted of 15 planting dates with three replications at each planting date. Replicates were spatially separated on different greenhouse benches. The first planting series occurred in 1999 on 17 July, 26 July, 3 August, 13 August, 23 August, 17 September and 5 November. The second planting series occurred in 2000 on 26 January, 15 March, 10 April, 18 April, 28 April, 9 May, 20 May and 2 June. The planting dates were chosen to form a paired arrangement in which the seedlings emerged during similar photoperiods, with the difference that the 1999 dates represented shortening days and 2000 dates represented lengthening days. The longest day length including civil twilight (defined as the period prior to sunrise or following sunset when the centre of the sun is geometrically 6 degrees below the horizon) occurred on 21 June 1999 at 16 h 56 min. The shortest day length including civil twilight occurred on 21 December 1999 at 9 h 51 min. Days to the first opened flower were recorded for each plant.

Photosynthetically active radiation (PAR) was calculated using an estimate of the percent irradiance transmitted through the greenhouse and daily solar irradiance measured at a weather station approximately 500 m from the greenhouse. To estimate percent PAR transmission, an experiment was devised in April 2004 where irradiance was measured continuously on a quantum basis, within the greenhouse using five 1-m line quantum sensors (Model Li-191SA, LICOR, Lincoln, NE, USA) set at five locations in the greenhouse at plant height. Average hourly values were determined. Irradiance was also measured outside the greenhouse with a quantum sensor (Model Li-190SA, LICOR, Lincoln, NE, USA). Percent PAR transmission into the greenhouse was affected by cloudiness and solar elevation (S_E). Solar elevation was calculated from astronomical equations [44]. Cloudiness was represented by the fraction (F_1) where irradiance was divided by the quantity of a maximum 2500 μmol m^{-2} s^{-1}, multiplied by the SIN of the solar elevation (S_E). On a clear day in summer at solar noon this fraction would be close to one while on a cloudy day it could fall to as low as 0.1. Percent transmission (P_T) of radiation into the greenhouse was related to F_1 and S_E by:

$$P_T = (a_1 + a_2F_1) + (a_3 + a_4F_1)S_E \quad (1)$$

The empirical coefficients a_1 to a_4 were calculated by least squares from the values of P_T, F_1 and S_E generated from the greenhouse transmissivity experiment (Table 4).

Since irradiance was measured on an energy basis (W m^{-2}) at a nearby weather station during the phenology experiment (1999 to 2000), and measured on a quanta (μmol m^{-2} s^{-1}) basis during the transmissivity experiment, the cloudiness fraction during the phenological experiment (F_2) was calculated for each hour of this experiment by dividing irradiance by the quantity of 1250 W m^{-2} times the SIN of S_E. To relate the two irradiance measures, we assumed that approximately 1250 W m^{-2} of irradiance would be measured as 2500 μmol m^{-2} s^{-1} using a quantum sensor. This also assumed that F_1 (calculated on an energy basis during the percent transmissivity experiment) was equivalent to F_2 (calculated throughout the phenology experiment) since visible and solar radiation have the same transmissivity through glass. Percent transmission into the greenhouse was calculated for each hour of the phenology experiment using Equation (1) with F_2 substituted for F_1. P_T multiplied by the hourly values of irradiance recorded at the weather station during the phenology experiment predicted hourly values of irradiance at the plant canopy level which were summed each day. Daily irradiance sums from the supplemental lamps were also added to daily irradiance values.

3.2. Theoretical Considerations for Modeling

In formulating a new model to predict time to first flower, we began with the mathematical model developed by Stewart *et al.* [29] which simulates time from planting to first flower from the average daily temperature, and latitude. The rate of change of phenological development was expressed as:

$$dD/dt = b[(T - T_B) - C_FP_F] \quad (2)$$

where D was phenological development, t was time in days, T was average daily air temperature, T_B was a base temperature below which temperature had no effect on development, b was an empirical coefficient, C_F was a temperature-photoperiod interaction function and P_F was a photoperiod function. The photoperiod function was defined following Stewart *et al.* [29] and Cober *et al.* [14] as:

$$P_F = (P - 13.5) \quad P_F = 0 \text{ when } P < 13.5 \tag{3}$$

P_F is illustrated in Figure 5. The function C_F was defined in Stewart *et al.* [29] and Cober *et al.* [14] as:

$$C_F = c - d(T_D - T) \tag{4}$$

where c was a photoperiod coefficient for each isoline (Table 1), d was an interactive coefficient, and T_D was a daily mean temperature limit. C_F had the restriction that it must be greater than or equal to zero. In this model, the interactive term ($d(T_D\text{-}T)$) increased as T fell below T_D. When the temperature decreased to T_B, photoperiod sensitivity disappeared where:

$$T_B = T_D - d/c \tag{5}$$

From a previous study where values of T were changed systematically, T_D was set to 28 °C but could vary from 24 to 30 °C with compensating changes in d, an empirical coefficient [14].

Following Cober *et al.* [14], we assumed that D (phenological development) increased from zero to one from seeding to first flower. Using this assumption, Equation (2) can be expressed in integral form as:

$$\int_{i=1}^{N} b[(T - T_B) - C_F P_F]dt = 1 \tag{6}$$

where N was the number of days from planting to first flower. Dividing by b and integrating numerically resulted in a photothermal unit (G_{PDD}) expressed as:

$$G_{PDD} = \sum_{i=1}^{N}[(T - T_B) - C_F P_F)]\Delta t = 1/b \tag{7}$$

where t was a time step of one day.

In a multiplicative approach, the rate of phenological development was expressed as:

$$dD/dt = b^*[(T - T_B)(1 - c^*P_F)] \tag{8}$$

where b* and c* were the multiplicative equivalents to b and c. The multiplicative photo thermal equation was expressed as:

$$G_{PDD} = \sum_{i=1}^{N}[(T - T_B)(1 - c^* P_F)]\Delta t = 1/b^* \tag{9}$$

In the current paper, the above models were modified in three ways. The first modification was to change the photoperiod function to P_{FM} so that photoperiod sensitivity would start at lower photoperiods. P_F was defined as

$$P_{FM} = 0.13(P - 8.2) \quad P_{FM} = 0 \text{ when } P < 8.2 \tag{10}$$

$$P_{FM} = P_F \quad \text{when } P > 14.3$$

When the photoperiod was greater than 14.3 h, P_{FM} was set equal to P_F. In other words, a one line function for photoperiod response was replaced with a two line function (Figure 5). Values for the coefficients in Equation (10) (0.13 and 8.2 h) were determined by least squares which is explained more fully below when we discuss the evaluation of all coefficients.

The second modification was to change the photoperiod functions to account for effects of the rate of photoperiod change. For the additive equation this was expressed as:

$$C_{FM1} = c[1 + g(\Delta P/\Delta t)] - d(T_D - T) \tag{11}$$

where $\Delta P/\Delta t$ was the rate of photoperiod change, and g was the rate of photoperiod change coefficient. The equivalent multiplicative equation was:

$$C^*_{FM1} = c^*[1 + g^*(\Delta P/\Delta t)] \tag{12}$$

The final modification was to add an irradiance function (R_F) to the equations expressed as:

$$R_F = h\exp(-kR) \tag{13}$$

where R was the daily total irradiance. The hypothesis was that during some days, low amounts of irradiance lowered photosynthate production to a level where the rate of phenological development was reduced. The irradiance function has a maximum value h at zero daily irradiance but falls exponentially to near zero as irradiance increases. The C_F functions were changed to:

$$C_{FM2} = c[1 - g(\Delta P/\Delta t)] - d(T_D - T) - R_F \tag{14}$$

and

$$C^*_{FM2} = c^*[1 - g^*(\Delta P/\Delta t)] - R^*_F \tag{15}$$

where R^*_F was the corresponding equation for the irradiance function with h* and k*.

The final version of the additive model was expressed as:

$$G_{PDD} = \sum_{i=1}^{N}[(T - T_B) - C_{FM2}P_{FM})]\Delta t = 1/b \tag{16}$$

The final version of the multiplicative model was expressed as

$$G_{PDD} = \sum_{i=1}^{N}[(T - T_B)(1 - C^*_{FM2}P_{FM})]\Delta t = 1/b^* \tag{17}$$

To summarize, Equations (16) and (17) modeled flowering time using temperature, non-linear photoperiod, rate of photoperiod change and daily irradiance functions, and a photoperiod coefficient for each isoline. These two models are labeled complete models throughout this report compared to reduced models where functions are deleted from the complete models.

Following the methods of Stewart *et al.* [29], we assumed that c and c* varied with isoline but all other coefficients were constant with isoline. We assumed T_b and d were 5.78 and 0.196 respectively from Cober *et al.* [14] and Stewart *et al.* [29]. Values for b, b*, c, c*, f, f*, g, g*, h, h*, k and k* were calculated by least squares by fitting Equations (7) and (9) to observations of days to planting from the experimental data of this study using a non-linear least squares algorithm [45]. Two values each of g and g* were determined; one for increasing day length and one for decreasing day length. A value for c and c* was determined for each of the 26 isolines used in this study. Since Equation (10) was discontinuous and also had a relatively small effect on model calculations, we determined the values of 0.13 and 8.2 h by using simple iteration. That is, we ran the least squares algorithm for a range of values for the two coefficients. Values of 0.13 and 8.2 h resulted in the minimum sums of squares of model calculations and observation differences.

Models were evaluated using the coefficient of determination (R^2) and standard error of estimate (SEE) defined by:

$$SEE = \sqrt{\sum_{i=1}^{N}(X_C - X_O)^2 / N_{DF}} \tag{18}$$

where N was the number of observations, X_C calculated values, X_O observed values and N_{DF} the number of degrees of freedom. We also plotted differences between model calculations and observations of days from planting to first flower as a function of day of year to determine if systematic differences were evident.

The effects of the modifications were determined by deleting the irradiance function, the rate of change function, and the non-linear photoperiod response function individually in turn and determining R^2 and SEEs for each model where a function was deleted.

To address concerns for independent validation of the model, we tested the complete models further by dividing the data into two almost equal groups using alternating planting dates to sub divide the data set, for example, data from the 17 July and 3 August planting dates were in set 1 and data from 26 July and 13 August planting dates were in set 2. In one group we resolved for all the coefficients and then compared model calculations using the new set of coefficients with observations from the second independent data set. The parameters derived from the two data sets are shown in Table 1 as dependent (first dataset) and independent (second dataset). These datasets are truly independent since the environmental conditions experienced by the plants in each set are different.

4. Conclusions

The complete model (model with temperature, non-linear photoperiod, rate of photoperiod change and daily irradiance functions, and a photoperiod coefficient for each isoline) predicted days to first flower in agreement with observations, with R^2 and SEEs of 0.92 and 2.84 d for the multiplicative model and 0.91 and 2.88 d for the additive model (Figure 9a, Figure 10a, Table 2). These SEEs compared well with the average standard deviation of the observed flowering times which was 0.97 days. The model was able to account for wide ranges in photoperiod responses throughout the year (Figure 11). Our original soybean flowering time model which did not contain an irradiance function, a rate of photoperiod change function, nor a non-linear photoperiod function had a much higher standard error of estimate for flowering time (mean of 8.8 d for both multiplicative and additive models, Figures 9b,c and 10b,c) compared to our complete model with the addition of these three functions (mean of 2.9 d, Table 2). Rating the improvements generated by the individual functions showed that the largest improvement resulted from the addition of the irradiance function, followed by the rate of photoperiod change function, with the smallest improvement resulting from the addition of the non-linear photoperiod function (Table 2). The rate of photoperiod change can be a factor impacting soybean phenology, especially near the equinoxes, when photoperiod changes are the greatest. Surprisingly both increasing and decreasing photoperiods delayed flowering. Low (<10 MJ m^{-2} day^{-1}) daily irradiance also delayed flowering. Lowered solar irradiance resulting from cloudiness could be a minor factor in Maritime climates or in cloudy years in continental climates.

Figure 9. A comparison of calculated and observed days from planting to first flower for Harosoy and Clark soybean isolines for (**a**) the complete multiplicative model, (**b**) the complete model without the rate of photoperiod change function, (**c**) the complete model without the irradiance function.

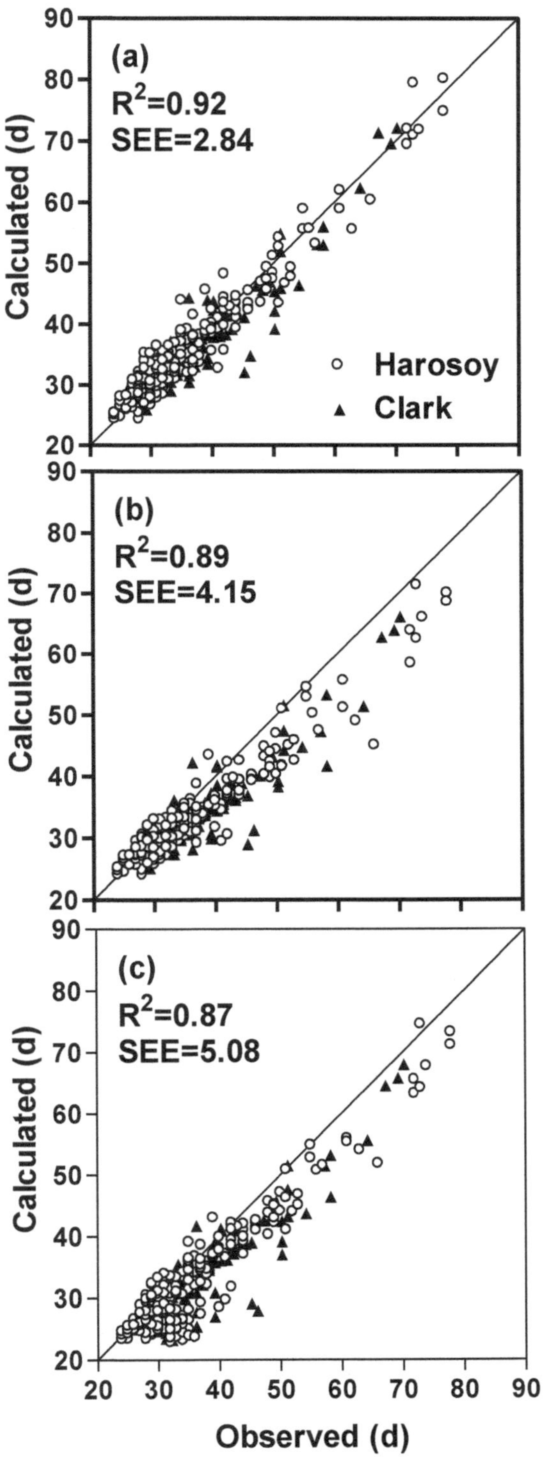

Figure 10. A comparison of calculated and observed days from planting to first flower for Harosoy and Clark soybean isolines for (**a**) the complete additive model, (**b**) the complete model without the rate of photoperiod change (**c**) the complete model without the irradiance function.

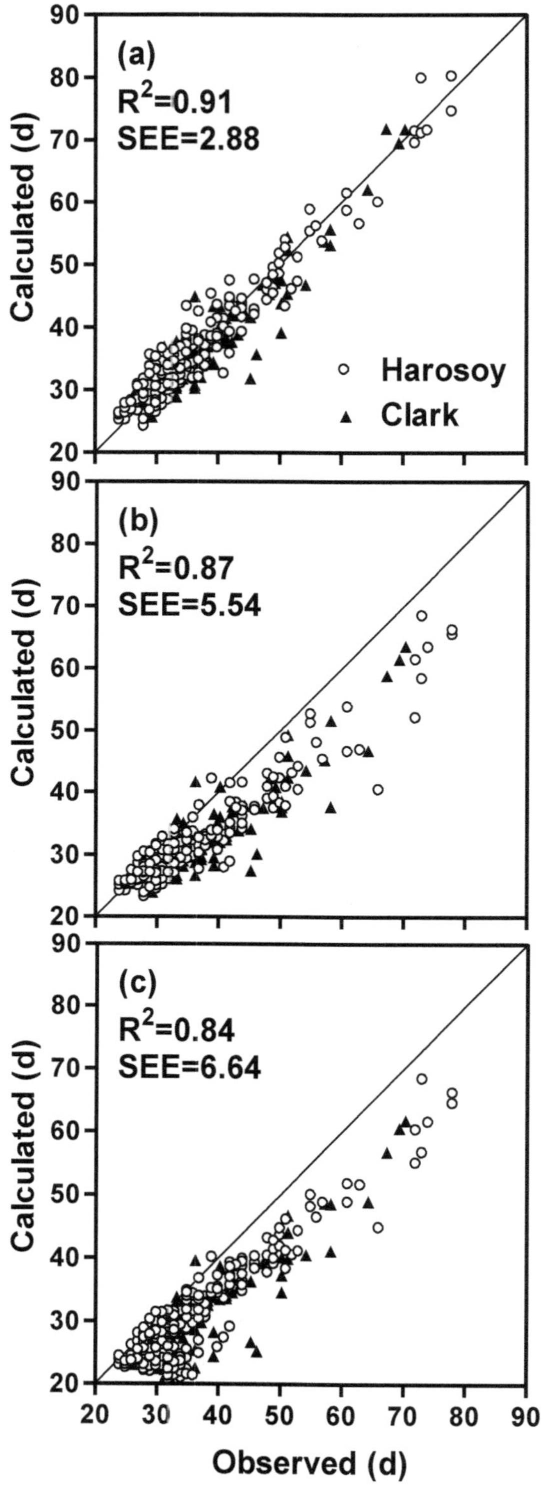

Figure 11. Days from planting to first flower *versus* the day of year when soybeans were planted based on the multiplicative model (lines) and observed values (symbols) for 12 soybean isolines varying from (**a**) least to (**d**) greatest photosensitivity.

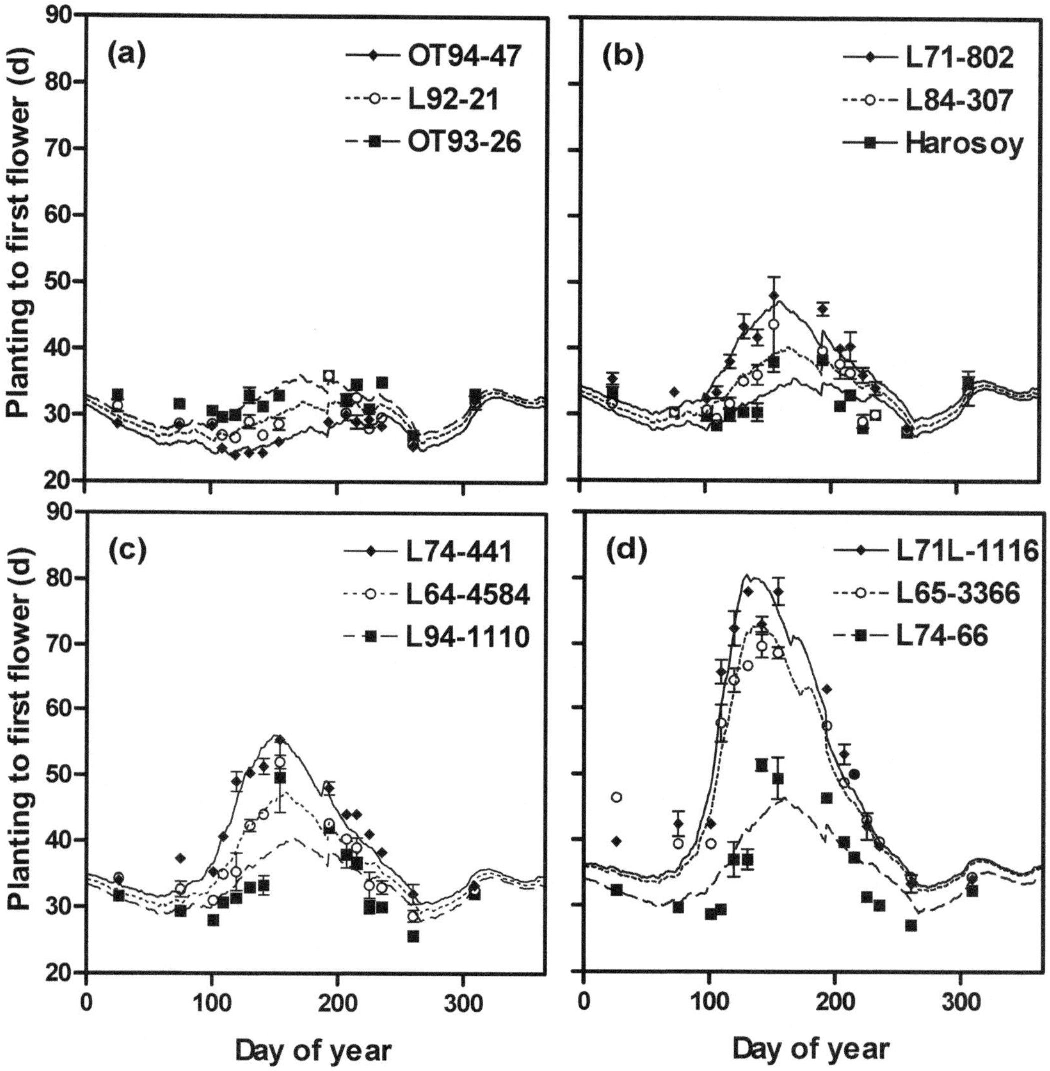

Acknowledgments

Funding was provided by Agriculture and Agri-Food Canada. We acknowledge the excellent technical assistance of Betty Wilson, and Doug Balchin, and plant care provided by George Hill and the phytotron staff.

Author Contributions

Elroy R. Cober and Daniel F. Curtis designed and carried out the plant experiment. Douglas W. Stewart developed the model equations, derived model parameters, and prepared figures. Elroy R. Cober and Malcolm J. Morrison managed interpretation and most of the manuscript preparation.

Conflicts of Interest

The authors declare no conflict of interest.

References

1. Bernard, R.L. Two major genes for time of flowering and maturity in soybeans. *Crop Sci.* **1971**, *11*, 242–244.
2. Buzzell, R.I. Inheritance of a soybean flowering response to fluorescent-daylength conditions. *Can. J. Genet. Cytol.* **1971**, *13*, 703–707.
3. Buzzell, R.I.; Voldeng, H.D. Inheritance of insensitivity to long daylength. *Soybean Genet. Newsl.* **1980**, *7*, 26–29.
4. Saindon, G.; Beversdorf, W.D.; Voldeng, H.D. Adjustment of the soybean phenology using the E4 locus. *Crop Sci.* **1989**, *29*, 1361–1365.
5. McBlain, B.A.; Bernard, R.L. A new gene affecting the time of flowering and maturity in soybeans. *J. Hered.* **1987**, *78*, 160–162.
6. Bonato, E.R.; Vello, N.A. E6, a dominant gene conditioning early flowering and maturity in soybeans. *Genet. Mol. Biol.* **1999**, *22*, 229–232.
7. Cober, E.R.; Voldeng, H.D. A new soybean maturity and photoperiod-sensitivity locus linked to E1 and T. *Crop Sci.* **2001**, *41*, 698–701.
8. Cober, E.R.; Molnar, S.J.; Charette, M.; Voldeng, H.D. A new locus for early maturity in soybean. *Crop Sci.* **2010**, *50*, 524–527.
9. Watanabe, S.; Hideshima, R.; Xia, Z.; Tsubokura, Y.; Sato, S.; Nakamoto, Y.; Yamanaka, N.; Takahashi, R.; Ishimoto, M.; Anai, T.; *et al.* Map-based cloning of the gene associated with the soybean maturity locus E3. *Genetics* **2009**, *182*, 1251–1262.
10. Liu, B.; Kanazawa, A.; Matsumura, H.; Takahashi, R.; Harada, K.; Abe, J. Genetic redundancy in soybean photoresponses associated with duplication of the phytochrome A gene. *Genetics* **2008**, *180*, 995–1007.
11. Watanabe, S.; Xia, Z.; Hideshima, R.; Tsubokura, Y.; Sato, S.; Yamanaka, N.; Takahashi, R.; Anai, T.; Tabata, S.; Kitamura, K.; *et al.* A map-based cloning strategy employing a residual heterozygous line reveals that the GIGANTEA gene is involved in soybean maturity and flowering. *Genetics* **2011**, *188*, 395–407.
12. Xia, Z.; Watanabe, S.; Yamada, T.; Tsubokura, Y.; Nakashima, H.; Zhai, H.; Anai, T.; Sato, S.; Yamazaki, T.; Lü, S.; *et al.* Positional cloning and characterization reveal the molecular basis for soybean maturity locus E1 that regulates photoperiodic flowering. *Proc. Natl. Acad. Sci. USA* **2012**, *109*, E2155–E2164.
13. Cober, E.R.; Tanner, J.W.; Voldeng, H.D. Genetic control of photoperiod response in early-maturing, near-isogenic soybean lines. *Crop Sci.* **1996**, *36*, 601–605.
14. Cober, E.R.; Stewart, D.W.; Voldeng, H.D. Photoperiod and temperature responses in early-maturing, near-isogenic soybean lines. *Crop Sci.* **2001**, *41*, 721–727.
15. Verma, V.D.; Sawaji, B.V. Screening of soybean germplasm for photoperiodic insensitivity under natural conditions in Vidarbha region of Maharashtra, India. *Soybean Genet. Newsl.* **1994**, *21*, 137–140.
16. Bhatia, V.S.; Manglik, P.; Bhatnagar, P.S.; Guruprasad, K.N. Variation in sensitivity of soybean genotypes to varying photoperiods in India. *Soybean Genet. Newsl.* **1997**, *24*, 99–100.

17. Mayers, J.; Lawn, R.; Byth, D. Agronomic studies on soybean (*Glycine max* (L.) Merrill) in the dry season of the tropics. III. Effect of artificial photoperiod extension on phenology, growth and seed yield. *Aust. J. Agric. Res.* **1991**, *42*, 1109–1121.
18. Hadley, P.; Roberts, E.H.; Summerfield, R.J.; Minchin, F.R. Effects of temperature and photoperiod on flowering in soya bean [*Glycine max* (L.) Merrill]: A quantitative model. *Ann. Bot.* **1984**, *53*, 669–681.
19. Constable, G.A.; Rose, I.A. Variability of soybean phenology response to temperature, daylength and rate of change in daylength. *Field Crops Res.* **1988**, *18*, 57–69.
20. Summerfield, R.J.; Lawn, R.J.; QI, A.; Ellis, R.H.; Roberts, E.H.; Chay, P.M.; Brouwer, J.B.; Rose, J.L.; Shanmugasundaram, S.; Yeates, S.J.; *et al.* Towards the reliable prediction of time to flowering in six annual crops. II. Soyabean (Glycine max). *Exp. Agric.* **1993**, *29*, 253–289.
21. Summerfield, R.J.; Roberts, E.H.; Ellis, R.H.; Lawn, R.J. Towards the reliable prediction of time to flowering in six annual crops. I. The development of simple models for fluctuating field environments. *Exp. Agric.* **1991**, *27*, 11–31.
22. Upadhyay, A.P.; Ellis, R.H.; Summerfield, R.J.; Roberts, E.H.; Qi, A. Characterization of photothermal flowering responses in maturity isolines of soyabean [*Glycine max* (L.) Merrill] cv. Clark. *Ann. Bot.* **1994**, *74*, 87–96.
23. Yan, W.; Wallace, D.H. Simulation and prediction of plant phenology for five crops based on photoperiod × temperature interaction. *Ann. Bot.* **1998**, *81*, 705–716.
24. Major, D.J.; Johnson, D.R.; Tanner, J.W.; Anderson, I.C. Effects of daylength and temperature on soybean development1. *Crop Sci.* **1975**, *15*, 174–179.
25. Grimm, S.S.; Jones, J.W.; Boote, K.J.; Hesketh, J.D. Parameter estimation for predicting flowering date of soybean cultivars. *Crop Sci.* **1993**, *33*, 137–144.
26. Grimm, S.S.; Jones, J.W.; Boote, K.J.; Herzog, D.C. Modeling the occurrence of reproductive stages after flowering for four soybean cultivars. *Agron. J.* **1994**, *86*, 31–38.
27. Piper, E.L.; Boote, K.J.; Jones, J.W.; Grimm, S.S. Comparison of two phenology models for predicting flowering and maturity date of soybean. *Crop Sci.* **1996**, *36*, 1606–1614.
28. Piper, E.L.; Smit, M.A.; Boote, K.J.; Jones, J.W. The role of daily minimum temperature in modulating the development rate to flowering in soybean. *Field Crops Res.* **1996**, *47*, 211–220.
29. Stewart, D.W.; Cober, E.R.; Bernard, R.L. Modeling genetic effects on the photothermal response of soybean phenological development. *Agron. J.* **2003**, *95*, 65–70.
30. Stewart, D.W. AAFC-ECORC, Ottawa, ON, Canada. Unpublished data, 2006.
31. Birch, C.J.; Vos, J.; Kiniry, J.; Bos, H.J.; Elings, A. Phyllochron responds to acclimation to temperature and irradiance in maize. *Field Crops Res.* **1998**, *59*, 187–200.
32. Tollenaar, M. Duration of the grain-filling period in maize is not affected by photoperiod and incident PPFD during the vegetative phase. *Field Crops Res.* **1999**, *62*, 15–21.
33. Bertero, H.D. Effects of photoperiod, temperature and radiation on the rate of leaf appearance in quinoa (*Chenopodium quinoa* Willd.) under field conditions. *Ann. Bot.* **2001**, *87*, 495–502.
34. Zhang, D.; Armitage, A.M.; Affolter, J.M.; Dirr, M.A. Environmental control of flowering and growth of *Achillea millefolium* L. "Summer Pastels". *HortScience* **1996**, *31*, 364–365.
35. Mattson, N.S.; Erwin, J.E. The impact of photoperiod and irradiance on flowering of several herbaceous ornamentals. *Sci. Hortic.* **2005**, *104*, 275–292.

36. Kirby, E.J.M.; Appleyard, M.; Fellowes, G. Effect of sowing date on the temperature response of leaf emergence and leaf size in barley. *Plant Cell Environ.* **1982**, *5*, 477–484.
37. Kernich, G.C.; Slafer, G.A.; Halloran, G.M. Barley development as affected by rate of change of photoperiod. *J. Agric. Sci.* **1995**, *124*, 379–388.
38. Baker, C.K.; Gallagher, J.N.; Monteith, J.L. Daylength change and leaf appearance in winter wheat. *Plant Cell Environ.* **1980**, *3*, 285–287.
39. Slafer, G.A.; Connor, D.J.; Halloran, G.M. Rate of leaf appearance and final number of leaves in wheat: Effects of duration and rate of change of photoperiod. *Ann. Bot.* **1994**, *74*, 427–436.
40. Slafer, G.A.; Halloran, G.M.; Connor, D.J. Development rate in wheat as affected by duration and rate of change of photoperiod. *Ann. Bot.* **1994**, *73*, 671–677.
41. Bonhomme, R.; Derieux, M.; Kiniry, J.R.; Edmeades, G.O.; Ozier-Lafontaine, H. Maize leaf number sensitivity in relation to photoperiod in multilocation field trials. *Agron. J.* **1991**, *83*, 153–157.
42. Clerget, B.; Dingkuhn, M.; Chantereau, J.; Hemberger, J.; Louarn, G.; Vaksmann, M. Does panicle initiation in tropical sorghum depend on day-to-day change in photoperiod? *Field Crops Res.* **2004**, *88*, 21–37.
43. Sinclair, T.R.; Neumaier, N.; Farias, J.R.B.; Nepomuceno, A.L. Comparison of vegetative development in soybean cultivars for low-latitude environments. *Field Crops Res.* **2005**, *92*, 53–59.
44. List, R.J. Smithsonian meteorological tables. *Smithson. Misc. Collect.* **1949**, *114*, 1–527.
45. Marquardt, D.W. An algorithm for least-squares estimation of nonlinear parameters. *J. Soc. Ind. Appl. Math.* **1963**, *11*, 431–441.

7

Live Cell Imaging During Germination Reveals Dynamic Tubular Structures Derived from Protein Storage Vacuoles of Barley Aleurone Cells

Verena Ibl and Eva Stoger *

Department for Applied Genetics and Cell Biology, Molecular Plant Physiology and Crop Biotechnology, University of Natural Resources and Life Sciences, Muthgasse 18, Vienna 1190, Austria; E-Mail: verena.ibl@boku.ac.at

* Author to whom correspondence should be addressed; E-Mail: eva.stoeger@boku.ac.at

---

Abstract: The germination of cereal seeds is a rapid developmental process in which the endomembrane system undergoes a series of dynamic morphological changes to mobilize storage compounds. The changing ultrastructure of protein storage vacuoles (PSVs) in the cells of the aleurone layer has been investigated in the past, but generally this involved inferences drawn from static pictures representing different developmental stages. We used live cell imaging in transgenic barley plants expressing a TIP3-GFP fusion protein as a fluorescent PSV marker to follow in real time the spatially and temporally regulated remodeling and reshaping of PSVs during germination. During late-stage germination, we observed thin, tubular structures extending from PSVs in an actin-dependent manner. No extensions were detected following the disruption of actin microfilaments, while microtubules did not appear to be involved in the process. The previously-undetected tubular PSV structures were characterized by complex movements, fusion events and a dynamic morphology. Their function during germination remains unknown, but might be related to the transport of solutes and metabolites.

Keywords: barley; aleurone; endosperm; PSV; germination

---

1. Introduction

Seeds produce several different classes of storage proteins (prolamins, albumins and globulins) and diverse storage organelles protect these proteins from degradation [1,2]. In most plant species, protein storage vacuoles (PSVs) are the main storage organelles found in seeds [3–6]. They contain mostly albumins and globulins which typically pass through the Golgi body and accumulate in the PSV directly [1,2,7]. In cereal species, prolamins are produced as an additional class of storage proteins. Prolamins are often the most prominent storage proteins in cereal endosperm and are typically deposited into specialized, membrane-bounded storage organelles known as protein bodies (PBs). Depending on the species, PBs may remain in the cytoplasm as distinct, ER-derived organelles or eventually they may be incorporated into the PSVs [2,7]. Whereas PSVs are most abundant in seeds, the storage of metabolites, the digestion of cytoplasmic constituents and turgor maintenance in vegetative tissues are accomplished by central lytic vacuoles (LV) [8,9]. Different sets of cargo proteins and tonoplast intrinsic proteins (TIPs) have been routinely used to distinguish between the functionally different types of vacuoles [8,10], although it should be noted that some overlap of TIP markers has been observed upon recombinant expression [4]. Vacuolar development and remodeling during seed germination may involve the conversion of PSVs into LVs through cell type-specific sets of transformation events including PSV fusion, storage protein degradation, and the gradual replacement of TIPs [11,12].

The mature cereal grain is desiccated and is packed with reserves of starch, protein, lipids and minerals, providing the germinating seeds with energy and nutrients. The outer region of the endosperm comprises between one and three layers of epidermal aleurone cells, within which the PSVs are the most prominent organelles [13]. Aleurone PSVs primarily contain globulins [14] but prolamin-type storage proteins can also accumulate in the aleurone PSVs of certain species, including maize [15]. Aleurone PSVs also store small amounts of non-starch carbohydrates [16] and mineral phytate complexes that provide a source of phosphorus, potassium and magnesium [17]. Unlike starchy endosperm, the aleurone is viable in mature seeds. After imbibition, the highly differentiated and specialized aleurone tissue uses its stored reserves to synthesize and secrete digestive enzymes, which in turn mobilize the insoluble reserves in the starchy endosperm [14,18].

Germination is initiated by the imbibition of water. This causes the embryo to release gibberellic acid (GA_3), which diffuses to the aleurone layer and induces the synthesis and secretion of hydrolases that degrade the starchy endosperm [18]. Expression profiling studies show that key transcripts required for the synthesis of GA_3 are already strongly expressed in the endosperm and aleurone during grain maturation, and remain abundant during early germination (24 h after imbibition) [19]. In barley, it takes 4 days for the endosperm storage reserves to be depleted, and then the aleurone cells die. The storage products are then transported to the embryo via the scutellum.

Germination is also characterized by changes to the fine structure of aleurone cells [20]. De-embryonated half-grains, isolated aleurone layers and aleurone protoplasts have been used to study how cereal aleurone cells respond to hormones and to understand the molecular and cellular basis of signaling and regulation in the aleurone [18,21]. Barley aleurone protoplasts treated with GA_3 secrete hydrolases that are synthesized *de novo* from amino acids released by the breakdown of storage proteins in PSVs [12,22]. The aleurone protoplasts then undergo dramatic morphological changes, in which the multiple small PSVs found in freshly isolated cells coalesce to form one large central vacuole after 4 days of GA_3 treatment [20,23–25]. Secondary lytic vacuoles are found in barley aleurone protoplasts at the same time, and are possibly aleurain-containing vacuoles [26,27]. The vacuolation process correlates strongly with the duration of GA_3 treatment and is complete after 5 days of incubation [21,28]. Vacuolation is linked to the dramatic conversion of nutrient-storing compartments into lytic organelles. Noninvasive measurements of the vacuolar pH in barley aleurone protoplasts has shown that the vacuole lumen pH in barley aleurone cells declines from 6.6 to 5.8 or below in a few hours in the presence of GA_3 [25,27].

TIPs fused to fluorescent proteins are widely used as tonoplast markers for live cell imaging [4,29,30]. TIP3 (α-TIP) was first characterized as TP25, a seed-specific aquaporin that is strongly expressed in seeds but decreases rapidly during germination [31]. TIP3 is synthesized on the rough ER and appears to reach the tonoplast without passing through the Golgi body [31–33]. Expression assays and microscopy in Arabidopsis, barley aleurone protoplasts, and pea cotyledons confirmed that TIP3 is predominantly associated with PSVs in the seeds [4,10,27,29,32,34,35] and TIP3 has recently been used as a PSV (PBII) marker in rice subaleurone cells [30]. The ultrastructural changes of PSVs during barley endosperm maturation have been followed in situ by live cell imaging using a TIP3-GFP fluorescent tag [36]. Labelled PSVs in the subaleurone and central starchy endosperm cells underwent remarkable morphological changes, but the spherical PSVs in the aleurone layer did not change significantly [36]. The PSVs in the aleurone layer would be expected to undergo more profound changes during germination, but time-lapse imaging studies of living cells during germination have not been reported. We therefore used our established transgenic TIP3-GFP line [36] for the three-dimensional reconstruction and time-lapse analysis of PSV remodeling in barley aleurone cells during germination. Sections of intact germinating seeds were observed by confocal laser scanning microscopy (CLSM), showing that PSVs in barley aleurone cells undergo continuous remodeling, including a transition from spherical to ellipsoid to elongated morphology and complex processes of fusion, and possibly also scission and collapse. After 4 days of germination, highly mobile, tubular structures emerged from the PSVs, and their formation and mobility was promoted by actin microfilaments rather than microtubules. Potential functions of these TIP3-labeled tubular PSV extensions are discussed.

2. Results

2.1. Aleurone PSVs Become Larger and More Ellipsoid during Germination

The morphological changes that characterize PSVs in barley aleurone cells were studied by *in vivo* confocal microscopy, using homozygous T_4 TIP3-GFP seeds up to 4 days after germination. At every time point, individual scans and 3D projections of z-series stacks were prepared for optimal visualization. In mature seeds, the PSVs in aleurone cells were strongly labeled with TIP3-GFP and appeared predominantly as spherical, discrete compartments (Figure 1A) although ellipsoid PSVs were occasionally observed (Figure 1A, arrow). This confirmed our previous study showing a similar labeling of aleurone PSVs during the late stages of endosperm development [36].

Figure 1. The dynamic morphology of tonoplast intrinsic proteins (TIP)3-GFP-labelled protein storage vacuoles (PSVs) during 4 days of germination. (**A**) TIP3-GFP labels spherical PSVs in the aleurone layer of mature kernels, plus occasional ellipsoid PSVs (arrows); (**B**) Homogeneous and tightly packed PSVs after one day of germination; (**C**) Arrows indicate PSVs starting to form buds at 2 days after germination; (**D,E**) PSVs become larger and more heterogeneous, with spherical, ellipsoid, small and large structures. Arrows indicate that small PSVs are still connected to large ones. Scale bar = 5 µm. (**F**) The average PSV size in µm during 4 days of germination. Standard error bars are indicated (n = 100).

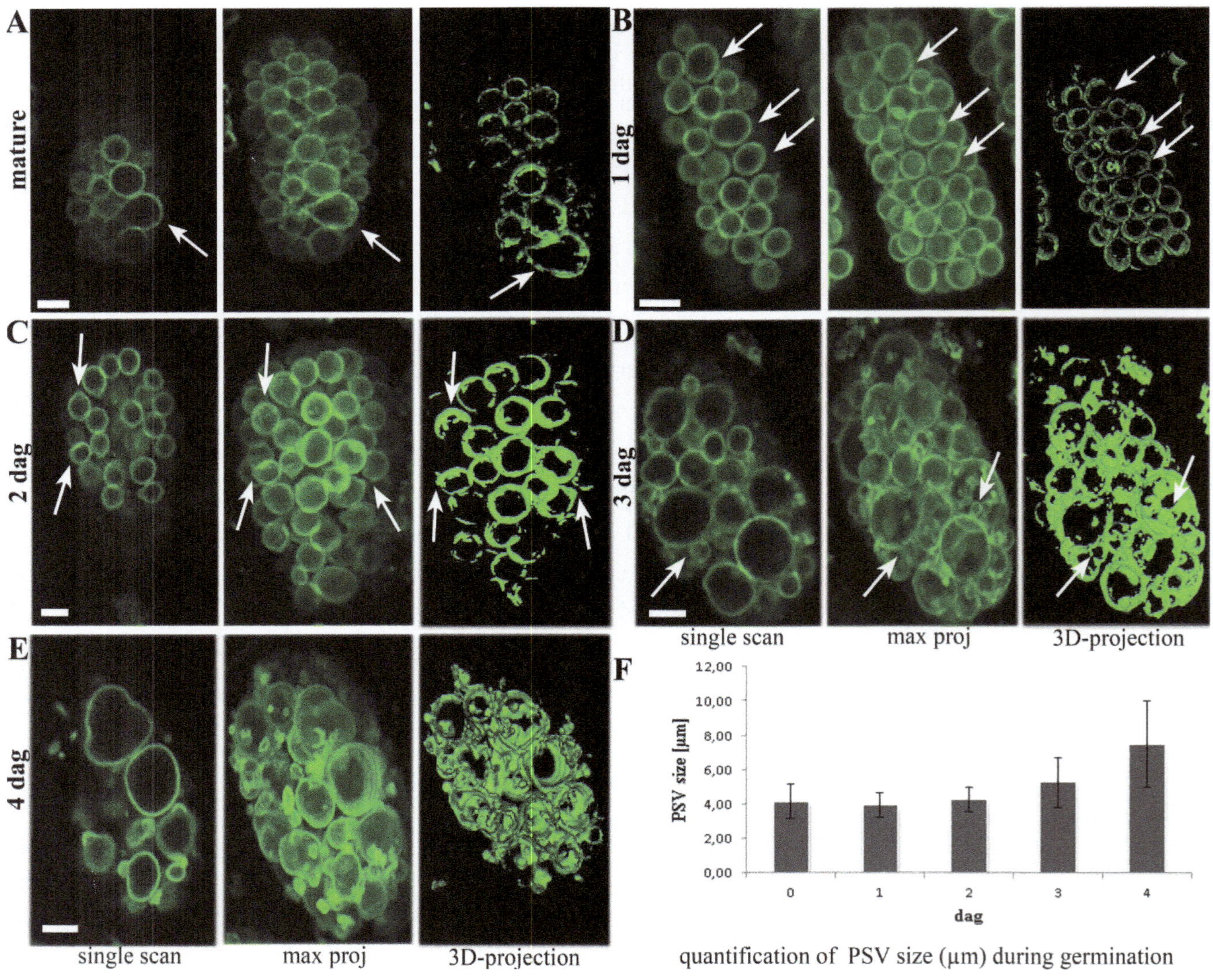

After one day of germination, the ellipsoid PSVs were more abundant, but the overall appearance of the PSVs remained homogenous and tightly packed (Figure 1B, arrows). Two days after germination, PSVs with buds were observed for the first time, indicating the beginning of the remodeling process (Figure 1C, arrows). Three days after germination, the morphological changes were more obvious, with the PSVs appearing more heterogeneous in shape and size (Figure 1D). Small PSVs appearing at this stage (Figure 1D, arrows) were shown to be connected to the larger ones in 3D projections, indicating that they probably result from reshaping and invagination processes (Figure 1D). Four days after germination, the TIP3-GFP-labeled PSVs underwent a more dramatic change in morphology, with few still spherical compartments but many more of the larger and more ellipsoid structures (Figure 1E). The average diameter of TIP3-labelled PSVs increased from 5 μm at 3 DAG (days after germination) to almost 8 μm a day later (Figure 1F). In addition, small TIP3-GFP-labeled PSVs appeared, only some of which were still connected to larger ones. In summary, live cell imaging provided a detailed overview of dynamic PSV morphology in the aleurone layer during 4 days of germination, showing that PSVs almost doubled in average size and changed from homogenous spherical to more diverse ellipsoid structures.

2.2. Real-Time Observations Indicate that Aleurone PSVs may Undergo Fusion, Scission and Collapse

Live cell imaging also allowed us to follow the dynamic behavior of aleurone PSVs *in situ*. The most striking morphological changes took place 3–4 days after germination so time series were collected on these days by confocal microscopy over periods of 5–20 min. Figure 2A shows two PSVs of different sizes on day 4, fusing to form a larger PSV in 16 min (see also Movie 1). Potential scission events were also observed, including a large, narrow PSV (Figure 2B, arrows) apparently splitting into two smaller, spherical PSVs (Figure 2B, asterisks) in 5–6 min (Movie 2). A TIP3-GFP-labeled vesicle was also observed budding off from another PSV (Figure 2B, arrowheads) in 12 s. Occasionally, our observations indicated that PSVs were also collapsing (Figure 2C and Movie 3), although we cannot exclude the possibility that these observations might be attributed to other events giving the impression of collapse. Remarkably, we also observed highly mobile TIP3-GFP-labeled tubular structures at later stages of germination (Figure 2D and Movie 4). These structures were shown to appear, reshape, expand and retract again (Figure 2D arrows), predominantly 3–4 days after germination and usually at the periphery of the cells. Germination is therefore characterized by several dynamic events that affect aleurone PSVs, including fusion, remodeling and the generation of TIP3-GFP-labeled extensions that have not been reported thus far.

2.3. Highly Mobile Tubular Extension Structures Originate from PSVs

The dynamic events leading to the formation of TIP3-GFP-labeled tubular extensions were investigated by following the dynamic behavior of PSVs 3–4 days after germination by time-lapse confocal microscopy. The labeled structured first appeared 3 days after germination but became more prominent a day later. Two different time series were constructed, confirming that the TIP3-GFP-labeled tubules originate from PSVs (Figure 3A, arrows; Movies 5 and 6). Interestingly, we not only observed tubular extensions arising directly from the PSV surface, but another was formed by remodeling a connecting tube between two PSVs (Figure 3A and Movie 6).

The movement of the PSV tubules was tracked by time lapse imaging in a single cell (Figure 3B and Movie 7). These images show that the tubular PSV structure elongates to double its original length, retracts again and then realigns (Figure 3B, arrowhead). An adjacent loop in the same cell contracts and aligns with another PSV membrane, where it is released. At the same time, the original PSV tube expands and contracts again (Figure 3B, arrowheads; Movie 7). Several tubules also expand, contact each other, separate, and new ones are generated (Movie 7). These data show that the PSV tubules are constantly remodeling, coming into contact with each other and with other PSVs. The speed of the movement was 0.1–0.2 μm/s.

Figure 2. Live cell imaging 4 days after germination, indicating PSV fusion, scission and collapse, and the extension of tubular structures. (**A**) PSV fusion over a duration of 16 min (images captured every 4 s); (**B**) PSV apparently splitting into two PSVs (images captured every 8 s) and a TIP3-GFP-labeled vesicle budding off from another PSV (images captured every 4 s). (**C**) Potential PSV collapse over a duration of 6 min (images captured every 8 s). (**D**) Dynamic behavior of tubular extension structures (images captured every 4 s). Scale bar = 2 μm.

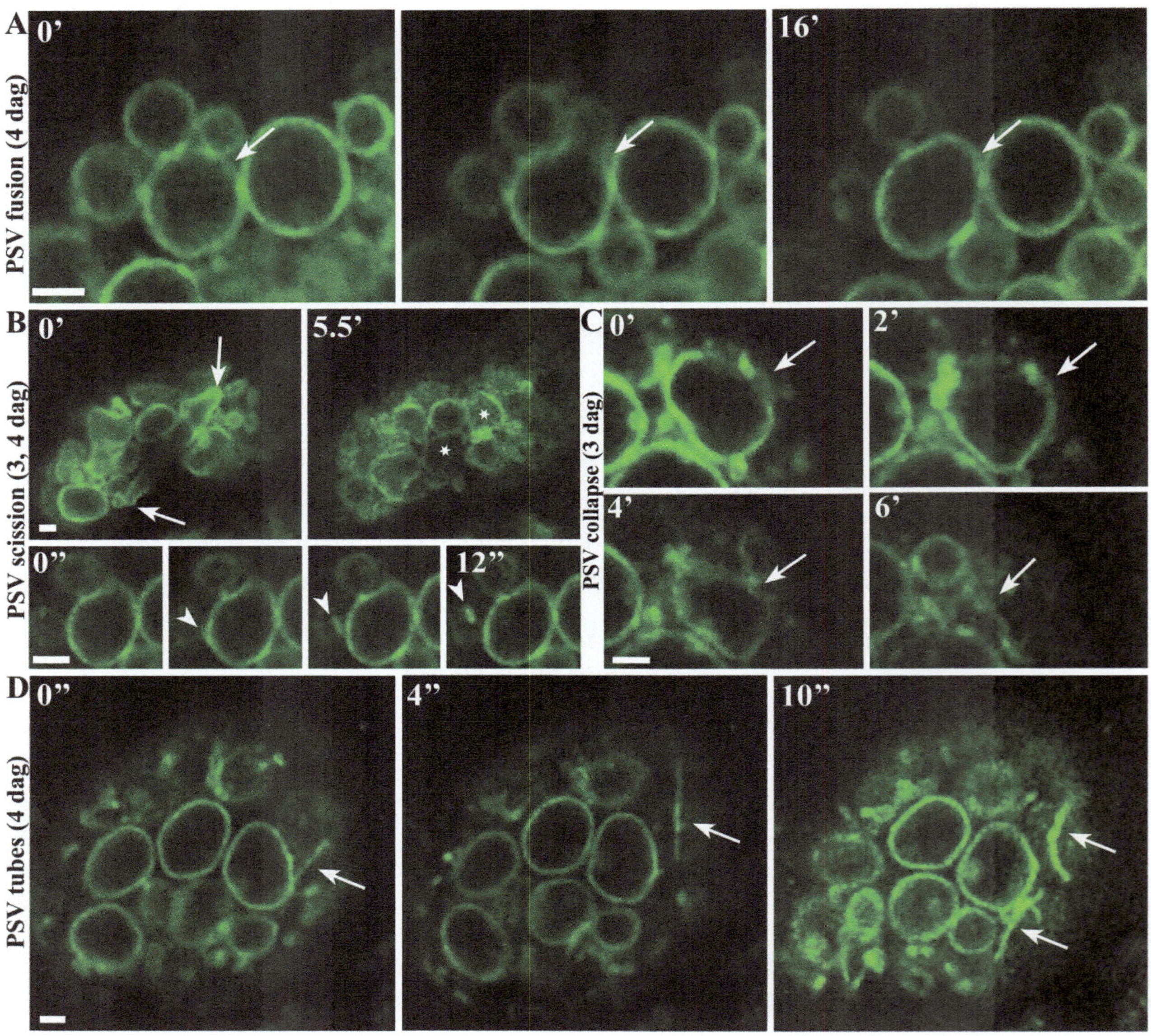

Figure 3. Confocal time series showing tubular structures arising from PSVs and undergoing continuous remodeling. (**A**) Time series of 7 and 2.5 min showing tubular structures arising from PSVs (images captured every 4 s). See the corresponding Movies 5 and 6 for more details; (**B**) Arrowheads and arrows show two different PSV extensions at the beginning of a 10-min time series (images captured every 4 s). See the corresponding Movie 7 for more details. Note the expansion and subsequent retraction of the structure, allowing the PSV extension to contact another PSV before release (arrow). Scale bar = 2 μm.

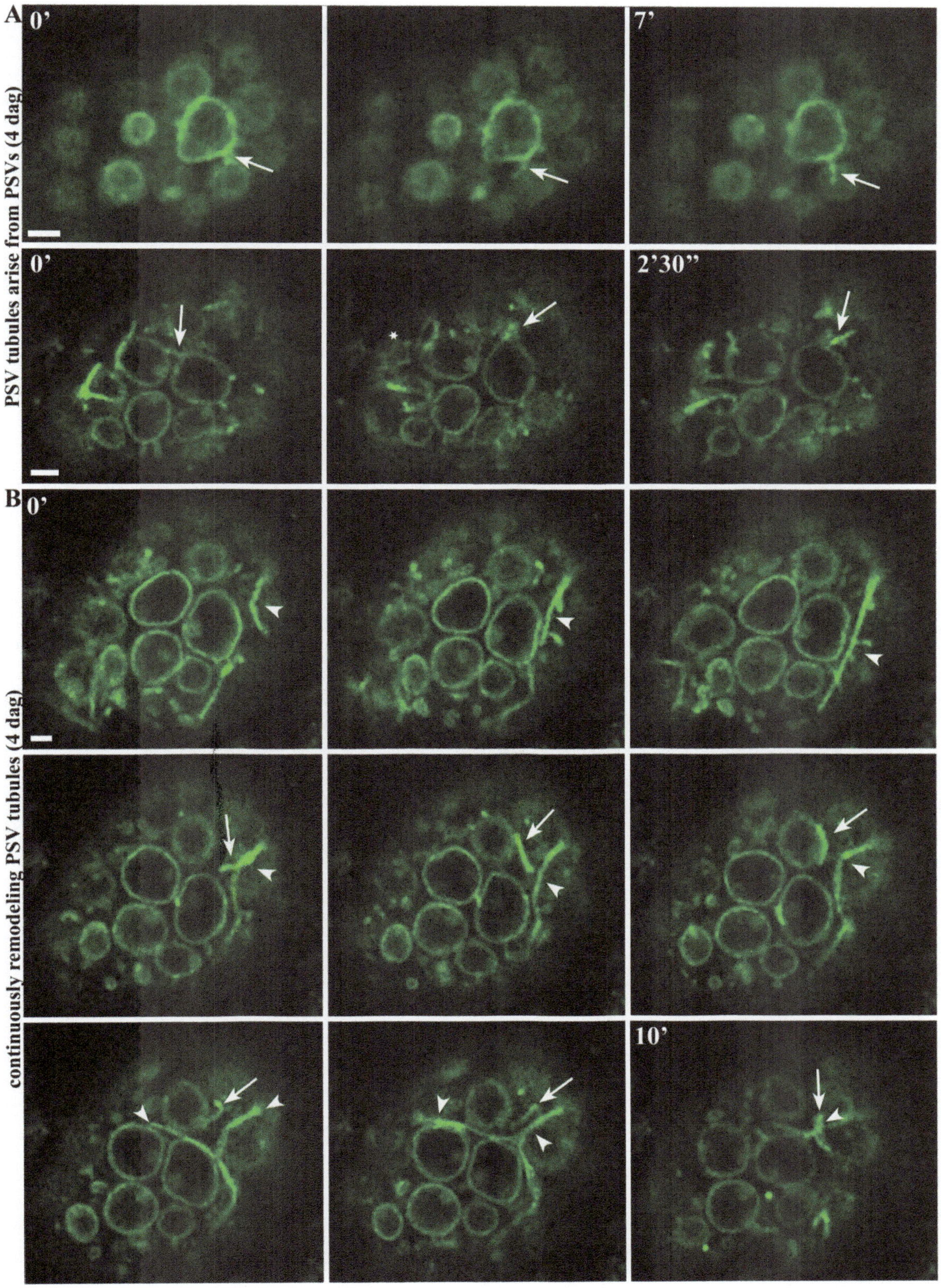

2.4. The Formation of PSV Tubules Depends on Microfilaments but not Microtubules

The cytoskeleton is generally important for vacuolar transport and remodeling, so we specifically inhibited the activity of microfilaments or microtubules during germination to determine their role in the dynamic behavior of PSVs. Seeds were soaked twice, first at the mature stage and then at 2 days after germination, with 10 μM latrunculin B (to inhibit microfilaments) or 10 μM oryzalin (to inhibit microtubules). Both chemicals inhibited germination substantially, whereas the equivalent working concentration of DMSO (the solvent used for the stock solutions) had no effect [37]. Observations by confocal microscopy 4 days after germination showed that seeds incubated with latrunculin B did not produce PSV tubules (Figure 4A) whereas seeds treated with oryzalin produced dynamic PSV extensions like those observed in untreated cells (Figure 4B). These data indicate that the formation of PSV tubules is actin-dependent. Notably, neither latrunculin B nor oryzalin had a significant effect on the underlying PSV morphology during germination.

Figure 4. The remodeling of tubular PSV extensions depends on microfilaments but not microtubules. (**A**) Representative pictures from movies showing PSV tubules in untreated cells (arrow) but no PSV tubules in cells treated with latrunculin B; (**B**) Representative pictures showing PSV tubules in untreated cells (arrow) and in cells treated with oryzalin. Scale bar = 2 μm.

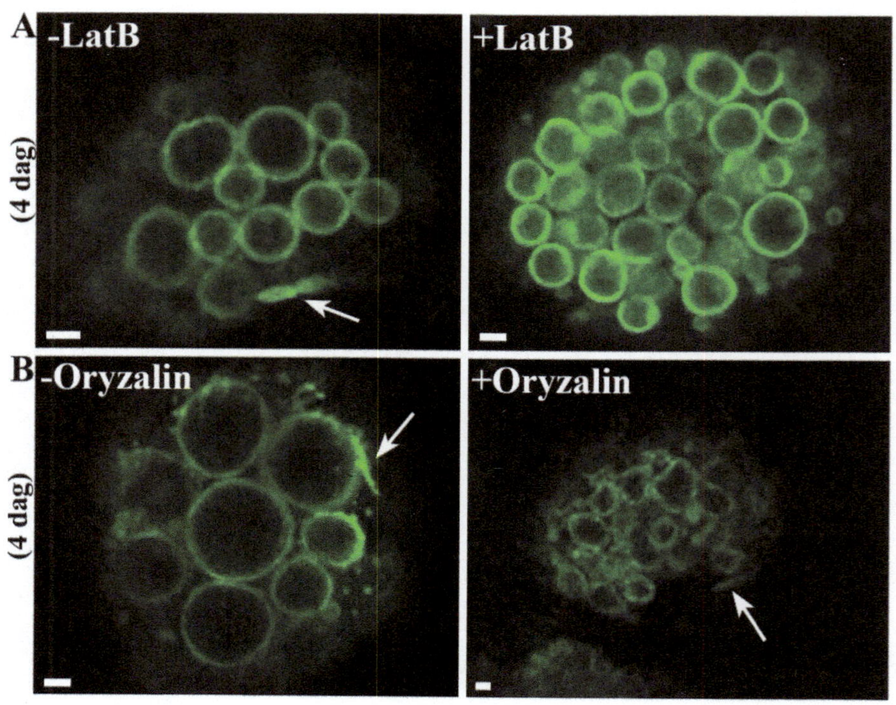

3. Discussion

The endomembrane system of cereal seeds undergoes a series of developmental changes during maturation and germination, first to accomplish the deposition of storage compounds and then their mobilization, both within a relatively short period of time. Germination takes only a few days, during which the cells of the aleurone layer deplete their reserves stored in PSVs. The ultrastructural changes of aleurone organelles during germination have been investigated previously mainly by static imaging techniques, often using fixed cells from de-embryonated half-grains, isolated aleurone layers or aleurone protoplasts treated with GA_3 [12,21]. There have been only few reports describing the analysis of endomembrane organelles in living cereal seeds, partly reflecting the limited availability of transgenic lines containing fluorescent organelle markers.

We therefore used a transgenic barley line expressing a fluorescent TIP3-GFP fusion protein to visualize the PSV and carried out detailed *in vivo* studies including time-lapse imaging, z-series and 3D projections. These studies revealed the remarkably dynamic morphology of PSVs during germination (Figure 5). The homogeneous spherical PSVs present in the mature seed changed progressively to larger, ellipsoid structures over 4 days of germination.

Figure 5. Schematic summary of the morphological changes of the PSVs and the formation of PSV tubules in barley aleurone cells during germination. dag = days after germination.

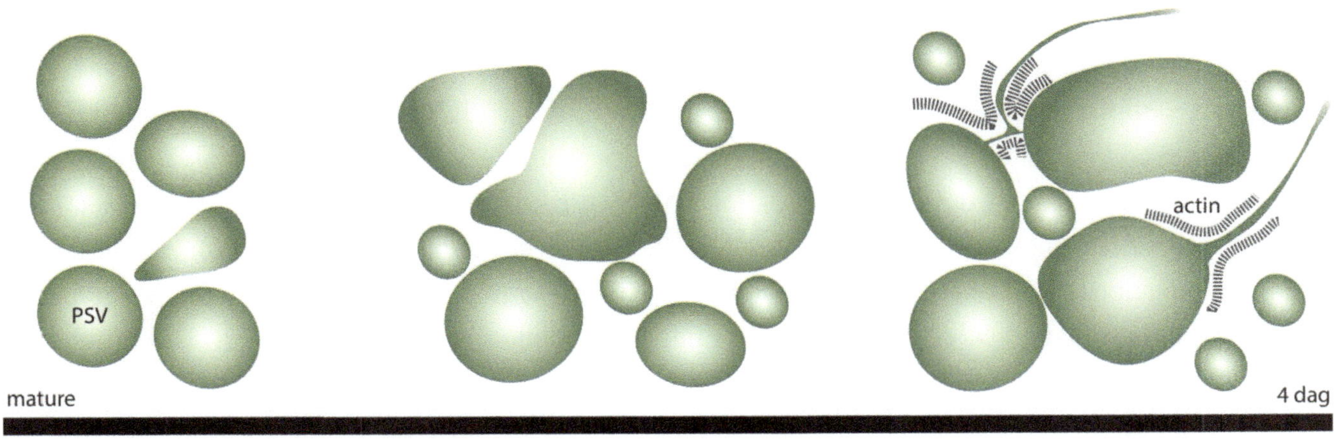

Unlike previous studies using isolated aleurone layers and protoplasts treated with GA_3 [20,23–25], we did not observe a single large central vacuole at the end of germination. It is therefore likely that interaction and communication among the aleurone, subaleurone and starchy endosperm cells contribute to the regulation of germination and particularly the behavior of storage organelles. The aleurone and scutellum are controlled by plant hormones such as GA_3 and ABA [19,25,27,38] but may also be influenced by feedback from the mobilized reserved released from starchy endosperm cells [18]. ABA antagonizes the events induced by GA_3 in the aleurone at many levels, suggesting there is a sensitive regulatory system based on the balance between these two hormones [18] as well as processes regulated by auxins and ethylene [19]. Although the same cellular events occur in the aleurone of whole imbibed grain and GA_3-treated aleurone layers and protoplasts, the events seem more rapid in isolated aleurone layers and protoplasts, perhaps because the direct application of single hormones overwhelms the endogenous interaction network [21] and eliminates interactions based on the arrangement of tissues and the macromolecules stored during development [18]. This is consistent with our observation that PSVs in aleurone cells near to the starchy endosperm behave in a different manner to PSVs near the testa [37]. The PSVs in the aleurone layer adjacent to the starchy endosperm are larger, indicating that PSV fusion is favored in these cells.

The fusion of the PSVs into larger vacuolar structures is a common feature that has been observed in both monocotyledonous and dicotyledonous plants with seed storage tissues that are still alive at the point of reserve mobilization [14]. This includes embryo cotyledon and root tissues, castor bean endosperm cells and cereal aleurone layers [11,12,39]. By focusing on living aleurone cells containing TIP3-GFP-labeled PSVs, we were able to monitor in real time the PSV remodeling processes throughout germination. Our time-lapse series revealed the dynamic processes underlying the observed morphological changes, including vacuole fusion and potential fission events and the deformation of initially spherical PSVs to produce ellipsoid or irregular structures that budded off smaller PSVs. Three-dimensional reconstructions revealed that connections were maintained between some small PSVs and the larger ones, whereas others were independent and spherical. Vacuoles also appeared to collapse occasionally, perhaps reflecting tonoplast destabilization at the onset of programed cell death [21].

Most conspicuously, on the third and fourth days of germination we observed TIP3-GFP-labeled, constantly remodeling tubular structures that arose directly from the PSVs. Tubular vacuoles in plant cells were first observed in 1929 in rose leaves, and were further characterized in onion epidermal and guard cells, and soybean root cells [40–42]. Such structures are often associated with specialized cell types including root hairs [43] and pollen tubes [44]. Tubular vacuoles extending from the large central vacuole were described in great detail in red onion epidermal cells [45]. More recently, unique tubular structures were observed using a photoconvertible mEosFP::2xFYVE probe, which labeled putative endosomes, prevacuoles and vacuoles [46]. The structures were described as aberrant oscillating or rotating vesicles that extended tubular projections 5–15 μm in length and ~0.6 μm in diameter. Time-lapse sequences showed that structures were generated constantly, extending from different vesicles, forming loops and snaring other vesicles into loose aggregates. The ends of the observed tubules were shown to be sealed rather than forming a continuous compartment [46]. The tubular structures we observed were similar in size and mobility, and likewise did not form a continuous network. Membrane protrusions have also been observed to various degrees for other organelles. For example, in plastids, dynamic tubular extrusions of the inner and outer envelopes, termed stromules, have been described [47,48]. Since stromules increase the surface area of plastids it has been speculated that they might facilitate the interaction and the exchange of metabolites with other plastids, mitochondria, peroxisomes and the endoplasmic reticulum [49–52].

The functional role of vacuolar tubules remains unclear. They may play a role in the transport of sugars, other nutrients and metabolites in fungi and yeast cells [53,54]. Therefore, the dynamic PSV tubules we observed in barley aleurone cells during germination may also transfer reserves or other materials between PSVs. It is also possible that the role of the PSV extensions may be linked to an increased rate of transport between the cytoplasm and vacuolar lumen due to the expansion of the tonoplast surface.

Alternatively, the PSV tubules may lack a specific function and may instead represent a stage in vacuolar development [55]. Tubular intermediate structures are involved in the transformation of PSVs into lytic vacuoles in the root cells of tobacco seedlings during germination and early root development [11]. Based on TEM studies combined with high-pressure freezing and freeze-substitution techniques, it has been suggested that mobilization of storage molecules induces the gradual osmotic collapse of the vacuolar membranes upon themselves, thereby squeezing the residual vacuolar contents into the remaining bulging vacuolar regions [11]. However, the PSV tubules we observed appear later in the process, following the bulging vacuolar regions that were pictured during the mobilization of storage reserves. However, the duration and individual steps of the vacuolar transition process are highly variable depending on the cell type [11]. Therefore, it will be necessary to determine the content and pH of the PSV tubules in order to clarify their role during barley germination. It is notable that we did not observe the tubular structures when tracking PSVs in the same barley line during endosperm maturation, suggesting they are specific for germination and do not represent artifacts induced by laser scanning microscopy or the expression of TIP3 as a vacuolar marker [36].

Time-lapse imaging after the treatment of germinating seeds with the microfilament inhibitor latrunculin B or the microtubule inhibitor oryzalin indicated that the formation of PSV extensions depends on microfilaments but not on microtubules. Similar results have been reported in lily pollen tubes, where vacuolar morphology and the movement of long tubular vacuolar structures are significantly affected by latrunculin B but not oryzalin [56]. The dependence of higher plant vacuolar tubules on actin and myosin has also been reported in other systems [43,57] and is consistent with the actin-based motility of vesicle trafficking to the central vacuole [58,59]. In contrast to previous studies showing the inhibition of vacuolar tubule movement, latrunculin B prevented the formation of PSV tubules in barley aleurone in the first place, agreeing with reports indicating that actin-interacting proteins may be involved in the formation of intravacuolar structures and in vesicle formation and/or release [58,60]. In the absence of actin filaments, proteins needed for vesicle formation and/or release (e.g., dynamins, Rab GTPase) may not be efficiently recruited, thus inhibiting reshaping events [45,61]. Actin filaments might also affect the generation and dynamic behavior of the PSV tubules indirectly by inhibiting cytoplasmic streaming, if the induction and dynamic behavior of the PSV extensions were influenced by the bulk flow of the cytoplasm [45]. Further studies are needed to clarify the role of actin filaments in the formation and expansion of PSV tubules and such studies will also help to determine the function of these novel structures.

4. Experimental

4.1. Plant Material

Barley wild-type (Golden Promise) and T4 homozygous TIP-GFP transgenic seeds [36] were plated on filter paper soaked in tap water and incubated for 4 days in Petri dishes at room temperature. Mature seeds were soaked for a maximum of 2 h in tap water.

4.2. Confocal Microcopy

Samples were mounted within a small border of Vaseline to avoid floating, and silicon glue was used to stabilize the cover slips. GFP images were captured using a Leica SP5 CLSM with excitation and emission wavelengths of 488 and 500–530 nm. Transgenic homozygous T_4 TIP3-GFP barley seeds at different germination stages (mature to 4 days after initiating germination) were sectioned, washed, mounted in tap water and analyzed immediately. Images were processed using ImageJ and arranged using Photoshop CS5. Movies are displayed at 17 frames per second.

4.3. Calculation of PSV Size

The average diameter of PSVs (n = 100) was calculated at different stages ranging from mature seeds to 4 days after germination, by measuring the length of the maximum distance between opposite membranes of the spherical and irregular PSVs.

4.4. Inhibitor Studies

Homozygous T_4 TIP3-GFP seeds were germinated in 24-well plates for 4 days. Dry seeds were placed in 400 μL of solution containing 10 μM latrunculin B or 10 μM oryzalin in tap water. The solutions were replaced with fresh ones after two days of germination. Each solution was prepared from a 1 mM stock solution in DMSO, and the control was therefore prepared by diluting pure DMSO to the same working concentration in tap water. Sections were mounted in tap water and analyzed immediately for a maximum of 1 h by CLSM as above. For each experiment, at least four seeds were sectioned at the appropriate germination stage to determine the effect of the inhibitor.

5. Conclusions

Live cell imaging has allowed us to follow in real time the morphological changes of PSVs in the aleurone cells of germinating transgenic barley seeds expressing a fluorescent PSV marker. This process has been studied before, but only by reconstructing static images, mostly based on fixed tissues and cells. Our live cell imaging method confirmed that PSVs undergo spatiotemporal remodeling during barley germination. Remarkably, during the late stages of germination we observed thin tubules extending from PSVs in an actin dependent mode. These previously unreported structures underwent complex movements, comprising fusion events and dynamic changes in morphology. The function of the PSV tubules remains unclear but may be related to the transport of reserves between PSVs or from the PSV to the cytoplasm. Alternatively, they may represent a stage in vacuolar development heralding the onset of programed cell death. Additional investigations are required to determine the content and pH of the PSV tubules and the precise mechanism by which actin filaments control their formation, extension, retraction and interactions, thus providing further insight into their function during germination.

Acknowledgments

This work was carried out in cooperation with the BOKU Leica Imaging Facility in Vienna. The authors would like to acknowledge financial support by the Austrian Science Fund FWF (P25736-B20 and I1461-B16).

Author Contributions

Verena Ibl performed the experiments and analyzed the data. Both authors designed and discussed the experiments and wrote the article.

Conflicts of Interest

The authors declare no conflict of interest.

References

1. Shewry, P.R.; Napier, J.A.; Tatham, A.S. Seed storage proteins: Structures and biosynthesis. *Plant Cell* **1995**, *7*, 945–956.

2. Galili, G. Er-derived compartments are formed by highly regulated processes and have special functions in plants. *Plant Physiol.* **2004**, *136*, 3411–3413.
3. Bassham, D.C.; Raikhel, N.V. Unique features of the plant vacuolar sorting machinery. *Curr. Opin. Cell Biol.* **2000**, *12*, 491–495.
4. Hunter, P.R.; Craddock, C.P.; Di Benedetto, S.; Roberts, L.M.; Frigerio, L. Fluorescent reporter proteins for the tonoplast and the vacuolar lumen identify a single vacuolar compartment in arabidopsis cells. *Plant Physiol.* **2007**, *145*, 1371–1382.
5. Herman, E.M.; Larkins, B.A. Protein storage bodies and vacuoles. *Plant Cell* **1999**, *11*, 601–614.
6. Jiang, L.; Phillips, T.E.; Hamm, C.A.; Drozdowicz, Y.M.; Rea, P.A.; Maeshima, M.; Rogers, S.W.; Rogers, J.C. The protein storage vacuole: A unique compound organelle. *J. Cell Biol.* **2001**, *155*, 991–1002.
7. Tosi, P.; Parker, M.; Gritsch, C.S.; Carzaniga, R.; Martin, B.; Shewry, P.R. Trafficking of storage proteins in developing grain of wheat. *J. Exp. Bot.* **2009**, *60*, 979–991.
8. Neuhaus, J.M.; Rogers, J.C. Sorting of proteins to vacuoles in plant cells. *Plant Mol. Biol.* **1998**, *38*, 127–144.
9. Martinoia, E.; Maeshima, M.; Neuhaus, H.E. Vacuolar transporters and their essential role in plant metabolism. *J. Exp. Bot.* **2007**, *58*, 83–102.
10. Jauh, G.Y.; Phillips, T.E.; Rogers, J.C. Tonoplast intrinsic protein isoforms as markers for vacuolar functions. *Plant Cell* **1999**, *11*, 1867–1882.
11. Zheng, H.; Staehelin, L.A. Protein storage vacuoles are transformed into lytic vacuoles in root meristematic cells of germinating seedlings by multiple, cell type-specific mechanisms. *Plant Physiol.* **2011**, *155*, 2023–2035.
12. Bethke, P.C.; Swanson, S.J.; Hillmer, S.; Jones, R.L. From storage compartment to lytic organelle: The metamorphosis of the aleurone protein storage vacuole. *Ann. Bot.* **1998**, *82*, 399–412.
13. Fath, A.; Bethke, P.; Lonsdale, J.; Meza-Romero, R.; Jones, R. Programmed cell death in cereal aleurone. *Plant Mol. Biol.* **2000**, *44*, 255–266.
14. Muntz, K. Protein dynamics and proteolysis in plant vacuoles. *J. Exp. Bot.* **2007**, *58*, 2391–2407.
15. Reyes, F.C.; Chung, T.; Holding, D.; Jung, R.; Vierstra, R.; Otegui, M.S. Delivery of prolamins to the protein storage vacuole in maize aleurone cells. *Plant Cell* **2011**, *23*, 769–784.
16. Jacobsen, J.V.; Knox, R.B.; Pyliotis, N.A. Structure and composition of aleurone grains in barley aleurone layer. *Planta* **1971**, *101*, 189–209.
17. Stewart, A.; Nield, H.; Lott, J.N. An investigation of the mineral content of barley grains and seedlings. *Plant Physiol.* **1988**, *86*, 93–97.
18. Ritchie, S.; Swanson, S.J.; Gilroy, S. Physiology of the aleurone layer and starchy endosperm during grain development and early seedling growth: New insights from cell and molecular biology. *Seed Sci. Res.* **2000**, *10*, 193–212.
19. Sreenivasulu, N.; Usadel, B.; Winter, A.; Radchuk, V.; Scholz, U.; Stein, N.; Weschke, W.; Strickert, M.; Close, T.J.; Stitt, M.; *et al.* Barley grain maturation and germination: Metabolic pathway and regulatory network commonalities and differences highlighted by new mapman/pageman profiling tools. *Plant Physiol.* **2008**, *146*, 1738–1758.
20. Jacobsen, J.V.; Zwar, J.A.; Chandler, P.M. Gibberellic-acid-responsive protoplasts from mature aleurone of himalaya barley. *Planta* **1985**, *163*, 430–438.

21. Bethke, P.C.; Jones, R.L. Cell death of barley aleurone protoplasts is mediated by reactive oxygen species. *Plant J.* **2001**, *25*, 19–29.
22. Jones, R.L.; Jacobsen, J.V. Regulation of synthesis and transport of secreted proteins in cereal aleurone. *Int. Rev. Cytol.* **1991**, *126*, 49–88.
23. Fath, A.; Bethke, P.C.; Jones, R.L. Barley aleurone cell death is not apoptotic: Characterization of nuclease activities and DNA degradation. *Plant J.* **1999**, *20*, 305–315.
24. Bush, D.S.; Cornejo, M.J.; Huang, C.N.; Jones, R.L. Ca-stimulated secretion of alpha-amylase during development in barley aleurone protoplasts. *Plant Physiol.* **1986**, *82*, 566–574.
25. Swanson, S.J.; Jones, R.L. Gibberellic acid induces vacuolar acidification in barley aleurone. *Plant Cell* **1996**, *8*, 2211–2221.
26. Holwerda, B.C.; Rogers, J.C. Purification and characterization of aleurain: A plant thiol protease functionally homologous to mammalian cathepsin H. *Plant Physiol.* **1992**, *99*, 848–855.
27. Swanson, S.J.; Bethke, P.C.; Jones, R.L. Barley aleurone cells contain two types of vacuoles: Characterization of lytic organelles by use of fluorescent probes. *Plant Cell* **1998**, *10*, 685–698.
28. Hwang, Y.S.; Bethke, P.C.; Cheong, Y.H.; Chang, H.S.; Zhu, T.; Jones, R.L. A gibberellin-regulated calcineurin b in rice localizes to the tonoplast and is implicated in vacuole function. *Plant Physiol.* **2005**, *138*, 1347–1358.
29. Gattolin, S.; Sorieul, M.; Hunter, P.R.; Khonsari, R.H.; Frigerio, L. *In vivo* imaging of the tonoplast intrinsic protein family in arabidopsis roots. *BMC Plant Biol.* **2009**, *9*, e133.
30. Onda, Y.; Kumamaru, T.; Kawagoe, Y. Er membrane-localized oxidoreductase ero1 is required for disulfide bond formation in the rice endosperm. *Proc. Natl. Acad. Sci. USA* **2009**, *106*, 14156–14161.
31. Johnson, K.D.; Herman, E.M.; Chrispeels, M.J. An abundant, highly conserved tonoplast protein in seeds. *Plant Physiol.* **1989**, *91*, 1006–1013.
32. Gattolin, S.; Sorieul, M.; Frigerio, L. Tonoplast intrinsic proteins and vacuolar identity. *Biochem. Soc. Trans.* **2010**, *38*, 769–773.
33. Hillmer, S.; Movafeghi, A.; Robinson, D.G.; Hinz, G. Vacuolar storage proteins are sorted in the cis-cisternae of the pea cotyledon golgi apparatus. *J. Cell Biol.* **2001**, *152*, 41–50.
34. Jiang, L.W.; Rogers, J.C. Sorting of membrane proteins to vacuoles in plant cells. *Plant Sci.* **1999**, *146*, 55–67.
35. Paris, N.; Stanley, C.M.; Jones, R.L.; Rogers, J.C. Plant cells contain two functionally distinct vacuolar compartments. *Cell* **1996**, *85*, 563–572.
36. Ibl, V.; Kapusi, E.; Arcalis, E.; Kawagoe, Y.; Stoger, E. Fusion, rupture, and degeneration: The fate of *in vivo*-labelled psvs in developing barley endosperm. *J. Exp. Bot.* **2014**, *65*, 3249–3261.
37. Ibl, V.; Stoger, E. University of Natural Resources and Life Sciences, Vienna, Austria. Unpublished data, 2014.
38. Sreenivasulu, N.; Radchuk, V.; Strickert, M.; Miersch, O.; Weschke, W.; Wobus, U. Gene expression patterns reveal tissue-specific signaling networks controlling programmed cell death and ABA-regulated maturation in developing barley seeds. *Plant J.* **2006**, *47*, 310–327.
39. Greenwood, J.S.; Helm, M.; Gietl, C. Ricinosomes and endosperm transfer cell structure in programmed cell death of the nucellus during ricinus seed development. *Proc. Natl. Acad. Sci. USA* **2005**, *102*, 2238–2243.

40. Guilliermond, A. The recent development of our idea of the vacuome of plant cells. *Am. J. Bot.* **1929**, *16*, 1–22.
41. Palevitz, B.A.; Okane, D.J. Epifluorescence and video analysis of vacuole motility and development in stomatal cells of allium. *Science* **1981**, *214*, 443–445.
42. Palevitz, B.A.; Okane, D.J.; Kobres, R.E.; Raikhel, N.V. The vacuole system in stomatal cells of allium—Vacuole movements and changes in morphology in differentiating cells as revealed by epifluorescence, video and electron-microscopy. *Protoplasma* **1981**, *109*, 23–55.
43. Ovecka, M.; Lang, I.; Baluska, F.; Ismail, A.; Illes, P.; Lichtscheidl, I.K. Endocytosis and vesicle trafficking during tip growth of root hairs. *Protoplasma* **2005**, *226*, 39–54.
44. Hicks, G.R.; Rojo, E.; Hong, S.; Carter, D.G.; Raikhel, N.V. Geminating pollen has tubular vacuoles, displays highly dynamic vacuole biogenesis, and requires vacuoless1 for proper function. *Plant Physiol.* **2004**, *134*, 1227–1239.
45. Wiltshire, E.J.; Collings, D.A. New dynamics in an old friend: Dynamic tubular vacuoles radiate through the cortical cytoplasm of red onion epidermal cells. *Plant Cell Physiol.* **2009**, *50*, 1826–1839.
46. Mathur, J.; Radhamony, R.; Sinclair, A.M.; Donoso, A.; Dunn, N.; Roach, E.; Radford, D.; Mohaghegh, P.S.; Logan, D.C.; Kokolic, K.; *et al.* Meosfp-based green-to-red photoconvertible subcellular probes for plants. *Plant Physiol.* **2010**, *154*, 1573–1587.
47. Natesan, S.K.; Sullivan, J.A.; Gray, J.C. Stromules: A characteristic cell-specific feature of plastid morphology. *J. Exp. Bot.* **2005**, *56*, 787–797.
48. Kwok, E.Y.; Hanson, M.R. Stromules and the dynamic nature of plastid morphology. *J. Microsc.* **2004**, *214*, 124–137.
49. Hanson, M.R.; Sattarzadeh, A. Stromules: Recent insights into a long neglected feature of plastid morphology and function. *Plant Physiol.* **2011**, *155*, 1486–1492.
50. Hanson, M.R.; Sattarzadeh, A. Trafficking of proteins through plastid stromules. *Plant Cell* **2013**, *25*, 2774–2782.
51. Itoh, R.; Fujiwara, M.T. Regulation of leucoplast morphology in roots: Interorganellar signaling from mitochondria? *Plant Signal. Behav.* **2010**, *5*, 856–859.
52. Schattat, M.; Barton, K.; Mathur, J. Correlated behavior implicates stromules in increasing the interactive surface between plastids and er tubules. *Plant Signal. Behav.* **2011**, *6*, 715–718.
53. Lazzaro, M.D.; Thomson, W.W. The vacuolar-tubular continuum in living trichomes of chickpea (*Cicer arietinum*) provides a rapid means of solute delivery from base to tip. *Protoplasma* **1996**, *193*, 181–190.
54. Ashford, A.E.; Allaway, W.G. The role of the motile tubular vacuole system in mycorrhizal fungi. *Plant Soil* **2002**, *244*, 177–187.
55. Newell, J.M.; Leigh, R.A.; Hall, J.L. Vacuole development in cultured evacuolated oat mesophyll protoplasts. *J. Exp. Bot.* **1998**, *49*, 817–827.
56. Lovy-Wheeler, A.; Cardenas, L.; Kunkel, J.G.; Hepler, P.K. Differential organelle movement on the actin cytoskeleton in lily pollen tubes. *Cell Motil. Cytoskeleton* **2007**, *64*, 217–232.
57. Higaki, T.; Kutsuna, N.; Okubo, E.; Sano, T.; Hasezawa, S. Actin microfilaments regulate vacuolar structures and dynamics: Dual observation of actin microfilaments and vacuolar membrane in living tobacco by-2 cells. *Plant Cell Physiol.* **2006**, *47*, 839–852.

58. Kim, H.; Park, M.; Kim, S.J.; Hwang, I. Actin filaments play a critical role in vacuolar trafficking at the golgi complex in plant cells. *Plant Cell* **2005**, *17*, 888–902.
59. Spector, I.; Shochet, N.R.; Kashman, Y.; Groweiss, A. Latrunculins: Novel marine toxins that disrupt microfilament organization in cultured cells. *Science* **1983**, *219*, 493–495.
60. Sheahan, M.B.; Rose, R.J.; McCurdy, D.W. Actin-filament-dependent remodeling of the vacuole in cultured mesophyll protoplasts. *Protoplasma* **2007**, *230*, 141–152.
61. Uemura, T.; Yoshimura, S.H.; Takeyasu, K.; Sato, M.H. Vacuolar membrane dynamics revealed by gfp-atvam3 fusion protein. *Genes Cells* **2002**, *7*, 743–753.

8

Shielding Flowers Developing under Stress: Translating Theory to Field Application

Noam Chayut [†], Shiri Sobol [†], Nahum Nave and Alon Samach *

The Institute for Plant Sciences and Genetics in Agriculture, The Robert H. Smith Faculty of Agriculture, Food And Environment, The Hebrew University of Jerusalem, Rehovot 76100, Israel;
E-Mails: noamchayut@gmail.com (N.C.); shiri.naft@gmail.com (S.S.);
nahum@dalton-winery.com (N.N.)

[†] These authors contributed equally to this work.

* Author to whom correspondence should be addressed; E-Mail: alon.samach@mail.huji.ac.il

---

Abstract: Developing reproductive organs within a flower are sensitive to environmental stress. A higher incidence of environmental stress during this stage of a crop plants' developmental cycle will lead to major breaches in food security. Clearly, we need to understand this sensitivity and try and overcome it, by agricultural practices and/or the breeding of more tolerant cultivars. Although passion fruit vines initiate flowers all year round, flower primordia abort during warm summers. This restricts the season of fruit production in regions with warm summers. Previously, using controlled chambers, stages in flower development that are sensitive to heat were identified. Based on genetic analysis and physiological experiments in controlled environments, gibberellin activity appeared to be a possible point of horticultural intervention. Here, we aimed to shield flowers of a commercial cultivar from end of summer conditions, thus allowing fruit production in new seasons. We conducted experiments over three years in different settings, and our findings consistently show that a single application of an inhibitor of gibberellin biosynthesis to vines in mid-August can cause precocious flowering of ~2–4 weeks, leading to earlier fruit production of ~1 month. In this case, knowledge obtained on phenology, environmental constraints and genetic variation, allowed us to reach a practical solution.

Keywords: flower development; flowering time; off-season flowering; *Passiflora edulis*; cytokinins; gibberellins; temperature

1. Introduction

Flowers, containing a plant's reproductive organs, are sensitive to hot ambient temperatures (HAT). The temperature that will actually cause irreparable damage to these organs depends on the species and the genotypes within. Most perennial species reach anthesis once a year, in the spring. Flower initiation and early stages of flower development may occur in the previous summer (apples) or towards the end of winter (citrus) [1]. Assuming that the highest ambient temperatures are reached in summer, HAT will affect initial stages of flower development in species that initiate flowers in summer. In species that initiate flowers after winter, HAT becomes a more serious threat at further stages in reproductive development (pollen viability, stigma receptivity) [2–4] or during fruitlet initial development [5].

Climate change may likely lead to even warmer temperatures at early stages of flower development in species with summer initiation and in species with late winter initiation. Thus, understanding the damage to reproductive organs caused by HAT and finding means to protect crops from HAT damage is of high importance [6]. Many botanical studies focus on describing and understanding processes, while other applied studies are more focused on providing agricultural solutions. Here, we provide an example of translational research, in which insights from basic studies on *Passiflora edulis* [7] were useful for developing field applications capable of protecting flowers during HAT, leading to out of season passion fruit.

Passiflora edulis is a woody perennial vine, originating in Brazil and neighboring countries. The species has quite strong variation in many traits, such as fruit color, self-incompatibility and resistance to climate extremes. The literature traditionally mentions two major edible groups (varieties or forma), the purple (*P. edulis* Sims var. *edulis*) and yellow (*P. edulis* Sims f. *flavicarpa* O. Deg.) passion fruit. Each group has been assigned certain traits and may have originated from different climates, and some hybrids have combined positive agricultural traits [8]. Since traits seem to segregate independently of each other, so that yellow-colored passion-fruit found today in the market might have many traits traditionally assigned to purple passion fruit, it seems simpler to define them all under the name *P. edulis* [9].

Cultivation outside of the region of origin likely began in Hawaii in the 1880s. The first published reports on the cultivation of *P. edulis* came in the 1930s [10,11]. In 2007, world production was estimated at 0.64 Million tons per annum [12], yet it is likely much higher, since most of the production is by small scale farmers for local consumption. In Brazil, the largest producer, the main commercial growth is of *flavicarpa* or its hybrids, producing yellow, relatively sour (high levels of acid), quite large fruit for the juice industry, locally named "Maracujá-Amarelo". Brazil, Colombia and Ecuador produced 70% of the world production, yet this commodity is rapidly increasing in climates, ranging from cool subtropical to warm tropical (Asia, Africa, Australia, North and Central America, as well as other South American countries). The main product is juice concentrate with the EU consuming 85% [12]; the second product is fresh fruit, for example the purple "Passion dream" (PD) cultivar grown in Israel [13].

Flower induction in passion fruit is independent of environment. All non-juvenile nodes contain a leaf, a tendril meristem that produces both a tendril and a flower [14,15] and a vegetative lateral meristem. In apices of healthy plants of the PD cultivar, a new node forms approximately every two days [7]. The rate of growth of the flower primordium and the transition between stages of development along the length of the shoot is more or less uniform. Thus, flowers reach similar sizes and developmental stages at specific, more or less conserved node positions, relative to the apex [7]. Flower primordium formation occurs at node Position Number 6 (P6), meaning that there are five younger leaf primordia till the apex. Sepals are formed at P9, and all four whorls are present at P13. Anthesis, lasting ~12 h, occurs when primordia reach P24–P31 (28.6 nodes on average), ~45 days after the flower primordium has formed.

Successful flower development requires certain environmental conditions [7,15,16]. Flower primordia of most varieties abort at an early stage of development if exposed to photoperiods shorter than 11 h [15,17]. In many species, photoperiod influences flower induction [18]. In passion fruit, long days (LDs) are necessary for intact flower development and not for flower induction. In farming with no irrigation, a dry season is likely to cause a reduction in bloom [19–21]. An additional restraint seems to be low irradiance [22].

Primordia of most *P. edulis* varieties are sensitive to hot ambient temperatures (HAT) and abort during warm summers typical for many growing regions, such as Japan, Hawaii, Florida, Southern Australia and Israel [7,11,16,21,23–26]. In Israel, flowers of the PD cultivar reach anthesis only during spring and autumn and abort during winter, due to short photoperiods [15], and summer (mid-June till October), due to HAT conditions [7]. Flowers reaching anthesis normally set fruit if pollinated, even under HAT conditions. In the field, during summer, PD flowers abort before reaching P17 [7]. During September, daytime field temperatures are normally still high, yet nights become slightly cooler, as measured by hours below 22 °C [7]. PD flower primordia growing during this transition period are exposed to decreasing inhibitory HAT conditions, and the first ones to survive end of summer conditions and reach anthesis do so in the first weeks of October. We identified a genotype, "Anthesis all year" (AAY), capable of developing flowers under summer HAT conditions, yet of no commercial value. In comparison to PD, AAY leaves apparently contain much higher levels of cytokinin [7]. Spraying net house-grown PD plants with a synthetic cytokinin (forchlorfenuron; FCF; commercial name "Guliver") towards the end of summer (August 25) significantly increased the survival rate of primordia reaching P12–P14 on September 20, enabling them to reach anthesis; yet, this had only a subtle influence on the time of flowering [7].

An understanding of the phenology of PD flower development allows us to predict the developmental state of a floral primordium at a certain node position at different time points in the past or future. We can subject plants to changing controlled environments and determine at which developmental stage flower primordia are sensitive to HAT conditions and when they become resistant.

Under controlled HAT conditions (a 34/22 °C day/night temperature regime), flowers aborted after making stamens and carpels. When PD plants were moved from optimal conditions for flowering (OCF; 22/16 °C LDs) to controlled HAT conditions, primordia that already reached P17 at day of transfer/treatment (DOT) managed to properly reach anthesis. This suggests that primordia reaching

P17 are no longer sensitive to this HAT treatment. When PD plants were moved from controlled HAT conditions to OCF conditions, primordia that passed P11 on DOT could not be saved and aborted before anthesis [7]. This suggests that damage caused by HAT is irreversible after primordia passed P11. Pre-treating PD plants with the gibberellin biosynthesis inhibitor, uniconazole (Uni), before transferring them from controlled HAT conditions to controlled OCF conditions, delayed the irreversible damage caused by HAT to primordia passed P14 [7]. It could be that the ability of AAY vines to flower in summer is at least partially due to higher cytokinin levels, which were shown in several systems to reduce gibberellin activity. This might explain why directly reducing gibberellin levels with Uni achieved a more efficient means to protect flowers under controlled HAT conditions.

Aim of This Manuscript

Currently, the local industry cannot produce ripe PD fruit between September and late December, because no flowers bloom between June and October. The information we recently gained on the phenology of passion fruit, the stages that are HAT sensitive and the effect of UNI under controlled conditions [7], prompted us in this study to try and translate our basic research to a field application. The heterogeneity in flower development within a vine at any specific time, together with the gradual changes in the environment, provide both a challenge and an opportunity to design practical interventions that cause out of season fruit production in passion fruit. Here, we hypothesized that applications of Uni with or without FCF at specific dates would effectively create a "shield" for flowers developing outdoors towards the end of summer, as temperatures gradually decrease towards the autumn. Here, we conducted a series of field trial experiments over a period of three years and found a treatment that consistently shielded flowers, allowing out of season fruit production.

2. Results and Discussion

In general, under local Rehovot (32°N) conditions (Supplementary Figure S1), fruit of PD vines ripen in the summer (July–August) and winter (January–March) [7]. In 2009, the first wave of ripe winter fruit (the last week of December to the beginning of January) was set around October 12 (October 5–20), when flowers reached anthesis. For these flowers, using the parameters introduced above [7], we calculated the time in which earlier stages of development were reached. These flowers were initiated (P6) ~45 days earlier on approximately August 28, passed the critical P11 stage on approximately September 9 and the P17 stage on approximately September 21 (Figure 1A,B), when temperatures began to decrease (Supplementary Figures S1 and S2). Between the previous spring and September 9, all flowers that reached P11 aborted due to summer HAT conditions. Uni treatment was shown to shield HAT-subjected PD flower primordia, if they are moved by P14 to OCF conditions [7]. We hypothesized that by applying Uni in the field, we could protect additional primordia; those that already reached P12–14 by September 9 and without Uni normally abort. This should lead to earlier flowering in the field and an earlier yield. To accomplish this, we decided to expose such (P14 on September 9) primordia to Uni from the time their node is formed (P0), which is ~28 days earlier, on August 13 (Figure 1C).

Figure 1. Mapping the fate of flower primordia. (**A**) Pictures of developing Passion dream (PD) flowers past Position Number 17 (P17) under optimal conditions, including the fruit set and the start of fruit development. For earlier stages, see [7]. Anthesis lasts around 12 h; (**B**) The fate of flower primordia during summer and fall. The upper row describes the fate of primordia that reach P6 (initiation of the flower primordium) at different dates. Primordia in red will abort before anthesis, while primordia in green will continue developing till anthesis. The first primordia that will complete development are at P6 on approximately August 28. The middle row describes the fate of primordia that reach P11 at different dates. The first primordia that will not abort (green, P6 on August 28) are at P11 on approximately September 9. The bottom row follows the fate of primordia that reach P17 on different dates. PD primordia do not reach this stage during summer. Normally, a primordium that reaches P17 intact will continue developing till anthesis, no matter the external conditions [7]; (**C**) The fate of flower primordia on vines treated with Uni on August 13 (black star). Similar to (**B**), the upper row describes the fate of primordia that reach P6 on different dates. Primordia in dark green are those saved by the treatment. Due to the treatment, the first primordia that will complete development are at P6 on approximately July 30. Not all of these primordia managed to reach anthesis. The treatment saved primordia that were not older than P11–P13 on day of transfer/treatment (DOT). The first treated primordia that reached P17 likely did so by August 21.

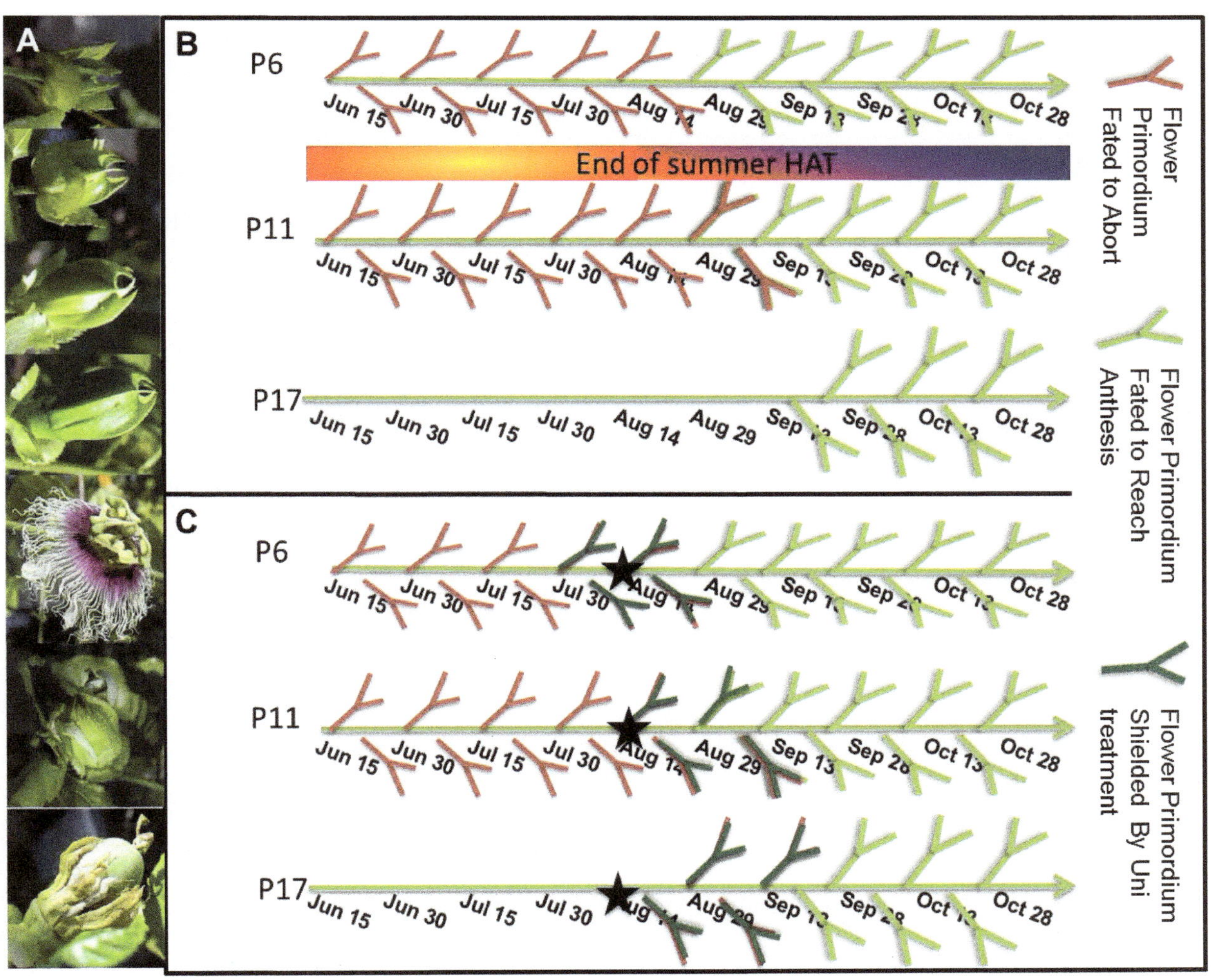

2.1. Summer 2009 Net House Trial

Our initial experiments outside of controlled growth rooms were performed in a net house (see Section 3.3 for the conditions) towards the end of summer. Potted plants were entrained under these conditions from the end of July. Similar to field conditions, the first control (untreated plants) flowers reached anthesis on October 5, and all control plants reached anthesis by October 25. We tested the effects of 200 ppm Uni (DOT: August 13) with or without FCF (10 ppm; applied on August 25). Since we marked node positions on DOT, we could follow and calculate the success rate (primordia reaching anthesis) per node position (on DOT), per treatment (Figure 2A). While we were expecting treated plants to reach anthesis ~1 week (3 nodes × 2 days per node) earlier than control plants, flowers from Uni-treated plants first reached anthesis already on approximately September 10, approximately one month before control plants. The first primordia of control plants that ended up reaching anthesis were at P-3 (position minus 3) on DOT. These primordia reached anthesis in the Uni-treated plants, as well. In addition, primordia that were already formed on DOT, but had not reached P14 (P13–P5), also reached anthesis due to the Uni treatment (Figures 1C and 2A). This might suggest that the shielding ability of Uni in the field is on existing primordia (not younger than P5–P6). Our field results also suggest that Uni shielding is much stronger in the end of summer conditions compared to the previously tested simulation of HAT under controlled conditions [7].

When the Uni treatment was followed by FCF 12 days later, the success rate of additional primordia, those at P14–P16 on DOT, increased (Figure 2A). We followed the fate of these particular primordia (P14–P16 on DOT). We recorded when these primordia reached anthesis (Figure 2B). Obviously, those of control plants aborted before reaching anthesis. The addition of FCF allowed most of these primordia to reach anthesis by September 12 (Figure 2B). An FCF treatment without previous Uni treatment did not protect these flowers [7]. This suggests that the early Uni treatment primed these primordia to respond to the FCF treatment, yet it was not sufficient, by itself, to shield these primordia till anthesis. We tried taking a closer look at these P14–P16 primordia, to better understand their fate (primed, but not capable of reaching anthesis) in Uni-treated plants. Although they did not reach anthesis, their size by August 27 in Uni-treated plants was much larger than untreated plants (Figure 2C).

In this experiment, there was a clear advantage to the additional FCF treatment. For farmers, considering the cost (chemicals, manpower) of this extra spray, the extra gain of 2–3 additional primordia is likely not worth the effort. In the field experiments described below, adding FCF to Uni never provided a substantial difference in the time of yield, so further experiments described here focus on the application of Uni alone.

Figure 2. Uniconazole (Uni) and forchlorfenuron (FCF) application on one-branched pot-grown PD plants (2009). The plants were grown under local end of summer conditions, in a net house (see Section 3.3). Treatments: plants sprayed with 200 ppm Uni on August 13 (DOT) or plants sprayed with 200 ppm Uni on August 13 and with 10 ppm FCF on August 25. Control plants were sprayed with water plus surfactant at DOT. Each treatment contained 5–6 plants. (**A**) Percent flower primordia reaching anthesis, per node position on DOT; bars represent the standard error of the mean. Node positions depicted as negative were not formed yet on DOT; (**B**) Anthesis date of P14–P16 (position on DOT) primordia. Control plants did not have developed flowers at these positions, and the first anthesis in control plants occurred on a later date (October 10). The additional FCF treatment caused a higher rate of flowers reaching anthesis at these positions, on an earlier date. For each treatment, the fate of 18 nodes (three nodes in six plants) was followed, and the percent of nodes reaching anthesis was calculated; (**C**) The short-term fate of nodes that were at P14–P16 on DOT, measured two weeks later (August 27). We measured the percent of primordia reaching a length of more than 24 mm, corresponding to normal primordia at P20. Reaching a length similar to P20 is not proof that the primordium is intact. $n = 18$ (one node in 18 plants).

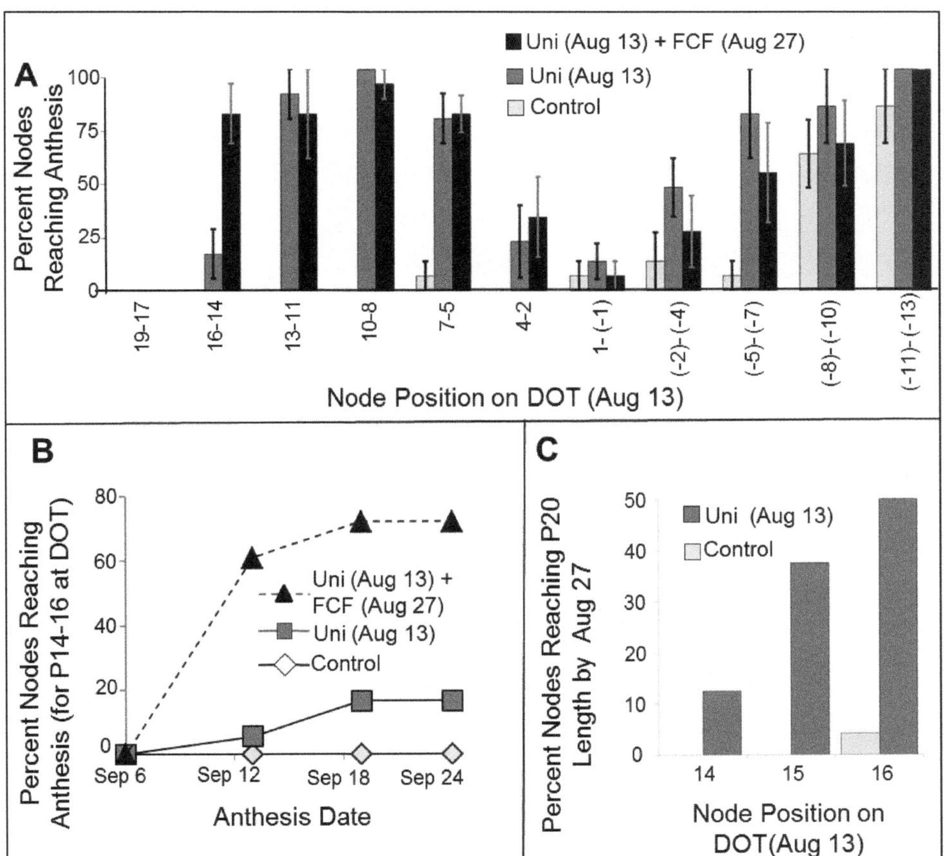

2.2. Summers of 2010–2011 Net House Trials

In the summer of 2010, we conducted a similar experiment using the same net house. First, we wanted to repeat the experiment, testing whether the Uni treatment will be effective again. In addition, we asked whether we could reach an even earlier date of anthesis by applying Uni earlier in the summer. Two hundred parts per million of Uni were applied on July 26 or August 10. Control plants (water plus surfactant sprayed at DOT) began anthesis on October 3, while treated plants (both Uni treatments) reached anthesis ~20 days earlier (Figure 3A). By October 3, all plants treated with Uni on July 26 reached anthesis, while in control plants, only 17% reached anthesis by this date (Figure 3A). The first fruit from treated plants ripened on December 16, more than two weeks before fruit from control plants.

Figure 3. Uni application on one-branched pot-grown PD plants (2010). The plants were grown under local end of summer conditions, in a net house (see Section 3.3). Treatments: plants sprayed with 200 ppm Uni on July 26 (DOT) or on August 10. Control plants were sprayed with water plus surfactant at DOT. Six plants were analyzed for each treatment. (**A**) The percent of plants reaching anthesis from September 12 (when treated plants reached anthesis) until October 6 (when control plants started reaching anthesis); (**B**) Average relative position on DOT (along the main branch) in which sequential primordia (more than one flower) reached a blooming stage. For control plants, these were primordia that were not yet formed (negative value) at DOT; (**C**) The effect of Uni treatments on the elongation of lateral meristems from nodes that were positioned at P15–P24 on DOT. The measurement was performed on August 19; (**D**) The effect of Uni on July 16 on the plastochron interval (see Section 3.4). For each plant, we counted the number of newly-formed nodes during the experiment and divided the experiment duration in days by the node number. Bars in (**B–D**) represent the standard error of the mean.

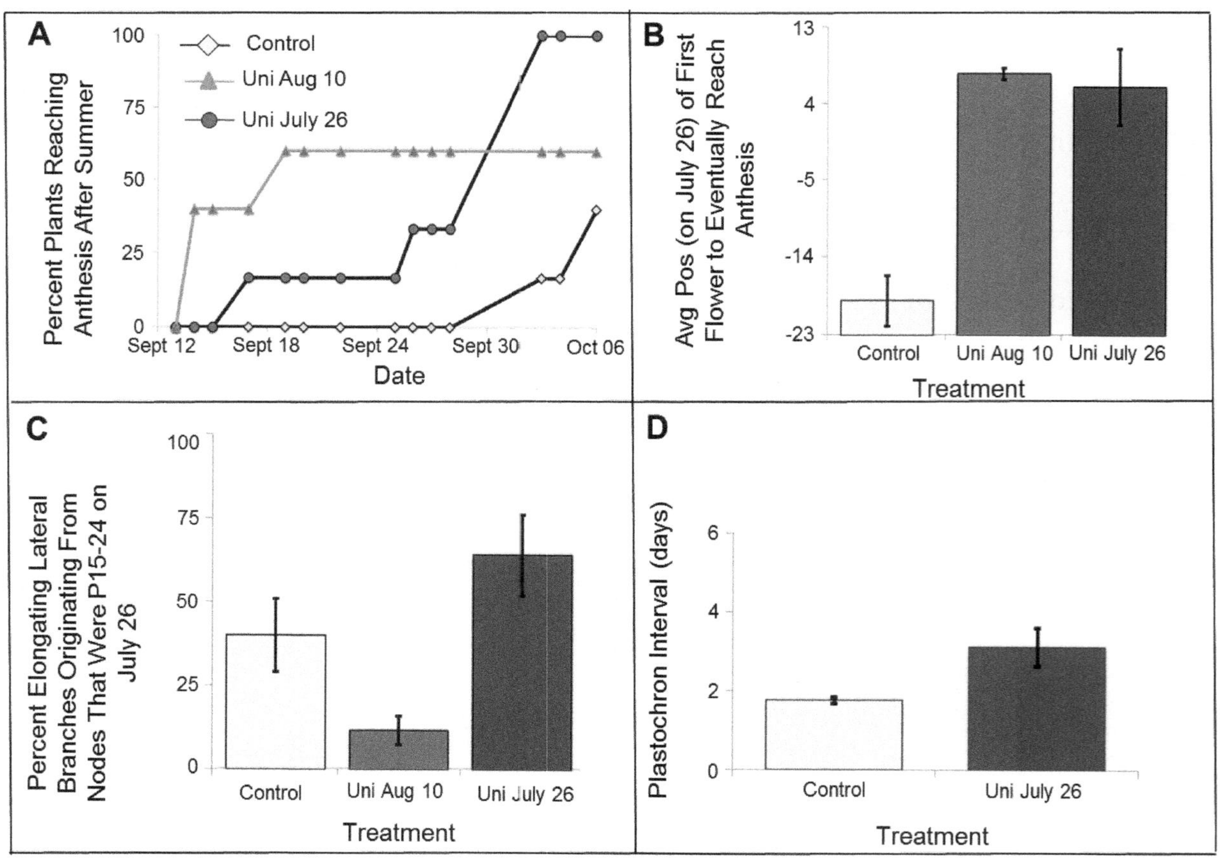

The first control flowers reaching anthesis after summer were, on average, at ~P-(minus)19 on July 26 (Figure 3B). The first intact August 10 Uni-treated flowers reaching anthesis were, on average, at ~P7.5 on July 26. Thus, Uni treatment in mid-August clearly shielded flowers one year after another. Spraying plants with Uni at an earlier date (July 26) did not shield extra flowers (Figure 3B). Thus, this specific Uni treatment, when provided at an earlier date, was not sufficient to enable flowers to reach anthesis before September under local conditions. It seems that our original calculated "window of intervention" on approximately August 12 was appropriate.

Uniconazole, as an inhibitor of gibberellin biosynthesis, can cause growth retardation in plants, including passion fruit [27]. In this experiment, we also followed the elongation of lateral branches in response to the treatment. Nine days after treatment, the elongation of lateral branches was inhibited compared to control plants, yet 24 days after treatment, elongation was restored (Figure 3C).

While the Uni treatment shielded flowers, the July 26 treatment also increased plastochron intervals by ~1.7-fold (Figure 3D), in other words, slowing down the production of new primordia and possibly flower development. This likely explains why a Uni-treated flower that reached P7.5 on July 26 required an additional 50 days to reach anthesis (September 14). It appears that the full agricultural potential of Uni shielding of primordia from HAT is partially masked by slower flower development. This effect of Uni on plastochron intervals may depend on environmental conditions at the time of application, since it was not noticed in other Uni treatments.

In the summer of 2011, we performed an additional one-branched potted PD plant trial in the same net house. Treatments were 200 ppm Uni on either July 28 or August 11 compared to untreated plants (control). In control plants, the first flowers appeared on October 7, and the average date of first anthesis was October 24. In this trial, an earlier date of application (end of July) led to an even earlier date of anthesis (Supplementary Figure S3). Uni treatment on July 28 caused first anthesis, in the beginning of September, while Uni treatment on August 11 caused early (less dramatic) flowering, on around September 21. This difference between years could be due to a cooler period before August 11 in the summer of 2011, compared to 2010 (Supplementary Figure S2).

2.3. Summer 2010 Experimental Field Trial

In the summer of 2010, we also tested the effect of Uni treatment at different dates (July 26, August 2 and August 10) on flowering in the faculty experimental orchard (no nets, field conditions). We conducted a "snapshot" survey by counting the number of open flowers per trellis running meter at a certain time point, every few days (Figure 4A). In this count, some of the flowers reaching anthesis are likely missed, and their distribution among branches is not documented. Thus, we also marked random branches in advance and recorded the date on which the first flower on them reached anthesis. Here, if we missed an open flower, we could still record its bloom a few days later, since it is still on the branch. This allowed us to calculate the accumulation of branches reaching anthesis (Figure 4B). Later on, we counted ripened fruit (fruit that detached) per trellis running meter (Figure 4C). Here, again, we could measure the accumulation of fruit, even when visiting every few days, since the fruit that fell remained on the ground till our next visit.

The first intact control flowers reached anthesis between October 4 and 7 (Figure 4A). The most effective Uni treatment in causing early flowering in this trial was the latest treatment on August 10.

Here, plants at anthesis were detected in the first date of observation, September 17 (Figure 4A). By October 13, ~50% of treated branches reached anthesis, while only ~6% of control branches reached anthesis (Figure 4B). Ripe control fruit began to abscise December 15, 72 days after sighting the first anthesis (Figure 4C). Ripe August 10 Uni-treated fruits began abscising December 2, 76 days after sighting the first anthesis (Figure 4C). By December 19, the treated vines produced nine-fold higher levels of ripe fruit (Figure 4C).

The two earlier treatments with Uni, on July 26 and August 2, did not cause a significant change in the flowering pattern of the plants, compared to control untreated plants. These treated plants had much shorter internode lengths compared to control plants, as is expected from Uni treatment [15]. Therefore, Uni treatment at earlier times definitely had an effect on plants, but did not cause obvious shielding of flowers.

Figure 4. Uni spray on PD vines in an experimental field (2010). Treatments: 200 ppm Uni sprayed on July 26/August 2/August 10 and control (not treated). We visited the plot seven times in a period of 22 days starting September 15. (**A**) The number of open flowers per running meter of trellis per side on that certain day; (**B**) For each treatment, we gathered information on 20 random shoots (10 per side, west or east) that were appointed and marked before treatments began. The accumulative percentage of blooming branches is presented; (**C**) Accumulated ripe fruit number per linear meter of trellised vines. Fruit was defined as ripe upon abscission. We collected and counted the number of fruit on the ground on each visit.

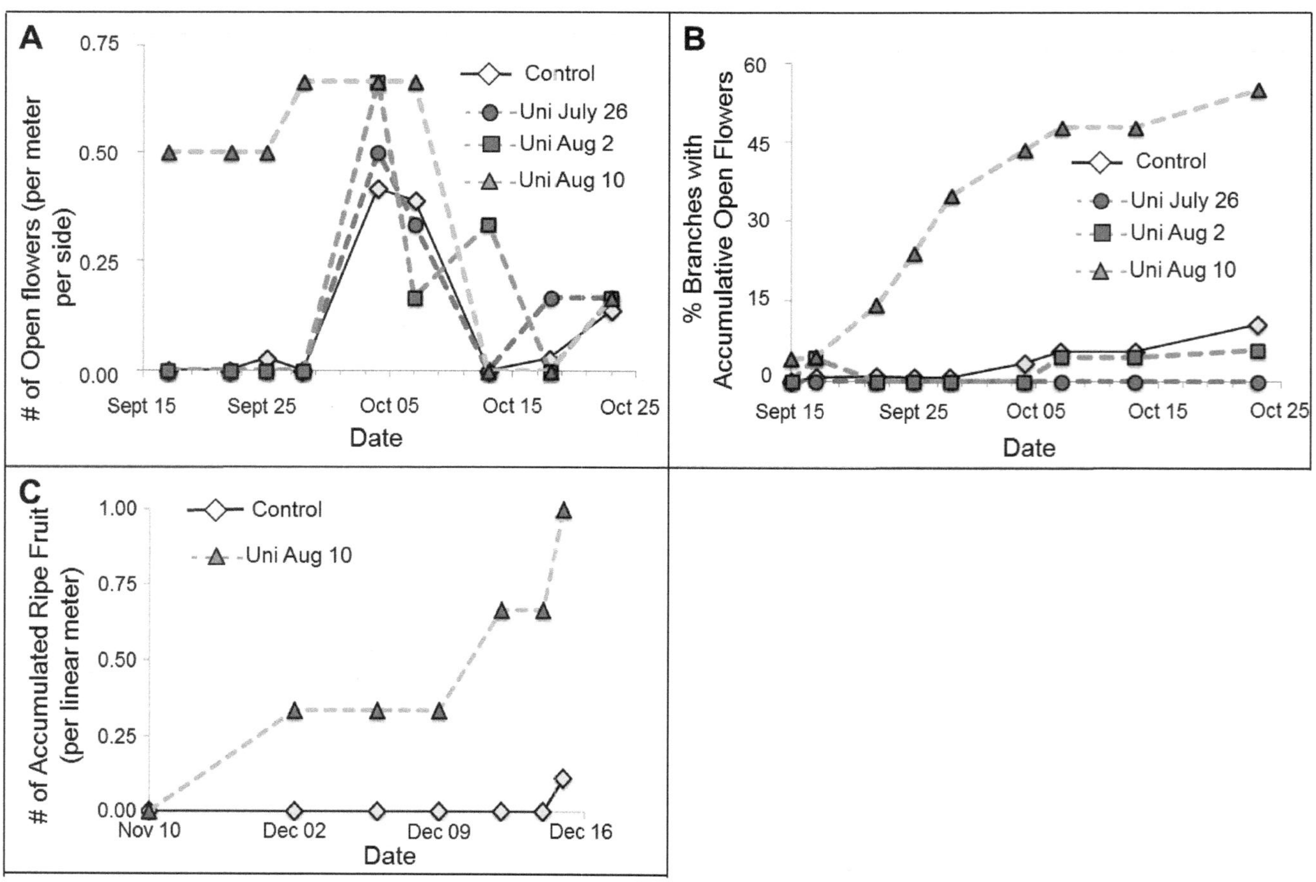

2.4. Commercial Field Trials 2010–2011

In the summer of 2010, we also made our first attempt to shield flowers in a commercial PD orchard (Moranim subtropical fruits LTD; Section 3.3.3). Vines were sprayed with 200 ppm Uni with a hand pump backpack sprayer on August 13, August 27 or on both dates. We also tried adding FCF (August 27) to a single Uni (August 13) treatment to see if adding FCF has a more pronounced affect in the field. Control vines (on each treated row) were sprayed with water and surfactant on August 13. We conducted a "snapshot" survey by counting the number of open flowers per trellis running meter at a certain time point, every few days, between September 15 and October 27. Passion fruit flowers remain open for ~12 h, so every flower was counted only once. Control flowers first reached anthesis at a rate of at least one flower per trellis-meter by October 27 (Figure 5A). The single Uni treatment on August 13 caused the earliest flowering, as flowers reached anthesis at a rate of at least one flower per trellis meter already by September 19. Adding an FCF treatment or a second Uni treatment on August 27 had no added benefit in causing precocious flowering. A single Uni treatment at a later date (August 27) was also less efficient (Figure 5A). It could be that the delay in flowering due to the second treatment of Uni is also because of it causing a slower plastochron interval (not measured).

Fruit of the PD cultivar reached final size ~20 days after fertilization. On November 1, we returned to the orchard to quantify fruitlet size. At this time, fruitlets would have reached their final size (~40 mm in diameter) if they developed from a flower that reached anthesis before October 13. We chose for each trellis meter and side the two largest fruits per treatment. Altogether, the fruit diameter was calculated for 20 fruits per treatment (Figure 5B). The largest control fruits from each row were, on average, 20 mm in diameter. Under ideal conditions, PD fruit reaches this size five days after pollination. Thus, these fruit likely came from flowers that reached anthesis on October 27, when we first observed the first uniform anthesis in control plants (Figure 5A). The largest fruit from the treated plants were all significantly larger, reaching a ~40 mm diameter, suggesting that they developed from flowers that reached anthesis before October 13. Finally, we collected data on initial ripe fruit in the orchard. We visited the orchard four times between November 22 and December 22 and counted the number of accumulated ripe (abscised with full color from the vine) fruit per meter of trellis per treatment (Figure 5C). Not surprisingly, fruit from the treatment that caused early flowering (Uni on August 13) ripened before the control fruit, thus extending the season of yield. The treatment affected ripening date more significantly than it affected anthesis date. One explanation for this may be that in the transition period from September to December, in which local temperatures are slowly declining, fruit maturation takes longer. Thus, the benefit of setting fruit in September *versus* October is that the earlier fruit will be exposed to warmer temperatures and will mature more rapidly. Thus, in this period, earlier anthesis could cause an even more pronounced earlier fruit ripening. A relative delay in fruit ripening was observed in Uni-treated plants that were later treated with FCF (Figure 5C). It is not clear at this point if this effect is significant and repeatable and, if so, what might be the reason for it.

Figure 5. Uni and FCF spray on PD vines in a commercial orchard (2010). Treatments: 200 ppm Uni on August 13/on August 13 and on August 27/200 ppm Uni on August 13 and 10 ppm FCF on August 27/Control (water and surfactant sprayed on August 13). (**A**) Counts of open flowers on several occasions between September 19 and October 27; the latter date is when flowers in the control plants began to reach anthesis. For each date, the average number of flowers reaching anthesis per running meter of trellis per side is presented; (**B**) Fruitlet size (diameter in mms) on November 1: the two largest developing fruits per running meter were documented in each treatment and in the control plants growing in the same orchard line; (**C**) Accumulated number of ripe (abscised) fruits (as in Figure 4C).

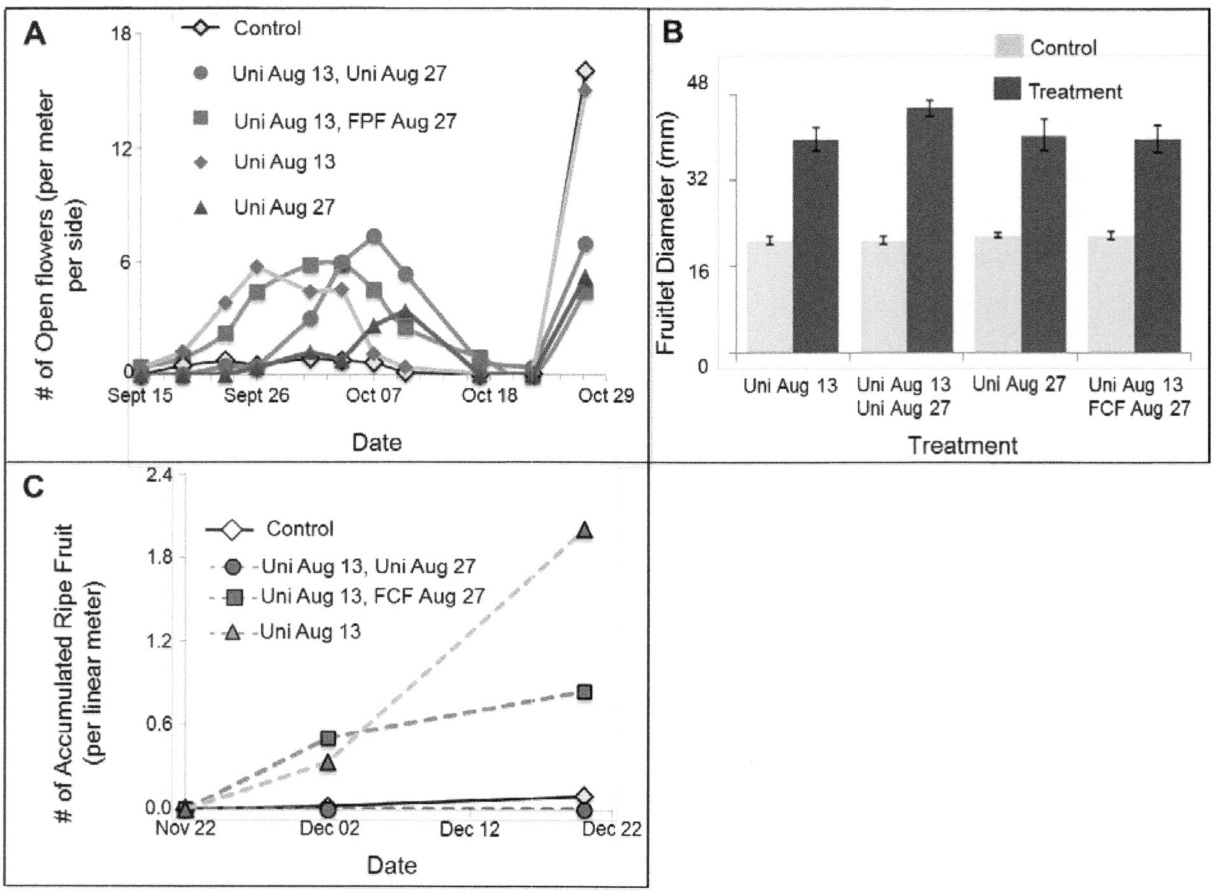

In the summer of 2011, we conducted additional trials in commercial orchards at two different locations: Moranim LTD and Mr. Isaac Tzipori's farm (see Section 3.3.3). At the Tzipori farm, three rows of trellised PD vines were left untreated, while three other rows were sprayed with 200 ppm Uni and surfactant on August 10, using a hand pump sprayer. In Moranim LTD commercial orchard trellised PD vines were left untreated or sprayed with 200 ppm Uni and surfactant on August 12 using an air blast sprayer. We visited the farms when notified that control blooming started (Tzipori's on September 22 and Moranim on September 27). We chose four branches from both sides of the treated and untreated rows (three rows in Tzipori's and six in Moranim). The branches chosen were those that seemed to have the most flowers and in which growth did not stop (apical meristem still active). We counted the number of flowers that already opened (since the end of summer) on each of the branches.

In both experiments, spraying 200 ppm Uni on approximately August 10 resulted in a substantially earlier bloom (Figures 6A and 7A,B). In Tzipori's control, untreated vines, no flowers reached anthesis on an earlier date, while three of the 24 branches contained an open flower on the day of measurement. In contrast, all Uni-treated branches contained previously and currently open flowers, with numbers of flowers per branch ranging from three to eight, with an average of six flowers that had already reached anthesis, per branch (Figure 6A). In Moranim, by September 27, Uni-treated vines contained on average five-fold more flowers that had already reached anthesis, compared to control vines (Figures 6B and 7C,D). We also measured the position of the most developed flower (or fruit) since the end of summer (Figure 6C). In Uni-treated vines, the average position was P33, and in control vines, it was P26. Assuming an average of 28.6 nodes for anthesis, the treated plants potentially had 4–5 set fruit per branch by this date. The presence of developing fruit in treated vines was obvious when studying the vines (Figure 7D).

Figure 6. Uni spray on PD vines in commercial orchards (2011). (**A**) Tzipori farm treatments: 200 ppm Uni and surfactant sprayed on 10 August and control untreated plants. The average number of flowers that reached anthesis by September 22 per branch is presented; (**B**) Moranim LTD Treatments: 200 ppm Uni and surfactant sprayed on August 12 and control-untreated plants. The average number of flowers that reached anthesis by September 27 per branch is presented; (**C**) The same experiment as in (**B**), calculating the average relative position from the apex (on September 27) of the first flower reaching anthesis on each branch. A higher number depicts the shielding of additional primordia. Bars in (**A–C**) represent the standard error of the mean.

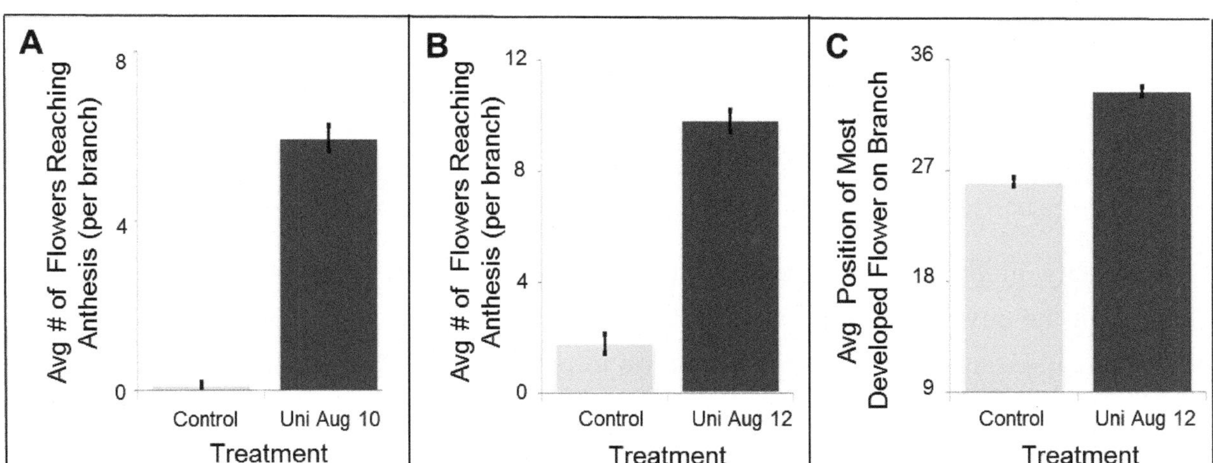

When conducting a series of field trials meant to overcome an environmental constraint, it is expected to find variation in environmental conditions between years and locations (Supplementary Figure S2). Nevertheless, we were able to repeat our main result in three consecutive years and in four different locations and growth conditions: earlier flowering and fruiting as a result of a single spray of 200 ppm of Uni on approximately August 12. The degree of damage caused to PD flowers by HAT conditions vary with temperature regime (Supplementary Figure S4). The ability of Uni treatment to shield PD flowers from HAT conditions depends on the scale of heat to which flowers are exposed. For example, Uni treatment, although promoting flower development up to a certain stage, did not bring about anthesis under controlled HAT conditions of 34 °C/22 °C day/night regime (Supplementary Figure S4b). Thus, it is expected that the success of earlier (before mid-August) applications of Uni in the field may vary with the conditions in every year, as shown here for the net house trials of 2010 (Figure 3) and 2011 (Supplementary Figure S3).

Figure 7. Pictures of PD vines in commercial orchards (2011). (**A**,**B**) Tzipori farm on September 22 with treatment as in Figure 6A. Most control plants (**A**) did not reach anthesis, while treated plants (**B**) are blooming. Red circles mark flowers at anthesis. (**C**,**D**) Moranim LTD farm on September 27 with treatment as in Figure 6B. Aborted flowers (orange circles) can be seen in control vines, while treated vines contain flowers at anthesis attracting pollinators and setting fruit (light blue circles).

2.5. Summary

The heterogeneity in flower development within a vine at any specific time, together with the gradual changes in the environment, provide both a challenge and an opportunity to design practical interventions that cause out of season fruit production in passion fruit. Our previous studies suggested that reducing gibberellin levels and increasing cytokinin levels (perhaps causing a reduction in gibberellin activity) might both be potentially useful in protecting flower development under HAT [7]. Here, an understanding of stages in which PD flowers are susceptible to HAT and comprehension of the species plastochron pace allowed us to reach the correct time of intervention, which enabled flower development, leading to out-of-season fruit in commercial field trials. Understanding the dynamics of the primordia development under changing environments helped to define a window in development in which a simple chemical treatment could shield flowers and extend the bloom season and the period in which fruit is available for marketing.

During three consecutive years, in four different locations, we have shown that by spraying PD vines in mid-August with 200 ppm Uni, we can extend the anthesis period by 3–6 weeks, allowing additional flowers to reach anthesis. These flowers can set fruit and ripen much before the untreated fruit, leading to a longer period of fruit production for farmers. In the northern hemisphere, this also allows producing fruit towards the holiday season (December–January), when demand is much higher. This treatment is not costly, suggesting a substantial gain in profit.

Since normal marketable fruits were indeed produced by the Uni-treated flowers, the treatment did not seem to interfere in fertilization, fruit development and maturation. Successful pollination in local orchards depends on honeybees, which are introduced by farmers during bloom, or by hand pollination. Clearly, earlier bloom will require these interventions at an earlier date.

While we have not tested this treatment in other cultivars or in other growing regions outside of Israel subjected to HAT conditions during summer, we think it is quite likely that it will cause a similar, positive affect. A promising approach will be to breed for cultivars that are less sensitive to environmental constraints and that still have high fruit quality. The local commercial practices are quite intense (irrigation, NPK fertilization, vine replacements every ~5 years). We cannot rule out that under less optimal conditions, other constraints may reduce the shielding effect of the Uni treatment.

Interestingly, previous reported field trials with gibberellin biosynthesis inhibitors in passion fruit provided no evidence for a change in the time of anthesis [27–29]. This could be due to the time of application. Here, we saw that in some of our trials, earlier application of Uni had no effect on flowering. In addition, the time in which the effect of a treatment is measured may affect the results recorded. In most other fruit crops, for example pome fruits, the sequence of events beginning with flower primordia induction until bloom can take ~260 days. In PD, and probably other passion fruit cultivars, anthesis occurs ~45 days after primordia initiation. The Uni treatment affects primordia that are already developing, so that the effect of the treatment on flowering can be seen in a relatively short time (~30 days) and will remain unnoticed if the orchard is revisited after two or more months and only the flowers are monitored.

Since reducing levels of gibberellin enables further PD flower development under HAT conditions, it appears that gibberellin inhibits flower development under HAT conditions. Gibberellin may be diverting photosynthetic assimilates from developing flowers to other organs, as was suggested in bougainvillea [30].

3. Experimental

3.1. Plant Material

The cv. "Passion dream" (PD) is the most common commercially grown passion fruit in Israel. It was previously described [15] as an F_1 hybrid of yellow and purple passion fruit. Unlike seed propagated cultivars commonly used in many other countries, the PD highly heterozygous cultivar is propagated by vegetative rooted cutting. Thus, both commercial and experimental orchards vines described in this paper are genetically identical. Plants used for the net house experiments were rooted from cuttings taken from vines of the experimental faculty orchard.

3.2. Measuring Size of Flower Primordia

The flower primordium size was determined by measuring the length of bracts, sepals and complete primordium, including pedicel, as was described previously [7,15].

3.3. Growth Conditions

In the manuscript, when discussing plants in pots, we use the term "plants", and when discussing plants in the field, we use the term "vines".

3.3.1. Net House

The faculty net house is a cubic white net shed, which reduces light intensity by around 10%, keeps birds and insects away, but maintains the seasonal outdoor temperature and day length conditions. In the summer of 2009, 2010 and 2011, PD plants were grown in 18-cm diameter, well-irrigated pots under net-house conditions at the Faculty of Agriculture. While plants were growing and acclimatizing to the conditions, we pruned emerging secondary branches, allowing only one branch to continue its growth. These plants were termed: "One-branched pot-grown PD plants". After DOT, we continued the removal of emerging side branches in 2009. In 2010–2011, in order to measure the treatment effect on side-branching, we ceased pruning the plants on DOT. In all experiments, we followed the fate of the developing flowers of the main branch.

3.3.2. Experimental Field Trial

The Faculty of Agriculture experimental field is in Rehovot, Israel (31°54'12"N; 34°47'47"E). Rooted vegetative cuttings of PD plants were planted in the field in June, 2007. Eight irrigated trellised rows (north to south) of PD vines, each row divided into two (north and south) stretches of 6 meters, were given different treatments

3.3.3. Commercial Orchards

Moranim Subtropical fruit LTD: trellised irrigated PD plot planted northeast to southwest. Location: 31°52'56"N, 34°50'46"E. The experiments were held in summer 2010 and 2011 at different locations of the farm.

Isaac Tzipori farm LTD: Location: 31°11'52"N, 34°22'38"E. The experiment was held in summer 2011.

3.4. Calculating Plastochron Intervals at the Shoot Apex of PD

We calculated plastochron intervals [31,32] of successive leaf primordia at the shoot apex of PD. The unfolding of the PD leaf is easily visible and occurs independent of environment, on average, around (~)15 nodes (varies 14–16) from the apex [15]. In other words, if we follow the fate of a leaf primordium formed in the apex, once 14 additional nodes are formed, this leaf will likely unfold. All older nodes will contain open leaves, while all younger nodes contain folded leaves. To calculate the plastochron interval, we marked the main branch twice at the node containing the youngest unfolded leaf and divided the number of days between the two marking time points by the number of newly formed nodes between the two marks. Having a visible marker (leaf unfolding) that can easily tell us at any time the distance (in nodes) from the meristem, together with knowing the plastochron, allowed us to predict for every visible node its current location relative to the apex where it will be, relative to the apex, in the future, and where it was, relative to the apex, in a past date of interest.

3.5. Hormone Treatments

To reduce GA biosynthesis, we used the commercial product "Magic" (Sumitomo Chemicals, Hyogo-Ken, Japan), which is 50 g/L uniconazole (Uni), a triazole capable of inhibiting P450 *ent*-kaurene oxidase (CYP701). The concentration of Uni in spray was 200 ppm. To increase CK levels, we used the commercial product "Guliver" (Jiangsu institute of ecomones, Jiangsu, China), imported by Tarsis Agricultural & Industrial Chemicals Co. Ltd., Petah Tikva, Israel. "Guliver" is an aqueous solution of forchlorfenuron (FCF), a synthetic CK. The applied concentration was 10 g/L. The concentration of FCF in spray was 10 ppm. In all sprays, we added 0.025% Triton X-100 as a surfactant. For net house trials, we used small hand plastic sprayers. In the experimental field trial, we used hand pump backpack sprayers. In the commercial orchards, a hand pump backpack sprayer or an air blast sprayer was used, as stated in the text. Spraying of plants was performed in the early morning.

4. Conclusions

Agriculture is in need of constant improvement due to growing human needs and increasing, often man-made, environmental constraints, such as a warmer environment [6]. Many improved agricultural practices originated from a chance set of circumstances leading to a phenomenon noticed by a keen eye, while others were developed from methodical, lengthy trials ending with a winning combination. The use of plant growth regulators to overcome a developmental constraint in plants has provided enormous benefits to agriculture. One common method for identifying the right cocktail is called "spray and pray". In such a trial, a wide matrix of chemicals, concentrations and schedules is used in an attempt to find a winning combination. This could take a very long time and many resources. While we do not challenge here the added benefit of prayers, our study implies that in some cases, a better comprehension of the underlying developmental events might fast-forward the process.

Acknowledgments

We would like to thank Barak Cohen for helpful horticultural assistance with the passion fruit vines, Israel (Ruli) Moran for the use of his commercial orchard at Moranim subtropical fruits LTD and Isaac Tzipori for the use of his commercial orchard. This research was funded by the Israeli Science Foundation—the Charles H. Revson Foundation (Grant No. 1464/07), the Chief Scientist Ministry of Agriculture and Rural Development, Israel (Grant No. 837-0086-10) and by the German-Israeli Project Cooperation (Project H 3.1).

Author Contributions

Noam Chayut and Shiri Sobol conducted the experiments and analyzed the data. Nahum Nave contributed data. Alon Samach wrote the paper with Noam Chayut and supervised all work.

Conflicts of Interest

The authors declare no conflict of interest.

References

1. Samach, A.; Smith, H.M. Constraints to obtaining consistent annual yields in perennials II: Environment and fruit load affect flowering induction. *Plant Sci.* **2013**, *207*, 168–176.
2. Barnabas, B.; Jager, K.; Feher, A. The effect of drought and heat stress on reproductive processes in cereals. *Plant Cell Environ.* **2008**, *31*, 11–38.
3. Zinn, K.E.; Tunc-Ozdemir, M.; Harper, J.F. Temperature stress and plant sexual reproduction: Uncovering the weakest links. *J. Exp. Bot.* **2010**, *61*, 1959–1968.
4. Sato, S.; Peet, M.M.; Thomas, J.F. Physiological factors limit fruit set of tomato (*Lycopersicon esculentum* Mill.) under chronic, mild heat stress. *Plant Cell Environ.* **2000**, *23*, 719–726.
5. Racsko, J.; Leite, G.B.; Petri, J.L.; Zhongfu, S.; Wang, Y.; Szabo, Z.; Soltesz, M.; Nyeki, J. Fruit drop: The role of inner agents and environmental factors in the drop of flowers and fruits. *Int. J. Hortic. Sci.* **2007**, *13*, 13–23.
6. IPCC. Climate change 2014: Impacts, adaptation, and vulnerability. Contribution to the Fifth Assessment Report of the Intergovernmental Panel on Climate Change, 2014. Available online: http://www.ipcc.ch/report/ar5/wg2/ (accessed on 1 April 2014).
7. Sobol, S.; Chayut, N.; Nave, N.; Kafle, D.; Hegele, M.; Kaminetsky, R.; Wunsche, J.N.; Samach, A. Genetic variation in yield under hot ambient temperatures spotlights a role for cytokinin in protection of developing floral primordia. *Plant Cell Environ.* **2014**, *37*, 643–657.
8. Nakasone, H.Y.; Hirano, R.; Ito, P. Preliminary observations on the inheritance of several factors in the passion fruit (*Passiflora edulis* L. and forma flavi-carpa). In *Technical Progress Report Hawaii Agricultural Experiment Station College of Ropical*; Agriculture University of Hawaii: Honolulu, HI, USA, 1967; pp. 1–12.
9. Bernacci, L.C.; Soares-Scott, M.D.; Junqueira, N.T.V.; Passos, I.R.D.S.; Meletti, L.M.M. *Passiflora edulis* sims: The correct taxonomic way to cite the yellow passion fruit (and of others colors). *Rev Bras. Frutic. Jaboticabal-SP* **2008**, *30*, 566–576.
10. Barnes, H. Passion fruit culture in queensland. *Qld. Agric. J.* **1934**, *41*, 667–675.
11. Pope, W.T. *The Edible Passion Fruit in Hawzaii*; Bulletin of the Hawaii Agricultural Experiment Station: Honolulu, HI, USA, 1935; Volume 74, pp. 1–22.
12. Morey, P. Report on Passion Fruit Demand Study Indonesia. Prepared for: IFC World Bank SADI 2007. Available online: http://capas.unpad.ac.id/penelitian/file/3bf79737242973f720135a606599e41f/Final%20Report%20MARKISSA%20Demand%20Study-Englsih%20Version.pdf (accessed on 1 April 2014).
13. Blumenfeld, A.; Shaya, F. Sweet dream. *Yevul See* **2004**, *6*, 40–41.
14. Cutri, L.; Nave, N.; Ben-Ami, M.; Chayut, N.; Samach, A.; Dornelas, M.C. Evolutionary, genetic, environmental and hormonal-induced plasticity in the fate of organs arising from axillary meristems in *Passiflora* spp. *Mech. Dev.* **2012**, *130*, 61–69.

15. Nave, N.; Katz, E.; Chayut, N.; Gazit, S.; Samach, A. Flower development in the passion fruit passiflora edulis requires a photoperiod-induced systemic graft-transmissible signal. *Plant Cell Environ.* **2010**, *33*, 2065–2083.
16. Menzel, C.M.; Simpson, D.R. Passionfruit. In *Handbook of Environmental Physiology of Fruit Crops*; Schaffer, B., Andersen, P.C., Eds.; CRC Press, Inc.: Boca Raton, FL, USA, 1994; Volume II: Sub-tropical and Tropical Crops, pp. 225–242.
17. Watson, D.P.; Bowers, F.A.I. Long days produce flowers on passion fruit. *Hawaii Farm Sci.* **1965**, *14*, 3–5.
18. Samach, A.; Coupland, G. Time measurement and the control of flowering in plants. *Bioessays* **2000**, *22*, 38–47.
19. Gachanja, S.P.; Gurnah, A.M. Pruning and trellising purple passion fruit. I. Yields and seasonal trends. *J. Hortic. Sci.* **1980**, *55*, 345–349.
20. Menzel, C.M.; Simpson, D.R.; Dowling, A.J. Water relations in passionfruit: Effect of moisture stress on growth, flowering and nutrient uptake. *Sci. Hortic.* **1986**, *29*, 239–249.
21. Menzel, C.M.; Simpson, D.R.; Winks, C.W. Effect of temperature on growth, flowering and nutrient uptake of three passionfruit cultivars under low irradiance. *Sci. Hortic.* **1987**, *31*, 259–268.
22. Menzel, C.M.; Simpson, D.R. Effect of continuous shading on growth, flowering and nutrient uptake of passionfruit. *Sci. Hortic.* **1988**, *35*, 77–88.
23. Ishihata, K.; Shindo, T.; Iwahori, S. Flower-bud differentiation and development in purple passionfruit, *Passiflore edulis* sims. *Bull. Fac. Agric. Kagoshima Univ.* **1989**, *39*, 103–119.
24. Yonemoto, Y.; Kozai, N.; Ogata, T. Suitable night temperature conditions for reducing the acidity content in winter-harvested passion fruit. Japan International Research Center for Agricultural Sciences, Rearch Highlights, 2006. Available online: http://www.jircas.affrc.go.jp/english/publication/highlights/2006/2006_24.html (accessed on 1 April 2014).
25. Fisch, M.B. Passionfruit the world over. In *California Rare Fruit Growers Yearbook*; California Rare Fruit Growers, Inc. Fullerton, CA, USA, 1975; Volume 7, pp. 13–55.
26. Gurnah, A.M.; Gachanja, S.P. Spacing and pruning of purple passion fruit. *Trop. Agric. (Trinidad)* **1983**, *61*, 143–147.
27. Tzipori, I.; Shuker, S.; Tomer, E. Use of growth regulators in order to retard growth in passion fruit. *Alon Hanotea* **2004**, *58*, 450–465.
28. Ataide, E.M.; Ruggiero, C.; de Oliveira, J.C.; Rodrigues, J.D.; de Oliveira, H.J. Effect of paclobutrazol and gibberellic acid in floral induction of yellow passion-fruit in intercrop condition. *Rev. Bras. Frutic. Jaboticabal-SP* **2006**, *28*, 160–163.
29. Ataide, E.M.; Ruggiero, C.; Rodrigues, J.D.; de Oliveira, J.C.; Rodrigues, T.D.J.D.; da Silva, J.R. Plant regulator and biostimulant in yellow passion-fruit floral induction under non inductive condition. *Rev Bras. Frutic. Jaboticabal-SP* **2006**, *28*, 347–350.
30. Tse, A.; Ramina, A.; Hackett, W.P.; Sachs, R.M. Enhanced inflorescence development in *Bougainvillea* “san diego red” by removal of young leaves and cytokinin treatments. *Plant Physiol.* **1974**, *54*, 407–414.
31. Michelini, F.J. The plastochron index in developmental studies of *Xanthium italicum* moretti. *Am. J. Bot.* **1958**, *45*, 525–533.

32. Lamoreaux, R.J.; Chaney, W.R.; Brown, K.M. The plastochron index: A review after two decades of use. *Am. J. Bot.* **1978**, *65*, 586–593.

9

Regulation of Pollen Tube Growth by Transglutaminase

Giampiero Cai [1], Donatella Serafini-Fracassini [2] and Stefano Del Duca [2,*]

[1] Dipartimento di Scienze della Vita, Università degli Studi di Siena, via Mattioli 4, Siena 53100, Italy; E-Mail: giampiero.cai@unisi.it

[2] Dipartimento di Scienze Biologiche, Geologiche e Ambientali, Università degli Studi di Bologna, via Irnerio, Bologna 40126, Italy; E-Mail: donatella.serafini@unibo.it

* Author to whom correspondence should be addressed; E-Mail: stefano.delduca@unibo.it

---

Abstract: In pollen tubes, cytoskeleton proteins are involved in many aspects of pollen germination and growth, from the transport of sperm cells to the asymmetrical distribution of organelles to the deposition of cell wall material. These activities are based on the dynamics of the cytoskeleton. Changes to both actin filaments and microtubules are triggered by specific proteins, resulting in different organization levels suitable for the different functions of the cytoskeleton. Transglutaminases are enzymes ubiquitous in all plant organs and cell compartments. They catalyze the post-translational conjugation of polyamines to different protein targets, such as the cytoskeleton. Transglutaminases are suggested to have a general role in the interaction between pollen tubes and the extracellular matrix during fertilization and a specific role during the self-incompatibility response. In such processes, the activity of transglutaminases is enhanced, leading to the formation of cross-linked products (including aggregates of tubulin and actin). Consequently, transglutaminases are suggested to act as regulators of cytoskeleton dynamics. The distribution of transglutaminases in pollen tubes is affected by both membrane dynamics and the cytoskeleton. Transglutaminases are also secreted in the extracellular matrix, where they may take part in the assembly and/or strengthening of the pollen tube cell wall.

Keywords: pollen; pollen tube; transglutaminase; cytoskeleton; cell wall; polyamines; protein transport; protein secretion

---

1. The Pollen and Its Germination

The reproductive success of seed plants is performed and based on the invention of the pollen tube, an extension of the pollen grain that is produced under favorable conditions and that allows the sperm cells to move from the male gametophyte towards the female one. The invention of this biological road made it possible for land plants to reproduce in either the absence or the scarcity of water, thus increasing the reproductive success and dispersion of plants on earth. From a pure biological point of view, the pollen tube is an atypical plant cell, because it grows by a tip-growing mechanism, rather than by the typical diffuse way of somatic cells. The fact that pollen tubes grow exclusively in a specific domain (the tip) following a precise direction suggested that pollen tubes might be compared not only to rhizodermis trichoblastic cells, but also to nerve cells of animals [1]. Such a comparison is clearly forced and may simply stress the similarities between two very different cell types. The pollen tube grows by accumulating secretory vesicles in the tip domain. Vesicles are transported, together with many other organelles, along the cytoskeleton. The driving force is mainly based on the dynamic interaction between actin filaments (AFs) and myosins, although microtubules (MTs) and related motor proteins may have a regulatory function. Vesicles, which are produced by Golgi bodies, contain components required for the assembly of the cell wall and for maintaining the growth process. Two types of cell wall components are delivered to the tip domain: pectins/hemicelluloses and enzymes required for the membrane-localized synthesis of callose and cellulose. Pectins are used to form the primary wall of pollen tubes, a viscoelastic cell wall that allows expansion of the pollen tube by turgor pressure [2]. On the other hand, cellulose and callose are necessary for the strengthening of the cell wall [3], as well as for stabilizing the growth direction of pollen tubes [4]. Although the anatomy of pollen tubes is different from that of somatic cells, the growth of pollen tubes is likely to be based on the balance between the force exerted by turgor pressure and mechanical resistance due to the relative content of esterified/acid pectins. By interchanging the pectin composition of the apical cell wall, pollen tubes can grow by expanding such a wall through the force exerted by turgor pressure. Close to the growth domain, callose and (to a lesser extent) cellulose are deposited by their plasma membrane-localized enzymes [5]. This process reinforces and stabilizes the pollen tube cell wall. These preliminary concepts suggest that the composition of the cell wall is critical to determine the growth rate of pollen tubes. Therefore, every protein that can modify either the structure of the cell wall or the regulation of the growth process is elected as a key factor for the tip-growing mechanism.

Secretory vesicles are also likely to deliver proteins and other molecules required for regulating the growth process. These include receptors, ion channels, plasma membrane-associated G proteins and lipid-modifying enzymes. This network of proteins is necessary to maintain the molecular interplay that finely regulates growth. Regulation is critical, because pollen tube growth must be synchronized with external signals from the extracellular matrix (ECM) of surrounding cells (the pollen-pistil extracellular matrix—psECM—in the case of the female environment) [6]. It is obvious that the synchronization mechanism is linked to the assembly process of the cell wall (although such a correlation is largely unknown). Currently, accepted models indicate that the balance between pectin composition and turgor pressure constitutes the "motor" that promotes tube growth, while the molecular network in the tip domain represents the "driver" that either guides the growth or regulates its rate.

2. Structure and Function of TGases

Fifty years ago, a transamidating activity was described in liver and attributed to transglutaminases (TGs; EC 2.3.2.13) (TGase) [7]. Thereafter, many reports showed that the enzyme family responsible for this activity is ubiquitous in microorganisms, plants, invertebrates and vertebrates [8,9]. Twenty five years ago, the presence of a TGase activity was in fact reported in higher plants, and its characteristics were reported in some reviews [10,11].

Since its first description as an enzyme capable of calcium-dependent transamidation (7) (Figure 1), these enzymes were shown to catalyze the post-translational modification of proteins via either transamidation of selected glutamine residues to lysine ones or through covalently binding to primary amines. Because they form stable intra- and inter-molecular bridges, these enzymes are often referred to as "biological glues" [12]. TGases display strict specificity in recognition of glutamine protein substrates; however, the rules, which govern selection of only a few peptidyl glutamine residues, are unclear [13] and have poor specificity for the acyl-acceptor amine group. The transamidation of glutamine residues occurs to a variety of primary amines, which can be the ε-amino group of peptidyl lysine (Figure 1A) or the primary amine of a low molecular mass (Figure 1B) [14], such as aliphatic polyamines (PAs). In the absence of a suitable amine, water can act as an alternative nucleophile, leading to deamidation of the glutamine residue to glutamate (Figure 1C).

Figure 1. Transglutaminase could catalyze different biochemical reactions. Among them, (**A**) the Ca^{2+}-dependent acyl-transfer reaction between the γ-carboxamide group of a specific protein-bound glutamine and the ε-amino group of a distinct protein-bound lysine, residue giving rise to protein crosslinking. Second, the incorporation of substrates having a primary amino group, as polyamines, to the γ-carboxamide group of a specific protein-bound glutamine (**B**). Third, the deamidation of glutamyl-residue to a glutamic acid (**C**). This reaction occurs in the absence of a primary amine.

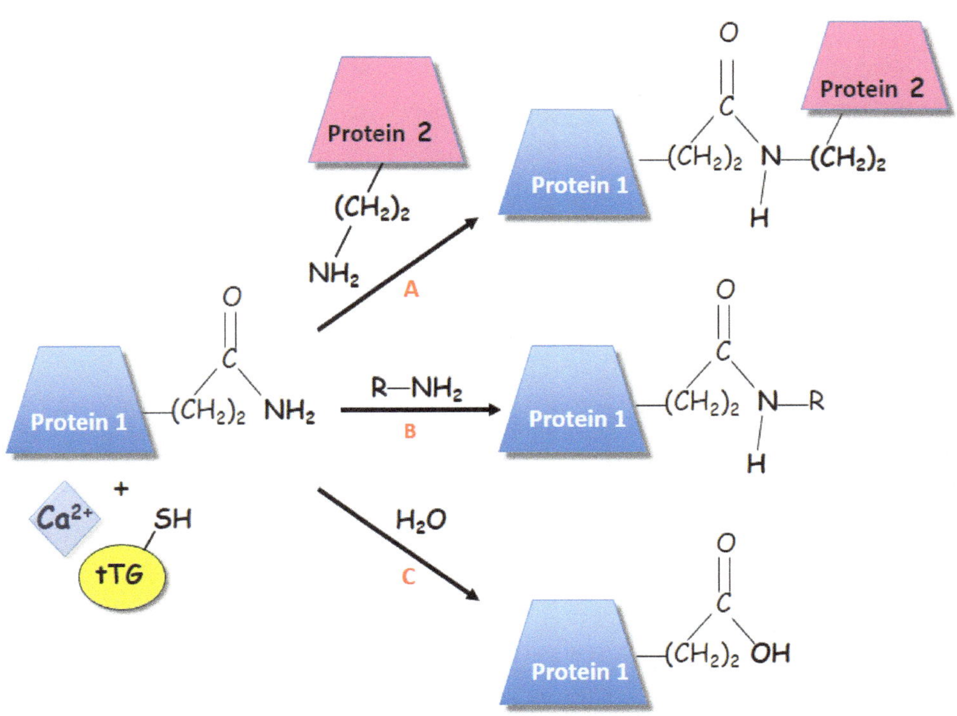

The catalytic mechanism is evolutionarily related to that of cysteine proteases. The active site thiol reacts with a glutamine side chain of a protein or peptide substrate to form a thioester intermediate from which the acyl group is transferred to an amine substrate (polyamines in Figure 2). As PAs have two terminal primary amino groups, one or both could be involved in TGase catalysis, leading to the formation of mono-γ-glutamyl or bis-γ-glutamyl derivatives, respectively. As PAs are polycations, these reactions could increase the electric charge of the modified proteins; in the case of bis-γ-glutamyl derivatives, the PA molecule could act as a bridge (with a different length according to the type of PA involved) between endo-glutamyl residues belonging to two different proteins or to the same proteins [7,15]. This catalytic activity, which requires Ca^{2+}, is inhibited by guanine nucleotides [16]. In the presence of guanosine triphosphate (GTP), TG2 lacks enzyme activity; instead, it may function as a G protein in the phospholipase C signaling pathway. Whereas the function of TGases is common to all the organisms studied until now, its function as a G protein has not been examined in plants.

Figure 2. The two-steps transamidase reaction of transglutaminase. All mammalian transglutaminases (TGases) belong to a superfamily of cysteine proteases, have structural homology and possess the catalytic triad of Cys-His-Asp/Asn; the reactivity of this Cys is activated by Ca^{2+}, which causes a conformational change in the enzyme, allowing the access of the substrate to the binding site. Step 1: the active Ca^{2+}-stabilized conformation of the enzyme forms a covalent intermediate between the active site thiol residue and a glutamyl residue in the protein substrate, releasing ammonia and activating the glutamine acyl moiety. Step 2: the active thioester undergoes an acyl transfer to a primary amine, in this case a polyamine, thus also introducing extra positive charges, as polyamines are protonated at physiological pH [or (not shown) to the lysyl residue of another protein]. A secondary cross-link might form between the free amine group of the bound polyamine and a glutamyl residue in another protein substrate, thus forming bis-(γ-glutamyl)-PA derivatives (not shown). E, Transglutaminase; C, cysteine residue; PA, polyamine; Q, glutamine residue.

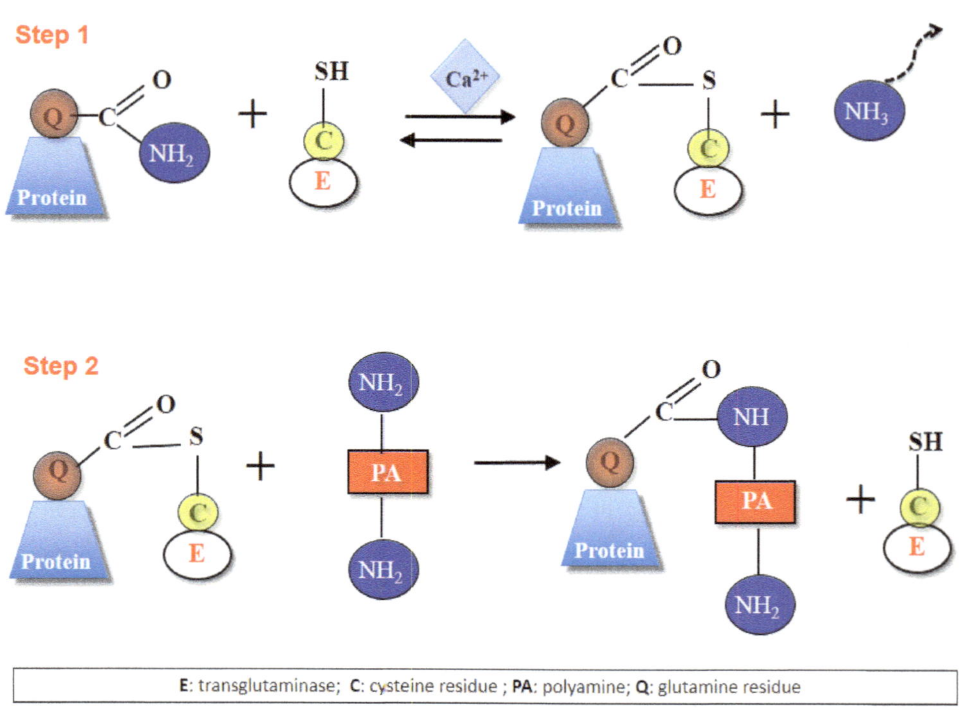

Both reactions involving protein cross-linking and polyamination are relevant *in vivo*, and competition between these amine substrates may take place within cells in a number of important physiological functions. The resulting cross-linked protein structures add strength to tissues and increase their resistance to chemical and proteolytic degradation [12,17].

In animals, nine TGase isoenzymes have been identified at the genomic level, but only six have been isolated and characterized at the protein level. The best-known are the ubiquitous type 2 tissue TGase (tTGase) and plasma Factor XIIIa, involved in the stabilization of fibrin clot. tTGase has been related to an increasing number of other biochemical capacities [9]. This enzyme is located both intra- (cytosol, mitochondria, nucleus) and extra-cellularly in the matrix, where it appears to be involved in differentiation, transmembrane signaling, cell adhesion organization of the extracellular matrix, motility, and pro- and anti-apoptotic roles [18,19]. The structural organization of TGase consists in four domains highly conserved during evolution among the mammalian TGase isoforms [12], as shown by the best known TGase, human type 2 tTGase.

3. Polyamines and Transglutaminases in Plants

Environmental and internal factors regulate plant growth. Among these, PAs are essential rejuvenation growth substances in all living organisms capable of regulating organogenesis and cell proliferation in higher plants and algae, apical growth of pollen, dormancy break, as well as senescence and homeostatic adjustments in response to external stimuli and stresses [20,21]. The first report on PA effects on plant growth was generated by Bertossi *et al.* [22] in *Helianthus tuberosus* dormant tubers and further confirmed in several other plants. The molecular mechanism of action of PAs is only partially known. These amines are present in free and bound form. Twenty-five years ago, interest on the possible role of TGase in plants arose from its known ability to covalently bind PAs to some animal proteins and, thus, possibly also to plant proteins. Despite several characteristics in common with the best known mammalian TGases, among which are the production of glutamyl-PA derivatives and the immunoreactivity by TGase animal antibodies and other characteristics, research on plant TGases has been hampered by difficulties encountered in their purification and in the lack of significant amino acid sequence homologies between animal TGases and the polypeptide sequences available in plant databases. A more recent computational analysis identified in *Arabidopsis thaliana* the presence of a gene, *AtPng1p*, which encodes a putative *N*-glycanase containing the Cys-His-Asp triad typical of the TGase catalytic domain. *AtPng1p* is a single gene expressed ubiquitously, but at low levels, as shown by nested real-time polymerase chain reaction RT-PCR undertaken in different organs of *Arabidopsis*, at all growth stages, in all light assayed conditions [16]. The overexpressed recombinant protein was purified, and an 86-kDa band was immuno-detected. Antibodies raised against this recombinant *AtPng1p* detected the same band in the *Arabidopsis* microsomal fraction and other bands of lower molecular mass in the cytosolic fraction, possibly proteolytic soluble fragments, as usually occur with animal TGases. This finding is in line with the presence of TGase activities and with the immunorecognition of different bands in extracts of several organs of the same plant. Analyses of the γ-glutamyl-derivatives confirmed the *AtPng1p* gene product acts as a TGase, having a Ca^{2+}- and GTP-dependent transamidase activity. This was the first plant protein, isolated and characterized at the molecular level, displaying a TGase activity with parameters typically exhibited by animal TGases. A structural comparison between this protein and the crystal of FXIII showed a considerable homology [23]. Despite having an amino acid sequence different from those of known animal TGases, with the exception of the active site triad, *AtPnglp* shares immunological and biochemical properties and possibly an overall similar conformation, probably being a TGase with a different, but convergent, phylogenetic history. The presence of the typical TGase active site in *AtPnglp* elucidates why this protein could exert the same catalytic activity of well-known TGases. AtPng1, which acts as a PNGase when expressed in *Saccharomyces*, possesses two residues for carbohydrate binding that may favor contact among the carbohydrate-PA-protein residues [24].

Recently, in *Helianthus tuberosus* immature cells of the primary apical meristem of the sprout, at least three immuno-recognized TGases were found to be expressed and active; these proteins are probably the same already found in differentiated cells of other plant systems, thus representing constitutive enzymes. Two of these TGases (namely the 75 and 85 kD) share a very high degree of molecular mass and amino acid composition similarity with two mammalian TGases, even though no information is available on their amino acid sequence. This suggests that these two TGases are phylogenetically conserved. The third protein of 58 kDa did not show significant similarity with other annotated TGases [25]. Enzymes of such molecular mass are active and distributed in several plants [26]. Two related cDNA, encoding for a putative TGase of 58 kDa, were cloned from *Helianthus* thylakoids; their transcripts were expressed mainly in young leaves and differentiated *Zea mays* callus under light exposure [27]. This TGase did not present the catalytic triad in the position of all annotated TGases and showed no similarity with the 58 kDa of sprout apices [25]. Other protein sequences from various plants can be found in database, where they are classified as TGases only on the basis of their homology with the TGase domain, but they were not experimentally assayed for enzymatic activity.

The cross-reactivity of the animal TGase antibodies with plant TGases has been reported in several organs and type of plants [28,29]. A similarity between plant and animal TGases is also supported by their catalytic activity, which can cross-recognize some of their respective substrates, suggesting a similarity in their specificity [30,31].

In plants, multiple TGase forms are active in the same organ, and many organs/tissues express these enzymes, like, for example, meristems, seeds, flower petals, pollen, leaves, shoots and roots. Most likely, these enzymes are ubiquitous, and some of them may exert similar effects in different cells, as they can occur with the modification of cytoskeleton substrates. On the contrary, other activities appear more specialized in relation to location and substrates, *i.e.*, those typical of plants, like the light-harvesting complexes of chloroplasts [32,33].

4. Transglutaminase in Pollen

In apple pollen, a Ca^{2+}-dependent TGase activity has been found, both intracellularly and extracellularly. TGase is widely distributed in the pollen tube of pear and apple, as well as of hazel (*Corylus avellana*) and pomelo (*Citrus grandis*), and it occurs as both a cytosolic and a membrane/cell wall form [34–37]. TGase monoclonal and polyclonal antibodies cross-react with a specific polypeptide of ~70 kD in apple pollen, whereas in pear pollen, antibodies cross-reacted with a 70-kD TGase-related polypeptide in the cytoplasmic protein pool, in the membrane (microsomal) fraction and in both

Triton- and sodium dodecyl sulfate (SDS)-extracted cell wall proteins. Analyses by two-dimensional gel electrophoresis (2-DE) indicated that each compartment contains distinct isoforms of TGase, suggesting that the pear genome might contain different TGase genes whose products are delivered to different cell regions [33]. Since preliminary observations suggest that the nuclear genomes of *Arabidopsis* and pear contain a single TGase gene [16,34] a single TGase protein seems to be post-translationally modified according to the final destination in the cell.

In pear [33] and apple pollen [35], TGase was visualized using immunofluorescence and immunogold electron microscopy in the growing half of pollen tubes, suggesting that the enzyme is actively constrained to accumulate in the growing segment. This effect might be possibly achieved by association of TGase with intracellular membranes, as association of gold particles with the plasma membrane of pollen tubes was often observed. This suggests that the enzyme might be either secreted or deposited in order to locally affect the structure of the cell wall, as also confirmed by the immunoblot assay of proteins extracted from different cell compartments.

4.1. Intracellular Targets of TGase Activity: The Cytoskeleton

Starting from initial reports on the binding of putrescine to rabbit skeletal muscle actin mediated by guinea pig liver TGase [38], amines were shown to bind actin covalently by TGase in cells undergoing apoptosis [39]. In addition, G-actin (but not F-actin) was cross-linked with myosin S1 by TGase [40]. Another cytoskeletal target of TGase is myosin [41]. On the tubulin side, MTs were shown to be the target of TGase activity in Alzheimer's disease brains [42]. The MT motor kinesin is also a target of TGase activity [43]. Altogether, these observations suggest that the cytoskeleton activity may be influenced and regulated by TGase activity, either by directly cross-linking specific amino acids or covalently linking PAs to the target surface.

Both AFs and MTs are aligned along the growth axis of the pollen tube and are involved in several aspects of the pollen tube growth mechanism. Generally, AFs are required for the transport of organelles and vesicles in the pollen tube (the so-called "cytoplasmic streaming") [44]. This process is necessary to deliver Golgi-derived secretory vesicles to the pollen tube apex, where vesicles fuse with the apical plasma membrane and provide new membrane and several components to the developing cell wall [45]. AFs are organized in at least three distinct ways, which are also spatially separated: a delicate meshwork of very short randomly organized AFs in the tip, an actin fringe in the subapex of pollen tubes (that separates the growing domain) and thick actin bundles that permeate the pollen tube to the grain. Bundles are required for the transport of organelles and for their relatively uniform distribution along the pollen tube [46]. The presence of three sub-domains of AFs suggests that an individual regulation of their organization is required at distinct cell locations, a process that definitely requires the presence of specific AF-binding proteins [47], Rac/Rop small GTPases [48] and oscillations of Ca^{2+} concentrations [49]. In summary, AFs are regulated by a complicated network of small molecules and proteins that finely tune AFs in order to promote the coordinated growth of pollen tubes.

Much less is known about the organization and dynamics of MTs. Immunocytochemical labeling allowed the visualization of cortical MT bundles in the pollen tube of several species [50]. As the use of green fluorescent protein (GFP)-labeled probes for visualizing MTs in living pollen tubes is still not standardized, with a few exceptions [51], the presence and organization of MTs in the apex and subapex of pollen tubes is uncertain. MTs are likely to be involved in the deposition of callose plugs [52], in the transport of sperm cells [53] and in the regulation of organelle/vesicle dynamics [44].

The crude extract of apple (*Malus domestica*) pollen contained a TGase activity that catalyzed the incorporation of polyamines into pollen proteins with molecular masses of 43 kD and 52–58 kD; the same proteins were also immunolabeled by mouse monoclonal antibodies to actin and tubulin [54]. TGase purified from apple pollen by hydrophobic interaction chromatography showed molecular and immunological characteristics very similar to those of the well-known TGase-2, which is widely expressed in many animal cells. The pollen enzyme catalyzed the binding of putrescine to actin and tubulin monomers purified from the same cell type. When tested on actin filaments, pollen TGase induces the formation of high-molecular-mass aggregates of actin. Use of fluorescein-cadaverine showed that the labeled polyamine was bound to actin by pollen TGase, as observed to occur with guinea pig liver TGase. The pollen TGase affected the binding of myosin and kinesin to actin filaments and microtubules, respectively [55], and also reduced the enzyme activity of myosin and the gliding activity of kinesin.

Either excessive concentrations or the absence of PAs disturbed the correct polymerization of actin in the presence of TGase: unordered bundles of actin appeared at high PA concentrations, suggesting that the actin binding sites are possibly saturated, giving rise to many mono-PA derivatives instead of forming the correct site-specific cross-links by bis-PA derivatives [41]. The natural PA concentrations (estimated to be approx. 50 μM in hydrated pollen) should decrease the competition for the protein-binding sites, allowing bis-derivatives to form. Consequently, pollen TGase may participate in actin regulation during pollen tube growth. Since the proper organization of the actin cytoskeleton is essential for tube growth and organelle movement [56], any disorganization of the motor apparatus generates critical changes in organelle motility and pollen tube growth. It was suggested that pollen TGase controls the transition between actin bundles and short filaments at the boundary between the apical and base domains of pollen tubes [54], favoring the assembly of more stable bundles.

TGase also changes the polymerization rate of tubulin in the presence of PAs. Putrescine at the presumptive natural concentration generated either filamentous or amorphous microtubule structures *in vitro*; motility assays show that TGase-modified microtubules moved with lower speed (at least a 50% reduction) compared with unmodified microtubules. Thus, pollen TGase can act as a "biological glue", because of its capacity to cross-link proteins and to reduce their movement ability, in agreement with the evidence that the motor activity of kinesin is also regulated by post-translational modification of tubulin [55,57]. The relevance of these findings is demonstrated by the fact that the functioning of pollen tube microtubules involves the control of vacuole positioning, the trafficking velocity of mitochondria and the focusing of secretory vesicles in the tip domain [58]. Since *in vitro* experiments showed that pollen TGase affects the organization and functioning of the cytoskeleton, the authors propose that the activity of TGase might be modulated also *in vivo* to exert a regulatory role on pollen tube growth.

Recent results suggest that TGase might be involved in the rearrangement of the cytoskeleton as it occurs during self-incompatibility (SI). In *Papaver rhoeas* (poppy), the SI response requires a specific recognition event between stigma and pollen S proteins. This event triggers a cascade of Ca^{2+}-dependent signals, which inhibit tip growth, produce critical changes to organelles, cause the depolymerization of AFs (with concurrent formation of actin foci) and determine a caspase-like protease activity [59]. Since the caspase activity also involves depolymerization of MTs [60], the cytoskeleton of pollen tubes is consequently part of the mechanism that regulates the development of the SI response. In Solanaceae, Rosaceae and Plantaginaceae, the SI response is based on the presence of S-RNases, small proteins with RNase activity that are produced by the pistil and are internalized into the pollen tube by either direct absorption or endocytosis [61]. The internalization process might involve the interaction between C2-domain proteins and arabinogalactan-proteins (AGPs), which might in turn bind to S-RNase [62]. Once internalized, S-RNase might degrade the RNA of incompatible pollen in order to prevent fertilization; however, it is likely that other processes, such as changes to Ca^{2+} concentration and modifications of the AF organization, are also involved [63]. Because of this evidence, the molecular mechanisms of rejection of self-pollen might share common features among different families. The molecules that mediate the S-RNase-based SI response are not known, but TGases might be considered a promising candidate. The involvement of TGases in the SI response is suggested by evidence that plant TGases participate to processes correlated to programmed-cell-death (PCD), such as the senescence of the flower corolla [29]. Free PAs declined continuously during the life of the tobacco corolla, but we have shown that the supply of 25 mM spermine for 3 h to flowers at early-mid stages of anthesis delayed senescence and caused diminished fresh weight loss and changes in shape, size and color [64]. Since TGase was immuno-localized in the epidermis and, at senescence, also in the mesophyll cell walls, this suggests the existence of a TGase-mediated mechanism by which PAs modulate senescence and cell death. In addition, the increase of PA content and TGase activity also occurs during incompatible pollinations in pear and citrus [36,65].

One more supporting piece of evidence comes from the finding that the cytoplasmic TGase of apple pollen can post-translationally modify actin and tubulin by conjugating them to PAs [54] and generating high molecular weight aggregates [55]. As mentioned above, such aggregates are capable of inhibiting the enzyme activity and the binding affinity of myosin and kinesin and, consequently, inhibit motor-dependent dynamic activities. The regulatory effect exerted by TGase on MTs may also be indirect; in fact, TGases have stimulatory effects on phospholipase A2 (PLA2) [37] that in turn might regulate the MT dynamics [65,66]. The presence of PLA2 in the pollen of *Arabidopsis thaliana* has been reported only recently [67]. Although plant PLA2s are likely involved in many biological functions (including senescence, wounding and stress responses), relatively little is known about plant PLA2s, and their genes remain essentially uncharacterized. In *Arabidopsis*, three of four PLA2 paralogs (PLA2 β, γ and δ) have been characterized; they are expressed in pollen, are localized in the endoplasmic reticulum and/or Golgi and are critical for pollen development and tube growth [68].

The picture arising from this evidence places TGase at the center of a mechanism that regulates the dynamics of both AFs and MTs under physiological and self-incompatibility conditions. TGase might regulate the dynamics of filamentous cytoskeleton, as well as single associated proteins (motor and non-motor). Such regulation may be direct or even indirect, involving the phospholipase pathways. Dramatic changes to TGase levels, as occurring during the SI response, may consequently lead to the loss of cytoskeleton integrity and function.

4.2. Secretion and Targeting of Extracellular TGase

The pollen tube TGase is also found in association with different membrane compartments, but mainly with Golgi membranes and the plasma membrane [33]. Secretion of TGase in eukaryotic cells might follow alternative routes other than the classic endoplasmic reticulum (ER)/Golgi pathway. This assumption is based on the evidence that TGase is not consistently associated with the Golgi cisternae, suggesting that secretion of extracellular TGase is not dependent on the ER/Golgi pathway [69]. In the pollen tube, TGase accumulates mainly in the subapical region, but not at the apex, where Golgi-derived secretory vesicles are usually present in large amounts [70]. On the other hand, the pollen tube apex is not the only region where secretion occurs [71], and secretion pathways alternative to the ER/Golgi route have been suggested to occur in plant cells [72]. In addition, novel intermediates in the secretory pathway have been identified in plant cells (the so-called "secretory vesicle cluster") and suggested to be involved in the transport of secretory proteins from the Golgi to the plasma membrane [73]. Consequently, the secretion of TGase might occur in regions out of the tip and involve modalities different from the standard vesicular pathway.

Association of TGase with the plasma membrane is likely to be dependent on AFs and, to a lesser extent, on MTs, the latter being important for the fine-tuning of membrane insertion [33]). When TGase is bound to the plasma membrane, it is likely to work as a G protein [74], whose GTP binding ability may regulate downstream signaling events involved in cell survival [75]. Since the plasma membrane TGase is likely to activate phospholipase C (PLC) [76], this suggests that the pollen tube TGase associates with PLC and regulate changes of Ca^{2+} concentration through production of inositol 1,4,5-trisphosphate(IP_3) [77,78].

The secretion process that should lead to the deposition of TGase in the extracellular matrix is still quite hypothetical, if not speculative. In fact, most of the steps are still not fully clarified, some steps are hypothetical; others, by contrast, are supported by experimental data. The model presented in Figure 3 is, therefore, theoretical, but it also aims to be an incentive for further debate and for future research. Data available in the literature suggests that pollen TGase might be distributed in two distinct locations and be involved in at least two general activities. Cytoplasmic TGase (1) is likely involved in the regulation of cytoskeleton activities by modulating the polymerization state of AFs (2) and MTs (3). This modification may consequently affect the activity of motor proteins (myosin and kinesin) and the motor-based organelle trafficking (4). Regulation of MT dynamics might also be achieved through modulation of phospholipase D (PLD) and A (PLA) activities, the latter enzyme being stimulated by cytoplasmic TGase (5). Cytosolic TGase is supposed to be also controlled by phospholipase C (PLC) through changes of Ca^{2+} concentration (6). Changes of apical Ca^{2+} concentration may affect the activity of actin-binding proteins (ABP), which in turn are likely to control the dynamics of AFs (7) leading eventually to the formation of aberrant structures (actin foci). The second form of TGase is predominantly associated with the membranes and with the cell wall. This form of TGase might be different from the cytosolic, probably being the result of posttranslational modifications. How the membrane TGase can be conveyed within secretory structures is not known. However, membrane-associated TGase is likely to be secreted and incorporated in the cell wall (8) where it participates in its structuring, possibly by interaction with pectins (9). Cell wall-associated TGase may be also the result of an alternative secretory pathway (10). TGase may be also secreted extracellularly (11) and take part in the pollen-pistil signaling interplay (Figure 3).

Figure 3. Hypothetical model of TGase trafficking in the pollen tube. Cytoplasmic TGase (**1**) might regulate both actin filaments (**2**) and microtubules (**3**), leading to changes in the cytoskeleton-based organelle transport (**4**). The dynamics of microtubules is also likely regulated by PLA and PLD, the former affected by TGase (**5**). TGase activity may also be controlled through changes of Ca^{2+} concentrations (**6**), which regulate the activity of ABP (**7**). Membrane TGase is secreted either through vesicular intermediates (**8**) or by alternative pathways (**10**), and it might take part in the assembly of the pollen tube cell wall (**9**). TGase may also be secreted in the extracellular matrix (**11**).

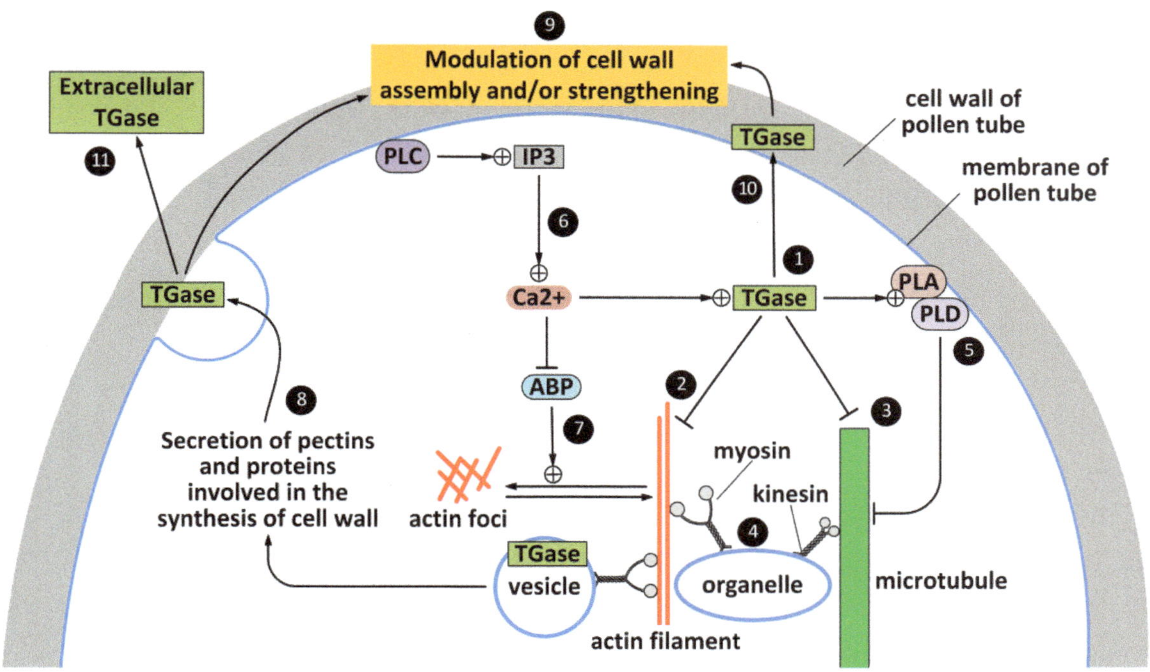

The presence of extracellular TGase (either incorporated in the cell wall or secreted outside of the cell) is an accepted feature in both animal [12] and plant cells (see below); extracellular TGase is likely to be involved in the stabilization of either the extracellular matrix or the cell wall through its protein cross-linking activity. The first indication of the presence of TGase products in the cell wall was provided by the digestion of cell wall polysaccharides of *Helianthus tuberosus* parenchyma, which caused the disaggregation of PA-conjugated high mass-proteins from polysaccharides [79]. In the alga, *Chlamydomonas reinhardtii*, TGase was involved in the formation of the cell wall, which allowed the zygote to survive desiccation. The TGase-directed formation of a soft protein envelope, which organizes the self-assembly of glycoproteins, was followed by oxidative cross-linking, which rendered the cell wall insoluble. The alga secretes an extracellular 72-kDa TGase, whose maximal activity precedes the insolubilization of the assembled glycoproteins [80]. In the fungus, *Phytophthora sojae*, a cell wall glycoprotein was identified as a Ca^{2+}-dependent TGase [81]. In *Nicotiana tabacum* corolla, an active 58-kDa TGase is present in the isolated cell wall fraction and might be responsible for the corolla strengthening finalized to the protection of the ovary containing the developing seeds [29]. The enzyme activity, as measured in isolated cell wall fractions, prevails in the distal part of the corolla and progressively increases during the flower lifespan, with the progression of differentiation and senescence [29]. TGase is immuno-localized almost exclusively in the epidermis, mainly in its cell walls, during corolla development, but at the onset of senescence, the enzyme spreads to the inner parenchymatous cells of the petals, whose cell walls appear rigid and straight, concomitantly with the rigid/papyraceous-like aspect of the corolla [64].

In apple pollen, Western blotting using anti-TGase antibodies also revealed two main immunoreactive bands of 70 and 75 kDa, the latter more evident in crude extracts of germinating pollen, in concentrated germination medium and in the cell wall [35].

The presence of TGase in the cell wall has been studied also in pollen by immunolocalization with anti-TGase antibodies and by identification of a TGase cross-linking activity at the apical part of pollen tubes, in the tube region close to the grain and in the pollen grain. The cross-linked products may provide strength to the pollen tube as it migrates through the style *in planta* [82]. As the pollen tip is continuously elongating during pollen germination, released TGase could be deposited along the tube wall and be specifically embedded under the effect of local micro-environmental conditions into the same accumulation sites of the protein cross-links, as shown by immunofluorescence colocalization, both in *in vitro* and *in planta* germinating pollen. The pollen enzyme is also able to catalyze the cross-linking of both mammalian TGase substrates H6-Xpr-GFP (a specific glutamine and lysine rich TGase substrate) and dimethylcasein (the most used substrate for the *in vitro* TGase assay) [35]. In animal tissues, H6-Xpr-GFP revealed a wide distribution of TGase in the extracellular matrices, suggesting an important role for TGase in maintaining the integrity of the extracellular matrix by cross-linking its constituent proteins [83].

In pollen, the *in situ* cross-links of H6-Xpr-GFP and pollen proteins were revealed by 81D4, an antibody that recognizes either the gln-lys or gln-PA links and identified by the analysis of the glutamyl-PA derivatives [35]. *In situ* localization by confocal microscopy in non-digested and non-permeabilized pollen revealed that His6-Xpr-GFP was associated extracellularly with the pollen tube wall, especially within 20 μm of the tip. The extracellular pollen TGase co-localized within the same accumulation sites of His6-Xpr-GFP observed along the pollen tube surface. The ability of pollen extracellular TGase to incorporate the exogenously supplied His6-Xpr-GFP in the same sites of pollen protein-protein and protein-amine cross-links was demonstrated by their co-localization in confocal micrographs. In addition, the administration of His6-Xpr-GFP resulted in a significant increase in the percentage of germinated pollen and tube length. The extracellular pollen TGase activity, the percentage of germination and tube growth were inhibited in a dose-dependent manner by the monoclonal antibody ID10 and by two site-directed irreversible inhibitors of TGase activity. The presence of the inhibitors resulted in shorter and thicker pollen tubes with a decreased growth rate within the first period of germination, followed by an inability to extend further the tip and, eventually, tube-burst [35]. The fluorescent-tagged amine substrate (FITC-cadaverine), routinely used to visualize tTGase activity in animal cells [84], which is a competitor of endogenous amine-donor substrates (amino group of lysine residues or PAs), limited or blocked tube elongation of both pollens of *Malus domestica* (entomophilous) and *Corylus avellana* (anemophilous) [37]. This primary amine was essentially detected around the pollen grain of *M. domestica*, mainly in the cell wall and in the apical part of the tube; a similar incorporation of FITC-cadaverine was detected in the pollen tube of *C. avellana* [37].

The identification of the natural substrates of TGase is, thus, of primary importance. Outside the cell of animal tissues, TG2 shapes the extracellular matrix by binding tightly to both fibronectin in the extracellular matrix and integrins on the cell surface; thus, TG2 promotes cell adhesion, signaling and differentiation in a manner independent of its catalytic activity [12]. These proteins can also fulfill other enzymatic activities, according to ligand or substrate availability [9]. In the pollen tube, only a few adhesion molecules have been implicated in guidance [85], and integrin analogues have only been characterized preliminarily in the pollen tube [86]. Whether these proteins are a target of TGase activity is not known.

In pear pollen, TGase is related to the secretion and/or distribution of methyl-esterified pectins [33]. In fact, methyl-esterified pectins accumulate and are interspersed with respect to the TGase signal. TGase is therefore secreted in the same region and is hypothetically related to the distribution and accumulation of pectins. Results obtained with the inhibitors either of membrane and cytoskeleton mediated-transport, such as Brefeldin A (BFA) oryzalin and Latrunculin B (LatB), indicated that TGase and pectins change their distribution in BFA-treated pollen tubes. Specifically, both TGase and pectins accumulate prominently in older (but not identical) segments of pollen tubes, suggesting that the membrane trafficking inhibitor affects the secretion of both components. On the other hand, LatB induced an apparently different behavior, because TGase accumulates in older segments, but pectins are likely to accumulate in the tip domain. Thus, TGase and pectins follow different secretion pathways that are affected similarly by membrane trafficking inhibitors, but differently by actin inhibitors. While the association of TGase with pectins is an interesting prospective, it is known that PAs can directly interact with polygalacturonic acid or pectins by competing with Ca^{2+} [87]. It can be hypothesized that PAs could be non-covalently bound to pectins by one terminal amino group and covalently linked via TGase to a protein by the other amino terminal group. This binding might regulate the strengthening of pectin-based cell walls and, therefore, the rate of pollen tube growth, which is based on a tug-of-war between turgor pressure and pectin deposition [88]. This hypothesis is also supported by evidence that the TGase AtPng1p possesses two residues for carbohydrate binding that may favor the contact among the carbohydrate-PA-protein residues [24]. TGase also partially overlaps with arabinogalactan proteins (AGPs), complex macromolecules composed of a polypeptide skeleton and branched glycan chains; AGPs are markers of several developmental and interaction processes, such as xylem development, tip growth of pollen tubes, somatic embryogenesis and PCD [89], since AGPs can interact with S-RNase-binding protein during the SI response [90]. AGPs are likely to be secreted at the apex and to be critical in the polarization of new pollen tubes [91]; thus, externalized TGase might interact with AGPs in order to regulate their positioning and/or functioning. These data are consistent with a role of extracellular TGase as the modulator of cell wall building and strengthening. In conclusion, pollen TGase plays an essential role in successful apple pollen tube growth. Therefore, the enzyme could function by protein cross-linking and amine protein conjugation, thus strengthening the cell-wall scaffold of the extending pollen tube.

5. Extracellular TGase Released in the Pollen Germination Medium

In apple pollen, two main immunoreactive bands of 70 and 75 kDa were detected in concentrated germination medium in which germinated pollen had been removed by filtration [35]. The extracellular TGase activity is able to conjugate PAs to pollen proteins secreted during germination *in vitro*. *In planta*, it can be hypothesized that the released TGase might be involved in connecting the pollen tube to surrounding stylar cells as it progresses through the psECM [35].

By comparing apple and hazel pollen during germination, the pollen grain wall of the entomophilous *M. domestica* was more intensively stained with the anti-TGase antibody than that of the anemophilous *C. avellana* [37]. The enzyme activity increased under stressful treatments, mimicking those induced by climate changes (temperature humidity, copper and acid rain pollution), and was released outside the pollen, along with its products; externalization of TGase was predominant in *C. avellana*, whose grain cell wall is different with respect to that of *M. domestica*. This difference could justify an easier or earlier release of various molecules, including allergens, extracellular TGase and its products by the anemophilous pollens. Pollen TGase could be one of the mediators of pollen allergenicity, especially under environmental stress induced by climate changes.

Moreover, *M. domestica* and *C. avellana* pollens also differ in their incompatibility mechanisms. *M. domestica* is gametophytic, while *C. avellana* is sporophytic, with the pollen/style recognition times more rapid in hazel, whose incompatibility glycoproteins are exposed to and immediately recognized by stigma cells.

6. Conclusions and Future Perspectives

Recent studies suggest that TGase is a relatively critical element in the growth process of pollen tubes. This activity could be carried out at least at two different levels. The cytoplasmic TGase would be involved, not only as clearly established in the cell free system, but also *in vivo*, in the general organization of the cytoskeletal apparatus that would be made more or less dynamic and, therefore, adapted to the growth rate of pollen tubes. TGase of membranes (or of cell wall) may be somewhat involved in the process of cell wall structuring in order to make it more or less elastic, depending on the medium that the pollen tube has to cross. This possible function is very interesting, because the structure of the cell wall clearly determines the growth speed of pollen tubes, which is also affected by the extracellular TGase acting directly on the cell wall. In fact, the deposition rate of methyl-esterified pectins and their enzymatic transformation in acid pectins is the "timer" that determines how fast the pollen tube can grow. The progressive transformation of pectins counterbalances the turgor pressure inside the pollen tube, setting up a mechanism that is able to respond effectively to changes of the external medium in which the pollen tube grows. Therefore, the membrane/cell wall TGase could be part of the mechanism that makes the cell wall more or less soft. It will therefore be important to check this aspect, determining the role of TGase in counterbalancing the internal force generated by turgor pressure. As the stabilization of the cell wall contributes to determining the growth direction of the pollen tube, the question is whether TGase may have an additional role in this process. In this specific context, the role of TGase would certainly be more important *in vivo* than *in vitro*.

When the activity of cell wall TGase increases above physiological levels, this may unleash a series of events that would ultimately lead to blocking pollen tube growth. This is the case that might occur during pollen rejection in the SI process. The increase of the enzyme activity of TGase, as triggered by SI signals, may determine, on the one hand, profound changes of the cytoskeleton and, on the other hand, an abnormal process of cell wall restructuring, with consequent damage to the growth process. It will be therefore important to study and analyze the activity and the targets of TGase activity under the SI response, for example, by mimicking *in vitro* the conditions that induce the SI response *in vivo*.

Acknowledgements

Some of the data presented here comes from recent work in our labs that was supported by grants from the RFO 2011, the POR FESR 2007-2013 (project "My Darling Clementine") to S.D.D. and D.S.F. and from the PRIN 2008 to G.C. We thank Mauro Cresti (Dipartimento di Scienze della Vita, University of Siena) for frequent and supportive discussions on pollen tube growth.

Reference

1. Palanivelu, R.; Preuss, D. Pollen tube targeting and axon guidance: Parallels in tip growth mechanisms. *Trends Cell Biol.* **2000**, *10*, 517–524.
2. Kroeger, J.H.; Zerzour, R.; Geitmann, A. Regulator or driving force? The role of turgor pressure in oscillatory plant cell growth. *PLoS One* **2011**, *6*, e18549.
3. Parre, E.; Geitmann, A. More than a leak sealant. The mechanical properties of callose in pollen tubes. *Plant Physiol.* **2005**, *137*, 274–286.
4. Aouar, L.; Chebli, Y.; Geitmann, A. Morphogenesis of complex plant cell shapes: The mechanical role of crystalline cellulose in growing pollen tubes. *Sex. Plant Reprod.* **2010**, *23*, 15–27.
5. Cai, G.; Faleri, C.; Del Casino, C.; Emons, A.M.C.; Cresti, M. Distribution of callose synthase, cellulose synthase and sucrose synthase in tobacco pollen tube is controlled in dissimilar ways by actin filaments and microtubules. *Plant Physiol.* **2011**, *155*, 1169–1190.
6. Cheung, A.Y.; Wu, H.M. Structural and signaling networks for the polar cell growth machinery in pollen tubes. *Ann. Rev. Plant Biol.* **2008**, *59*, 547–572.
7. Folk, J.E. Transglutaminases. *Annu. Rev. Biochem.* **1980**, *49*, 517–531.
8. Beninati, S.; Bergamini, C.M.; Piacentini, M. An overview of the first 50 years of transglutaminase research. *Amino Acids* **2009**, *36*, 591–598.
9. Lorand, L.; Graham, R.M. Transglutaminases: Crosslinking enzymes with pleiotropic functions. *Nat. Rev. Mol. Cell Biol.* **2003**, *4*, 140–156.
10. Icekson, I.; Apelbaum, A. Evidence for transglutaminase activity in plant tissue. *Plant Physiol.* **1987**, *84*, 972–974.
11. Serafini-Fracassini, D.; Del Duca, S.; D'Orazi, D. First evidence for polyamine conjugation mediated by an enzymic activity in plants. *Plant Physiol.* **1988**, *87*, 757–761.
12. Griffin, M.; Casadio, R.; Bergamini, C.M. Transglutaminases: Nature's biological glues. *Biochem. J.* **2002**, *368*, 377–396.
13. Coussons, P.J.; Price, N.C.; Kelly, S.M.; Smith, B.; Sawyer, L. Factors that govern the specificity of transglutaminase-catalysed modification of proteins and peptides. *Biochem. J.* **1992**, *282*, 929930.

14. Folk, J.E.; Park, M.H.; Chung, S.I.; Schrode, J.; Lester, E.P.; Cooper, H.L. Polyamines as physiological substrates for transglutaminases. *J. Biol. Chem.* **1980**, *255*, 3695–3700.
15. Beninati, S.; Folk, J.E. Covalent polyamine-protein conjugates: Analysis and distribution. *Adv. Exp. Med. Biol.* **1988**, *250*, 411–422.
16. Della Mea, M.; Caparros-Ruiz, D.; Claparols, I.; Serafini-Fracassini, D.; Rigau, J. AtPng1p. The first plant transglutaminase. *Plant Physiol.* **2004**, *135*, 2046–2054.
17. Nicholas, B.; Smethurst, P.; Verderio, E.; Jones, R.; Griffin, M. Cross-linking of cellular proteins by tissue transglutaminase during necrotic cell death: A mechanism for maintaining tissue integrity. *Biochem. J.* **2003**, *371*, 413–422.
18. Mukherjee, B.B.; Nemir, M.; Beninati, S.; Cordella-Miele, E.; Singh, K.; Chackalaparampil, I.; Shanmugam, V.; DeVouge, M.W.; Mukherjee, A.B. Interaction of osteopontin with fibronectin and other extracellular matrix molecules. *Ann. NY Acad. Sci.* **1995**, *760*, 201–212.
19. Lentini, A.; Abbruzzese, A.; Caraglia, M.; Marra, M.; Beninati, S. Protein-polyamine conjugation by transglutaminase in cancer cell differentiation: Review article. *Amino Acids* **2004**, *26*, 331–337.
20. Bagni, N.; Tassoni, A. Biosynthesis, oxidation and conjugation of aliphatic polyamines in higher plants. *Amino Acids* **2001**, *20*, 301–317.
21. Special Issue: Polyamines—Dedicated to the memory of Nello Bagni. *Plant Physiol. Biochem.* **2010**, *48*, 487–634.
22. Bertossi, F.; Bagni, N.; Moruzzi, G.; Caldarera, C.M. Spermine as a new growth-promoting substance for *Helianthus tuberosus* (Jerusalem Artichoke) *in vitro*. *Experientia* **1965**, *21*, 80–81.
23. Tasco, G.; Della Mea, M.; Serafini-Fracassini, D.; Casadio, R. Building a low resolution model of a transglutaminase domain of an hypothetical *N*-glycanase from Arabidopsis thaliana. *Amino Acids* **2003**, *25*, 197.
24. Masahara-Negishi, Y.; Hosomi, A.; Della Mea, M.; Serafini-Fracassini, D.; Suzuki, T. A plant peptide: *N*-glycanase orthologue facilitates glycoprotein ER-associated degradation in yeast. *Biochim. Biophys. Acta* **2012**, *1820*, 1457–1462.
25. Beninati, S.; Iorio, R.A.; Tasco, G.; Serafini-Fracassini, D.; Casadio, R.; Del Duca, S. Expression of different forms of transglutaminases by immature cells of *Helianthus tuberosus* sprout apices. *Amino Acids* **2012**, *44*, 271–283.
26. Serafini-Fracassini, D.; Del Duca, S. Transglutaminases: Widespread cross-linking enzymes in plants. *Ann. Bot.* **2008**, *102*, 145–152.
27. Villalobos, E.; Santos, M.; Talavera, D.; Rodriguez-Falcon, M.; Torne, J.M. Molecular cloning and characterization of a maize transglutaminase complementary DNA. *Gene* **2004**, *336*, 93–104.
28. Del Duca, S.; Tidu, V.; Bassi, R.; Esposito, C.; Serafini-Fracassini, D. Identification of chlorophyll-a/b proteins as substrates of transglutaminase activity in isolated chloroplasts of *Helianthus tuberosus* L. *Planta* **1994**, *193*, 283–289.
29. Della Mea, M.; de Filippis, F.; Genovesi, V.; Serafini Fracassini, D.; Del Duca, S. The acropetal wave of developmental cell death of tobacco corolla is preceded by activation of transglutaminase in different cell compartments. *Plant Physiol.* **2007**, *144*, 1211–1222.
30. Del Duca, S.; Serafini-Fracassini, D. Transglutaminases of higher, lower plants and fungi. *Prog. Exp. Tumor Res.* **2005**, *38*, 223–247.

31. Serafini-Fracassini, D.; Della Mea, M.; Tasco, G.; Casadio, R.; Del Duca, S. Plant and animal transglutaminases: Do similar functions imply similar structures? *Amino Acids* **2009**, *36*, 643657.
32. Della Mea, M.; di Sandro, A.; Dondini, L.; Del Duca, S.; Vantini, F.; Bergamini, C.; Bassi, R.; SerafiniFracassini, D. A Zea mays 39-kDa thylakoid transglutaminase catalyses the modification by polyamines of light-harvesting complex II in a light-dependent way. *Planta* **2004**, *219*, 754–764.
33. Del Duca, S.; Faleri, C.; Iorio, R.A.; Cresti, M.; Serafini-Fracassini, D.; Cai, G. Distribution of transglutaminase in pear pollen tubes in relation to cytoskeleton and membrane dynamics. *Plant Physiol.* **2013**, doi:10.1104/pp.112.212225.
34. Di Sandro, A.; Serafini-Fracassini, D.; Del Duca, S.; Della Mea, M.; de Franceschi, P.; Dondini, L.; Faleri, C.; Cai, G.; Sansavini, S. Pollen transglutaminase in pear self incompatibility and relationship with *S*-RNases and *S*-allele variability. *Acta Hort.* **2008**, *800*, 423–429.
35. Di Sandro, A.; Del Duca, S.; Verderio, E.; Hargreaves, A.J.; Scarpellini, A.; Cai, G.; Cresti, M.; Faleri, C.; Iorio, R.A.; Shigehisa, H.; *et al.* An extracellular transglutaminase is required for apple pollen tube growth. *Biochem. J.* **2010**, *429*, 261–271.
36. Gentile, A.; Antognoni, F.; Iorio, R.A.; Distefano, G.; Las, C.G.; La, M.S.; Serafini-Fracassini, D.; Del Duca, S. Polyamines and transglutaminase activity are involved in compatible and selfincompatible pollination of Citrus grandis. *Amino Acids* **2012**, *42*, 1025–1035.
37. Iorio, R.A.; di Sandro, A.; Paris, R.; Pagliarani, G.; Tartarini, S.; Ricci, G.; Serafini-Fracassini, D.; Verderio, E.; Del Duca, S. Simulated environmental criticalities affect transglutaminase of Malus and Corylus pollens having different allergenic potential. *Amino Acids* **2012**, *42*, 1007–1024.
38. Hegyi, G.; Michel, H.; Shabanowitz, J.; Hunt, D.F.; Chatterjie, N.; Healy-Louie, G.; Elzinga, M. Gln-41 is intermolecularly cross-linked to Lys-113 in F-actin by *N*-(4-azidobenzoyl)-putrescine. *Protein Sci.* **1992**, *1*, 132–144.
39. Nemes, Z., Jr.; Adany, R.; Balazs, M.; Boross, P.; Fesus, L. Identification of cytoplasmic actin as an abundant glutaminyl substrate for tissue transglutaminase in HL-60 and U937 cells undergoing apoptosis. *J. Biol. Chem.* **1997**, *272*, 20577–20583.
40. Eligula, L.; Chuang, L.; Phillips, M.L.; Motoki, M.; Seguro, K.; Muhlrad, A. Transglutaminase-induced cross-linking between subdomain 2 of G-actin and the 636–642 lysine-rich loop of myosin subfragment 1. *Biophys. J.* **1998**, *74*, 953–963.
41. Chowdhury, Z.A.; Barsigian, C.; Chalupowicz, G.D.; Bach, T.L.; Garcia-Manero, G.; Martinez, J. Colocalization of tissue transglutaminase and stress fibers in human vascular smooth muscle cells and human umbilical vein endothelial cells. *Exp. Cell Res.* **1997**, *231*, 38–49.
42. Miller, C.C.; Anderton, B.H. Transglutaminase and the neuronal cytoskeleton in Alzheimer's disease. *J. Neurochem.* **1986**, *46*, 1912–1922.
43. Esposito, C.; Mariniello, L.; Cozzolino, A.; Amoresano, A.; Orru, S.; Porta, R. Rat coagulating gland secretion contains a kinesin heavy chain-like protein acting as a type IV transglutaminase substrate. *Biochemistry* **2001**, *40*, 4966–4971.
44. Cai, G.; Cresti, M. Organelle motility in the pollen tube: A tale of 20 years. *J. Exp. Bot.* **2009**, *60*, 495–508.
45. Daher, F.B.; Geitmann, A. Actin is involved in pollen tube tropism through redefining the spatial targeting of secretory vesicles. *Traffic* **2011**, *12*, 1537–1551.

46. Cardenas, L.; Lovy-Wheeler, A.; Wilsen, K.L.; Hepler, P.K. Actin polymerization promotes the reversal of streaming in the apex of pollen tubes. *Cell Motil. Cytoskeleton* **2005**, *61*, 112–127.
47. Staiger, C.J.; Poulter, N.S.; Henty, J.L.; Franklin-Tong, V.E.; Blanchoin, L. Regulation of actin dynamics by actin-binding proteins in pollen. *J. Exp. Bot.* **2010**, *61*, 1969–1986.
48. Kost, B. Spatial control of Rho (Rac-Rop) signaling in tip-growing plant cells. *Trends Cell Biol.* **2008**, *18*, 119–127.
49. Cardenas, L.; Lovy-Wheeler, A.; Kunkel, J.G.; Hepler, P.K. Pollen tube growth oscillations and intracellular calcium levels are reversibly modulated by actin polymerization. *Plant Physiol.* **2008**, *146*, 1611–1621.
50. Del Casino, C.; Li, Y.; Moscatelli, A.; Scali, M.; Tiezzi, A.; Cresti, M. Distribution of microtubules during the growth of tobacco pollen tubes. *Biol. Cell* **1993**, *79*, 125–132.
51. Cheung, A.Y.; Duan, Q.H.; Costa, S.S.; de Graaf, B.H.J.; di Stilio, V.S.; Feijo, J.; Wu, H.M. The dynamic pollen tube cytoskeleton: Live cell studies using actin-binding and microtubule-binding reporter proteins. *Mol. Plant* **2008**, *1*, 686–702.
52. Laitiainen, E.; Nieminen, K.M.; Vihinen, H.; Raudaskoski, M. Movement of generative cell and vegetative nucleus in tobacco pollen tubes is dependent on microtubule cytoskeleton but independent of the synthesis of callose plugs. *Sex. Plant Reprod.* **2002**, *15*, 195–204.
53. Astrom, H.; Sorri, O.; Raudaskoski, M. Role of microtubules in the movement of the vegetative nucleus and generative cell in tobacco pollen tubes. *Sex. Plant Reprod.* **1995**, *8*, 61–69.
54. Del Duca, S.; Bregoli, A.M.; Bergamini, C.; Serafini-Fracassini, D. Transglutaminase-catalyzed modification of cytoskeletal proteins by polyamines during the germination of *Malus domestica* pollen. *Sex. Plant Reprod.* **1997**, *10*, 89–95.
55. Del Duca, S.; Serafini-Fracassini, D.; Bonner, P.L.; Cresti, M.; Cai, G. Effects of post-translational modifications catalyzed by pollen transglutaminase on the functional properties of microtubules and actin filaments. *Biochem. J.* **2009**, *418*, 651–664.
56. Xue, X.; Du, F.; Zhu, J.; Ren, H. Actin organization and regulation during pollen tube growth. *Front. Biol.* **2011**, *6*, 40–51.
57. Reed, N.A.; Cai, D.; Blasius, T.L.; Jih, G.T.; Meyhofer, E.; Gaertig, J.; Verhey, K.J. Microtubule acetylation promotes kinesin-1 binding and transport. *Curr. Biol.* **2006**, *16*, 2166–2172.
58. Cai, G.; Cresti, M. Microtubule motors and pollen tube growth—Still an open question. *Protoplasma* **2010**, *247*, 131–143.
59. Bosch, M.; Franklin-Tong, V.E. Self-incompatibility in Papaver: Signalling to trigger PCD in incompatible pollen. *J. Exp. Bot.* **2008**, *59*, 481–490.
60. Poulter, N.S.; Vatovec, S.; Franklin-Tong, V.E. Microtubules are a target for self-incompatibility signaling in *Papaver* pollen. *Plant Physiol.* **2008**, *146*, 1358–1367.
61. McClure, B.; Cruz-Garcia, F.; Romero, C. Compatibility and incompatibility in S-RNase-based systems. *Ann. Bot.* **2011**, *108*, 647–658.
62. Cruz-Garcia, F.; Nathan, H.C.; Kim, D.; McClure, B. Stylar glycoproteins bind to S-RNase *in vitro*. *Plant J.* **2005**, *42*, 295–304.
63. Liu, Z.Q.; Xu, G.H.; Zhang, S.L. Pyrus pyrifolia stylar S-RNase induces alterations in the actin cytoskeleton in self-pollen and tubes *in vitro*. *Protoplasma* **2007**, *232*, 61–67.

64. Cai, G.; Della Mea, M.; Faleri, C.; Serafini-Fracassini, D.; Del Duca, S. Spermine regulates the development and programmed cell death of *Nicotiana tabacum* corolla. University of Bologna, Bologna, Italy. Personal communication, 2013.
65. Del Duca, S.; Cai, G.; di Sandro, A.; Serafini-Fracassini, D. Compatible and self-incompatible pollination in *Pyrus communis* displays different polyamine levels and transglutaminase activity. *Amino Acids* **2010**, *38*, 659–667.
66. Dhonukshe, P.; Laxalt, A.M.; Goedhart, J.; Gadella, T.W.J.; Munnik, T. Phospholipase D activation correlates with microtubule reorganization in living plant cells. *Plant Cell* **2003**, *15*, 2666–2679.
67. Gardiner, J.; Andreeva, Z.; Barton, D.; Ritchie, A.; Overall, R.; Marc, J. The phospholipase A inhibitor, aristolochic acid, disrupts cortical microtubule arrays and root growth in Arabidopsis. *Plant Biol.* **2008**, *10*, 725–731.
68. Kim, H.J.; Ok, S.H.; Bahn, S.C.; Jang, J.; Oh, S.A.; Park, S.K.; Twell, D.; Ryu, S.B.; Shin, J.S. Endoplasmic reticulum- and Golgi-localized phospholipase A2 plays critical roles in Arabidopsis pollen development and germination. *Plant Cell* **2011**, *23*, 94–110.
69. Zemskov, E.A.; Mikhailenko, I.; Hsia, R.C.; Zaritskaya, L.; Belkin, A.M. Unconventional secretion of tissue transglutaminase involves phospholipid-dependent delivery into recycling endosomes. *PLoS One* **2011**, *6*, e19414.
70. Parton, R.M.; Fischer-Parton, S.; Watahiki, M.K.; Trewavas, A.J. Dynamics of the apical vesicle accumulation and the rate of growth are related in individual pollen tubes. *J. Cell Sci.* **2001**, *114*, 2685–2695.
71. Certal, A.C.; Almeida, R.B.; Carvalho, L.M.; Wong, E.; Moreno, N.; Michard, E.; Carneiro, J.; RodriguezLeon, J.; Wu, H.M.; Cheung, A.Y.; *et al*. Exclusion of a proton ATPase from the apical membrane is associated with cell polarity and tip growth in *Nicotiana tabacum* pollen tubes. *Plant Cell* **2008**, *20*, 614–634.
72. Ding, Y.; Wang, J.; Wang, J.; Stierhof, Y.D.; Robinson, D.G.; Jiang, L. Unconventional protein secretion. *Trends Plant Sci.* **2012**, *17*, 606–615.
73. Toyooka, K.; Goto, Y.; Asatsuma, S.; Koizumi, M.; Mitsui, T.; Matsuoka, K. A mobile secretory vesicle cluster involved in mass transport from the Golgi to the plant cell exterior. *Plant Cell* **2009**, *21*, 1212–1229.
74. Nakaoka, H.; Perez, D.M.; Baek, K.J.; Das, T.; Husain, A.; Misono, K.; Im, M.J.; Graham, R.M. Gh: A GTPbinding protein with transglutaminase activity and receptor signaling function. *Science* **1994**, *264*, 1593–1596.
75. Antonyak, M.A.; McNeill, C.J.; Wakshlag, J.J.; Boehm, J.E.; Cerione, R.A. Activation of the Ras-ERK pathway inhibits retinoic acid-induced stimulation of tissue transglutaminase expression in NIH3T3 cells. *J. Biol. Chem.* **2003**, *278*, 15859–15866.
76. Baek, K.J.; Kang, S.; Damron, D.; Im, M. Phospholipase Cdelta1 is a guanine nucleotide exchanging factor for transglutaminase II (Galpha h) and promotes alpha 1B-adrenoreceptor-mediated GTP binding and intracellular calcium release. *J. Biol. Chem.* **2001**, *276*, 5591–5597.
77. Dowd, P.E.; Coursol, S.; Skirpan, A.L.; Kao, T.H.; Gilroy, S. Petunia phospholipase c1 is involved in pollen tube growth. *Plant Cell* **2006**, *18*, 1438–1453.

78. Helling, D.; Possart, A.; Cottier, S.; Klahre, U.; Kost, B. Pollen tube tip growth depends on plasma membrane polarization mediated by tobacco PLC3 activity and endocytic membrane recycling. *Plant Cell* **2006**, *18*, 3519–3534.
79. Dinnella, C.; Serafini-Fracassini, D.; Grandi, B.; Del Duca, S. The cell cycle in *Helianthus tuberosus*: Analysis of polyamine-endogenous protein conjugates by transglutaminase-like activity. *Plant Physiol. Biochem.* **1992**, *30*, 531–539.
80. Waffenschmidt, S.; Kusch, T.; Woessner, J.P. A transglutaminase immunologically related to tissue transglutaminase catalyzes cross-linking of cell wall proteins in *Chlamydomonas reinhardtii*. *Plant Physiol.* **1999**, *121*, 1003–1015.
81. Brunner, F.; Rosahl, S.; Lee, J.; Rudd, J.J.; Geiler, C.; Kauppinen, S.; Rasmussen, G.; Scheel, D.; Nurnberger, T. Pep-13, a plant defense-inducing pathogen-associated pattern from *Phytophthora* transglutaminases. *EMBO J.* **2002**, *21*, 6681–6688.
82. Iorio, R.A.; di Sandro, A.; Scarpellini, A.; Del Duca, S.; Serafini-Fracassini, D.; Verderio, E. Visualisation of transglutaminase-mediated cross-linking activity in germinating pollen by laser confocal microscopy. *Plant Biosyst.* **2008**, *142*, 360–365.
83. Furutani, Y.; Kato, A.; Notoya, M.; Ghoneim, M.A.; Hirose, S. A simple assay and histochemical localization of transglutaminase activity using a derivative of green fluorescent protein as substrate. *J. Histochem. Cytochem.* **2001**, *49*, 247–258.
84. Verderio, E.; Nicholas, B.; Gross, S.; Griffin, M. Regulated expression of tissue transglutaminase in Swiss 3T3 fibroblasts: Effects on the processing of fibronectin, cell attachment, and cell death. *Exp. Cell Res.* **1998**, *239*, 119–138.
85. Chae, K.; Lord, E.M. Pollen tube growth and guidance: Roles of small, secreted proteins. *Ann. Bot.* **2011**, *108*, 627–636.
86. Sun, Y.; Qian, H.; Xu, X.-D.; Han, Y.; Yen, L.-F.; Sun, D.-Y. Integrin-like proteins in the pollen tube: Detection, localization and function. *Plant Cell Physiol.* **2000**, *41*, 1136–1142.
87. D'Orazi, D.; Bagni, N. *In vitro* interactions between polyamines and pectic substances. *Biochem. Biophys. Res. Commun.* **1987**, *148*, 1259–1263.
88. Zonia, L.; Munnik, T. Understanding pollen tube growth: The hydrodynamic model *versus* the cell wall model. *Trends Plant Sci.* **2011**, *16*, 347–352.
89. Majewska-Sawka, A.; Nothnagel, E.A. The multiple roles of arabinogalactan proteins in plant development. *Plant Physiol.* **2000**, *122*, 3–10.
90. Lee, C.B.; Swatek, K.N.; McClure, B. Pollen proteins bind to the *C*-terminal domain of Nicotiana alata pistil arabinogalactan proteins. *J. Biol. Chem.* **2008**, *283*, 26965–26973.
91. Mollet, J.C.; Kim, S.; Jauh, G.Y.; Lord, E.M. Arabinogalactan proteins, pollen tube growth, and the reversible effects of Yariv phenylglycoside. *Protoplasma* **2002**, *219*, 89–98.

10

Senescence-Associated Vacuoles, a Specific Lytic Compartment for Degradation of Chloroplast Proteins?

Cristian A. Carrión, Dana E. Martínez, M. Lorenza Costa and Juan José Guiamet *

Instituto de Fisiología Vegetal (INFIVE, CONICET-UNLP), cc327, La Plata 1900, Argentina; E-Mails: ccarrion@agro.unlp.edu.ar (C.A.C.); danamartinez@conicet.gov.ar (D.E.M.); lorenzacosta@agro.unlp.edu.ar (M.L.C.)

* Author to whom correspondence should be addressed; E-Mail: jguiamet@fcnym.unlp.edu.ar

---

Abstract: Degradation of chloroplasts and chloroplast components is a distinctive feature of leaf senescence. In spite of its importance in the nutrient economy of plants, knowledge about the mechanism(s) involved in the breakdown of chloroplast proteins is incomplete. A novel class of vacuoles, "senescence-associated vacuoles" (SAVs), characterized by intense proteolytic activity appear during senescence in chloroplast-containing cells of leaves. Since SAVs contain some chloroplast proteins, they are candidate organelles to participate in chloroplast breakdown. In this review we discuss the characteristics of SAVs, and their possible involvement in the degradation of Rubisco, the most abundant chloroplast protein. Finally, SAVs are compared with other extra-plastidial protein degradation pathways operating in senescing leaves.

Keywords: senescence-associated vacuoles; leaf senescence; chloroplast degradation; proteases; Rubisco; SAG12

---

1. An Overview of Leaf Senescence

Senescence is typically described as a highly organized process characterized by degradation of macromolecules and cellular structures, ultimately leading to cell, organ and/or plant death [1,2]. Typically, the progression of senescence is assessed through parameters related to the loss of chloroplast integrity (e.g., chlorophyll and protein content) and functioning (e.g., photosynthesis, potential quantum yield of photosystem II). This reflects that the breakdown of chloroplasts is an important process during leaf senescence, and integral to the role of senescence in nutrient redistribution [3–5]. Leaves constitute an important store of nutrients, particularly nitrogen, which are mostly held in the photosynthetic apparatus [6]. Within leaves, most N resides in chloroplasts [6]. The degradation of photosynthetic proteins during senescence releases large amounts of N in a readily remobilizable form (*i.e.*, amino acids), therefore senescence represents a salvage mechanism to recycle chloroplast nitrogen to actively growing organs of the plant (e.g., seeds) or to N reserves in the bark of deciduous trees [3–5]. In cereals, up to 80% of N harvested in the grains may come from senescence-associated redistribution of leaf protein N [6]. Given the importance of the breakdown of chloroplast proteins in the nitrogen budget of leaves, crops, and, eventually, global nutrient cycles [7], it is not surprising that senescence-associated changes in other organelles or cell compartments have received much less attention [8–10].

2. Chloroplast Degradation in Senescing Leaf Cells

Although the main focus of this review is on Senescence-Associated Vacuoles, given their presumed role in chloroplast protein degradation, a brief summary of breakdown processes taking place in senescing leaves seems warranted to provide a context of the developmental scenario where these vacuoles occur. During senescence, massive breakdown of chloroplasts results in the almost complete disappearance of recognizable plastids at late stages of senescence [11,12]. Many studies on chloroplast ultrastructure have documented extensive disorganization of the thylakoid membranes, with a decrease in grana stacking and dilation of thylakoids [11,13]. As thylakoids break down, osmiophillic globuli (plastoglobuli) containing products of membrane degradation appear, reaching relatively large numbers in severely deteriorated chloroplasts [11–13]. Eventually, chloroplast structure becomes completely disrupted at late stages of senescence [11]. The ultrastructural changes described above are the consequence of wide-range macromolecule degradation processes. Ultrastructural changes in thylakoids parallel the degradation of photosystem proteins [12–14], thylakoid galactolipids [12,15] and loss of chlorophylls [12,16,17], which result in impaired photosynthetic electron transport [13,18]. The stromal proteins of plastids, including Rubisco and other components of the C3 photosynthetic pathway [12,15], glutamine sunthetase II [19], and even sulfur assimilating enzymes [20] are also degraded, and their breakdown may start even before changes in the photochemical apparatus (e.g., chlorophyll loss) can be detected. The relative onset and rate of degradation of stromal *vs.* thylakoid proteins and chlorophyll may vary (e.g., [21]), which may help adjust the remaining photosynthetic capacity of a leaf in the course of senescence to environmental conditions (e.g., irradiance). This highlights the possibility that different proteolytic mechanisms are responsible for the fate of these different photosynthetic proteins.

The degradation of the membrane-bound proteins of photosystem II has been well studied in the context of the repair cycle following photoinhibition of PSII [22–26] and acclimation of the light-harvesting antenna to increased irradiance [27–29]. There is evidence that chloroplast proteases of the Deg and FtsH families are involved in the degradation of photodamaged D1 protein [22–26], and because photoinhibition persists during senescence [30], this mechanism might account for D1 breakdown in senescing chloroplasts. The light-harvesting proteins of PSII are degraded during acclimation to high irradiance, possibly through the operation of a metalloprotease [27,28]. *In vitro* experiments implicated FtsH6 in the degradation of the light-harvesting Lhcb2 protein [29], however, this could not be substantiated *in vivo* [31].

In many species, stromal protein degradation is apparently responsible for the marked decrease in photosynthetic activity during senescence, but, perhaps more important, Rubisco also constitutes the single most important source of remobilizable nitrogen in leaves of most species. Thus, understanding chloroplast protein and, particularly Rubisco, degradation, holds the promise of extending photosynthetic capacity or providing a handle to manipulate nutrient redistribution. Unfortunately, data linking a particular protease or a proteolytic pathway to the breakdown of stromal chloroplast proteins is often only correlative, or otherwise inconclusive.

In spite of the intense search for within-the-chloroplast degradation of Rubisco and other stromal proteins, there is still no convincing evidence to implicate chloroplast proteases in Rubisco degradation (reviewed in [14,32]). Moreover, recent findings showing Rubisco and other chloroplast proteins in vesicular structures outside the plastid (*i.e.*, Rubisco-Containing Bodies and Senescence-Associated Vacuoles [33–35]) supports the idea that plastidial degradation of Rubisco may be a minor proteolytic pathway for the breakdown of chloroplast stromal proteins in senescing leaves, whereas extra-plastidial pathways may be more important. These extra-plastidial pathways will be discussed in detail below.

3. Involvement of the Central Vacuole

Several reports have suggested that protein degradation in senescing leaves might occur within the central vacuole [36–38]. This is consistent with the fact that the central vacuole is an acidic compartment with high endopeptidase and carboxipeptidase activity [39,40], and that a number of senescence-associated proteases localize to the central vacuole [41–43], supporting its potential involvement in chloroplast protein degradation.

There are reports that provided microscopic evidence for the engulfment of entire plastids within the central vacuole [44–46], whereas other studies failed to confirm these observations (e.g., [47]). Although the available evidence suggests that engulfment of whole chloroplasts may be a relatively infrequent phenomenon, this does not rule out the involvement of the central vacuole. There is an early report on the late accumulation of amino acids (presumably arising from protein degradation) in the central vacuole of senescing leaf cells [48], and more recently Ono *et al.* [49] showed the accumulation of GFP in the central vacuoles of Arabidopsis plants expressing Rubisco-GFP. This implies that the central vacuole may participate in chloroplast protein degradation, either by accomplishing the complete degradation of proteins to their constituent amino acids, or part of this process. In any event, the operation of the entire or partial breakdown of chloroplast proteins in the central vacuole would require transport mechanisms between both organelles.

4. Senescence-Associated Vacuoles: Contents, Characteristics, Markers

In recent years evidence has accumulated showing the co-existence of different classes of vacuoles in the same cells. For example, primary and secondary vacuoles can be detected in aleurone layer cells [50], whereas functionally specialized CAM-associated and sodium storage vacuoles co-exist in mesophyll cells of Mesembryanthemum [51]. In general, the occurrence of two different vacuole types in the same cell, although clearly not universal, is possible under certain physiological or developmental transitions [52]. This is the case in senescing leaf cells, where in addition to the central vacuole, there are smaller vacuoles, absent from non-senescing leaves, and therefore termed "senescence-associated vacuoles" (SAVs) [53]. SAVs have been detected in soybean, Arabidopsis and tobacco [35,53,54]. The number of SAVs increases dramatically as senescence progresses, reaching relatively large numbers (40–60 SAVs per cell in a 1 μm thick optical section) during the period when chloroplast protein loss is most intense [35,55]. Further strengthening the association of SAVs with senescence, the number of SAVs per cell section increases when senescence is accelerated by ethylene, and, conversely, SAVs are almost non-detectable in leaves treated with cytokinins [35,54]. It is noteworthy that SAVs are detected in chloroplast-containing cells, *i.e.*, mesophyll, and, to a lesser extent in guard cells, but they seem to be absent from the rest of the epidermis [53].

SAVs contain vacuolar H^+ pyrophosphatase in their limiting membrane, are acidic, and display no organized internal structure [53], which justifies their assignment as a novel class of vacuoles. However, there are a number of features distinguishing SAVs from the central vacuole. As mentioned before, SAVs are smaller (0.5–0.8 μm in diameter). Furthermore, SAVs are more acidic than the central vacuole. Direct measurements with the pH sensor probe Lysosensor Yellow/Blue shows SAVs to have a lumenal pH around 5.2, whereas the pH of the central vacuole is approximately 6.0 [53]. SAVs membranes also lack the γ-TIP aquaporin, which is diagnostic of the central vacuole and may be important for internal equilibration of water [40].

A distinctive feature of SAVs is their high peptidase activity. When senescing leaf cells are incubated in the presence of a fluorescent probe (R6502) for peptidase activity, most of the fluorescence resulting from hydrolysis of the probe is detected in SAVs [35,53,54]. Since R6502 has been reported to detect cysteine type proteases with high efficiency [55] we surmised that SAVs peptidase activity was mostly due to cysteine-proteases [53], which was later confirmed by the observation that *in vivo* peptidase activity in SAVs is almost completely abolished by pre-incubation with the diagnostic cysteine protease inhibitor E-64 [54] (Figure 1). Likewise, subcellular fractionation combined with the use of an activity-based probe for cysteine proteases, DCG-04 [56,57] detected a large part of the cysteine protease activity of senescing cells in a fraction enriched in SAVs [54]. Up-regulation of cysteine protease genes is a common observation in many transcriptomic studies of leaf senescence in several plant species (reviewed in [14]), which suggests that SAVs contain some of the senescence up-regulated proteases of the cell. In line with this, the senescence-specific cysteine protease SAG12 appears to be located preferentially in SAVs [53,54] (Figure 2), confirming fluorescent (R6502) and activity-based probe (DCG04) results about the prevalence of cysteine proteases in these lytic vacuoles. The localization of a SAG12-GFP fusion in SAVs is quite significant in view of the strict senescence-associated expression of SAG12 [58–60], and this makes SAG12-GFP a convenient and specific marker for SAVs. Consistent with this, increased expression of SAG12 parallels the accumulation of SAVs during senescence of tobacco leaves [54]. SAG12 may account for part of the proteolytic activity of SAVs, but it is clearly not required for SAVs biogenesis [53].

Summing up, the available evidence suggests that at least two types of vacuoles co-exist in senescing mesophyll cells, the central vacuole, and a large number of smaller, senescence-associated vacuoles loaded with high peptidase activity.

Figure 1. *In vivo* inhibition of cysteine protease activity of SAVs with the specific inhibitor E-64. Cells were isolated from senescing (3 days) tobacco leaves, and treated for 2 h with E-64 (100 μM, E through H), or left in buffer as controls (**A** through **D**). Cells were later stained with R6502, a probe for protease activity that becomes brightly fluorescent upon cleavage (**C** and **G**) and Lysotracker Red, an acidotropic marker for acid organelles (**B** and **F**). Panels A and E show chlorophyll autofluorescence, whereas panels D and H show a merged image of Chl, Lysotracker Red and R6502. Co-localization between Lysotracker Red and R6502 is pseudocolored white. Bars represent 5 μm. Redrawn from [54].

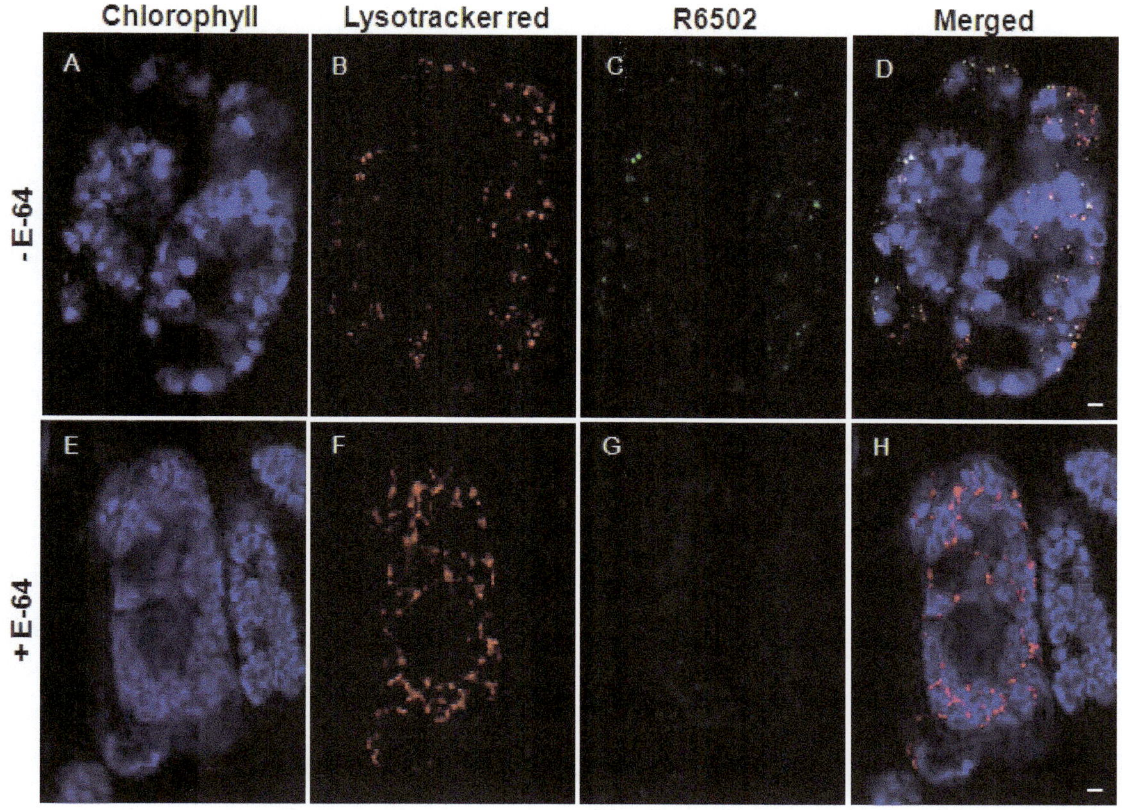

Figure 2. SAG12-GFP localization in senescing Arabidopsis leaves. Confocal images through the mesophyll of a senescing leaf from a SAG12-GFP transgenic plant incubated with the acidotropic dye Neutral Red. (**A**) Chlorophyll fluorescence (excitation 633 nm/emission > 650 nm), (**B**) SAG12-GFP (excitation 488 nm/ emission 505–550 nm), (**C**) Neutral Red (excitation 543 nm/emission 550–605 nm). Note that the GFP signal colocalizes with the fluorescence from Neutral Red, indicating that SAG12-GFP is located inside SAVs. Scale bar = 10 μm. Redrawn from [53].

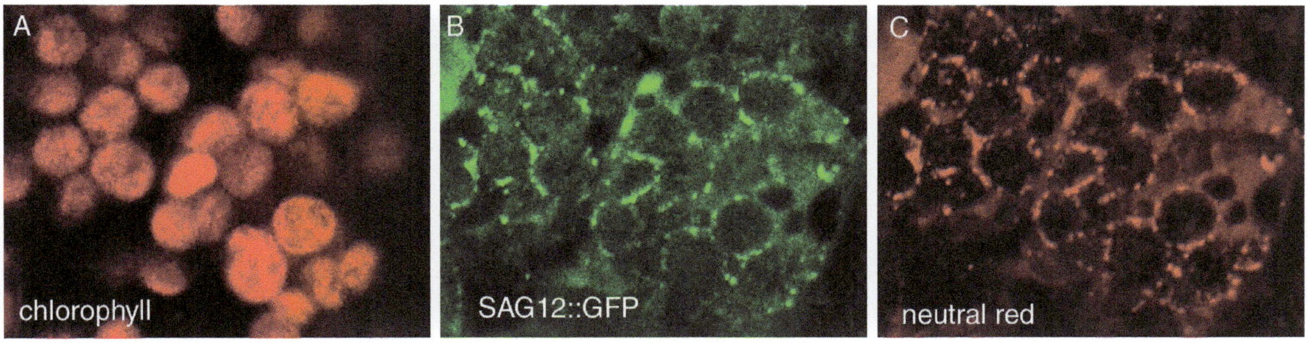

5. Involvement of SAVs in Chloroplast Protein Degradation

Their high peptidase activity and their occurrence in chloroplast-containing cells (*i.e.*, mesophyll and guard cells), led us to speculate that the breakdown of chloroplast proteins during senescence might partially occur in SAVs. Two lines of evidence point to the relocation of some chloroplast proteins to SAVs in senescing leaf cells. First, in a transgenic line of tobacco expressing GFP fused to the recA transit peptide for targeting to plastids, GFP fluorescence is restricted to the chloroplasts in mature leaves [35,61], while in senescing cells, in addition to the presence of GFP in chloroplasts, also SAVs (detected by their staining with the acidotropic marker Lysotracker Red) contain GFP (Figure 3) [35]. This suggests that some chloroplast proteins relocalize to SAVs in the course of senescence. A second evidence for the presence of chloroplast proteins in SAVs comes from subcellular fractionation and isolation of a SAVs-enriched fraction. The large subunit of Rubisco and glutamine synthetase II were readily detected in isolated SAVs confirming the relocation of chloroplast proteins to SAVs during senescence. The finding of the large subunit of Rubisco is significant, not only because this is the most abundant protein in C3 plants, but also because the large subunit is chloroplast-encoded and synthesized within the plastid, therefore the most likely explanation for its presence in SAVs is relocation. The lack of characteristic PSII proteins, such as D1 and Lhcb in this SAVs-enriched fraction indicates also that the presence of Rubisco is not due to contamination, and, perhaps more important, that PSII proteins are degraded by a different proteolytic pathway.

Since SAVs contain high peptidase activity, it is not surprising that isolated SAVs can degrade the chloroplast proteins (e.g., Rubisco) contained within them in autodigestion experiments [35,54]. About 25% of Rubisco contained in SAVs is lost in 4 h of incubation, and this moderate rate of degradation possibly explains why chloroplast-targeted GFP is relatively stable in SAVs, allowing for its detection through confocal microscopy [35,54]. The loss of Rubisco in autodigestion experiments is reduced to a large extent by addition of cysteine protease inhibitors, whereas serine protease inhibitors are largely ineffective [35,54]. Thus, *in vitro* proteolytic activity of SAVs seems to be due to cysteine proteases, which is consistent with all the observations indicating that cysteine proteases are an important component of SAVs. Correlative evidence suggests that Rubisco degradation by SAVs cysteine proteases also occurs *in vivo*. Incubation of wheat leaf segments [62] and tobacco leaf disks [54] with an inhibitor of cysteine proteases (E64) reduces the degradation of chloroplast proteins, including Rubisco. This correlates with severe inhibition of the cysteine protease activity of SAVs [54]. Although a direct genetic approach to test whether chloroplast proteins are degraded in SAVs is still lacking, the presence of chloroplast proteins in SAVs and the decreased protein degradation by treatments that block SAVs proteolytic activity lend support to the idea that SAVs may be part of a proteolytic pathway(s) involved in breakdown of the stromal proteins of chloroplasts in senescing leaves.

Figure 3. Z-stack (consecutive 1 μm thick optical slices) covering a depth of 4 μm in a mesophyll cell of a senescing tobacco leaf expressing GFP targeted to the plastid [61]. As an example, SAV detected in the middle planes (second or third optical slices) are indicated with white arrows (one SAV with LysoTracker Red and GFP and Chl fluorescence) or white arrowheads (two SAVs with LysoTracker Red, and GFP fluorescence). Empty arrows/arrowheads show fainter signals above or below the focal plane of selected SAVs. Note that most SAVs, stained by Lysotracker Red, also contain GFP, and that almost all GFP located outside chloroplasts is found in SAVs. Scale bar: 5 μm. Redrawn from [35].

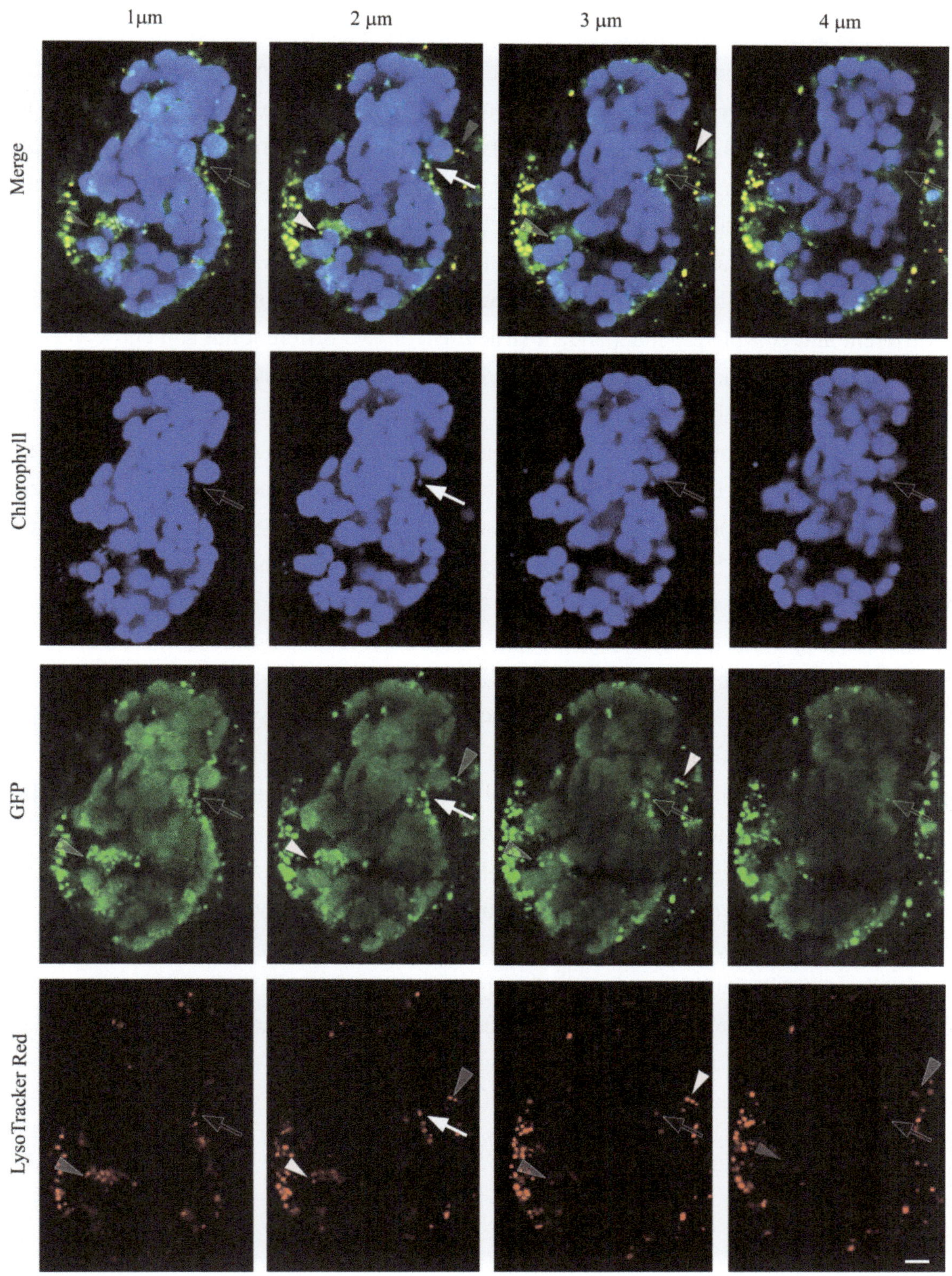

6. SAVs and Rubisco-Containing Bodies, Is There a Link?

Almost in parallel to the finding of SAVs, Rubisco-Containing Bodies (RCBs) were detected in senescing wheat leaves [33,34]. RCBs were first described as small (0.4–1.2 μm in diameter), double membrane bound vesicles detected in the cytosol of leaf cells through transmission electron microscopy [33]. Their numbers increased during senescence, and immunolocalization experiments showed that they contained chloroplast stromal proteins, such as Rubisco and glutamine synthetase II, but not thylakoid proteins [33]. This is similar to what we know about SAVs, although SAVs are apparently bound by a single membrane [53]. On the other hand, no evidences on the presence of proteases or peptidase activity within RCBs have been reported, which is an important difference with respect to SAVs. Testing for peptidase activity in RCBs would be important to confirm their role as a transport (not lytic) vesicle carrying stromal proteins to the central vacuole, and to differentiate RCBs functions from those of SAVS. Formation of RCBs depends on the operation of the autophagic pathway [34,46]. In transgenic lines expressing GFP or DsRed targeted to the plastid, RCBs are detected in the central vacuole of concanamycin-Atreated cells [34,46], possibly because concanamycin-A, an inhibitor of vacuolar H^+ ATPase, raises vacuolar pH thereby reducing its hydrolytic activity.

Since there seem to be a number of similarities between SAVs and RCBs, one may wonder whether they are two different and completely independent structures, or two different but somehow interrelated vesicles, or even whether they are the same structure given different names. Dependence of RCBs on the autophagic pathway and inhibition of SAVs by concanamycin-A treatment are two important differences which suggest these might be different structures representative of two different proteolytic pathways. We previously showed that small lytic vacuoles stained by the acidotropic dye Lysotracker Red and the cysteine protease probe R6502 accumulate normally in an *atg7* knock out mutant [53], indicating that formation of SAVs does not depend on the autophagic pathway. Regarding concanamycin-A, detection of RCBs is facilitated by pre-treatment of leaf disks for 20 h with concanamycin-A [34,46]. On the contrary, in our experiments with tobacco, concanamycin-A significantly reduced the staining of SAVs by R6502, *i.e.*, their proteolytic activity became almost non-detectable [63] (Figure 4). This might be simply explained on the basis of the operation of vacuolar H^+ ATPases in the acidification of SAVs lumen. Pre-treatment with concanamycin-A might render SAVs less acidic and, since many cysteine proteases have an acidic pH optimum, labeling by R6502 might be impaired by concanamycin-A. However, concanamycin-A treatment also decreases the number of SAVs detected through fluorescence of GFP in a line expressing a SAG12-GFP fusion [63] (Figure 5). The disappearance of SAG12-GFP is hard to attribute to increased SAVs pH, on the contrary, since GFP is reportedly less stable under low pH [64], an increase of the highly acidic pH of SAVs should increase GFP stability and, therefore, improve SAG12-GFP detection. Alternatively, inhibition of SAVs by concanamycin-A could be interpreted as evidence for the involvement of the Golgi apparatus in SAVs biogenesis, as concanamycin-A is known to disrupt trafficking through the trans-Golgi network [65], In any event, the scant evidence available so far, *i.e.*, dependence on the autophagic pathway and response to concanamycin-A, suggests that RCBs and SAVs may be different entities.

Figure 4. Inhibition of SAVs proteolytic activity by concanamycin-A. Detached tobacco leaves were induced to senesce by incubation in darkness for 3 days, then 5-mm diameter disks were excised and incubated in 1 µm concanamycin-A for 15 h. Cells were stained with Lysotracker Red and R6502 and observed through laser confocal microscopy. Note the almost complete disappearance of the characteristic punctuate pattern of R6502 (*i.e.*, proteolytic activity) corresponding to SAVs in cells from disks treated with concanamycin-A (lower panels).

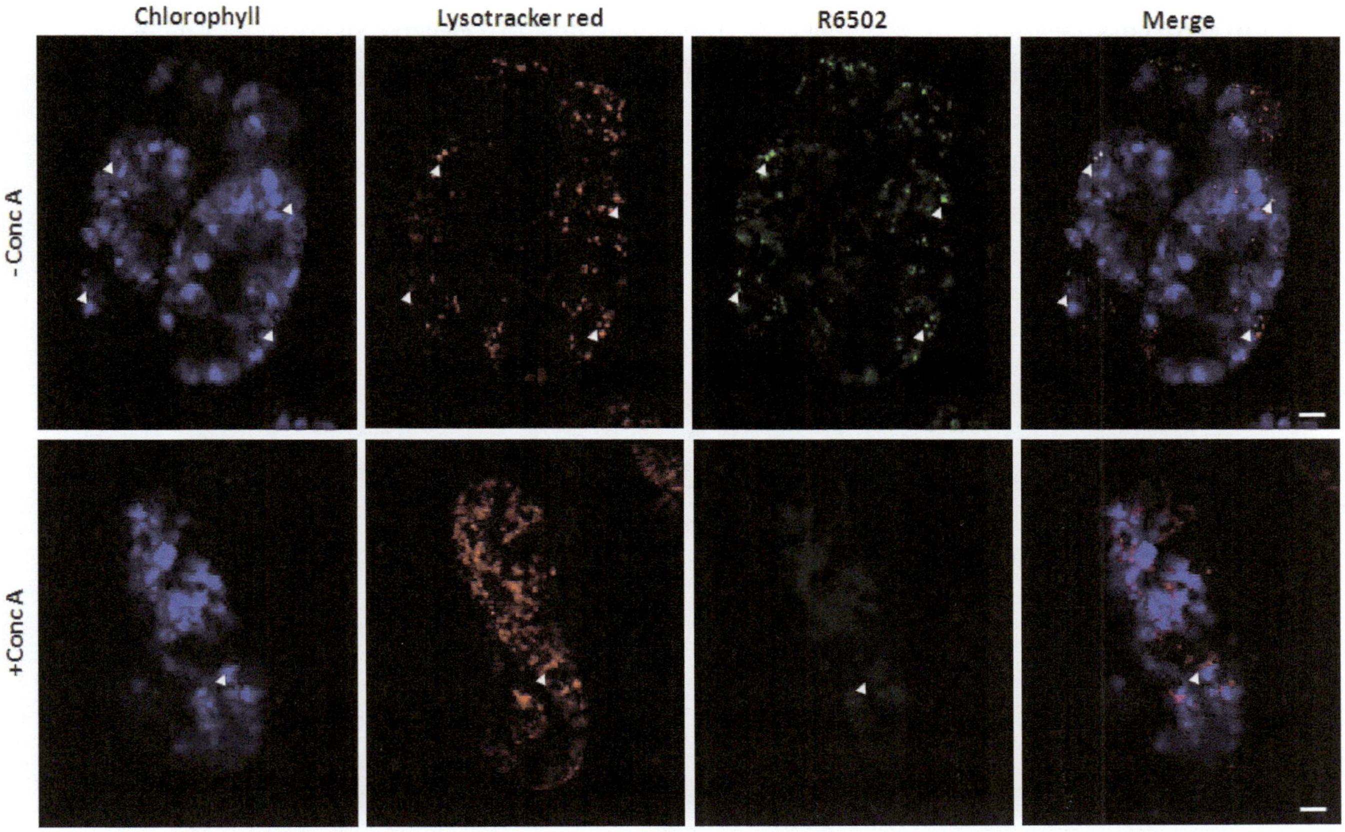

If SAVs and RCBs turned out to be different structures involved in stromal protein degradation, then a relevant question would be whether they co-exist in senescing cells or if they represent different breakdown pathways operating in specific developmental scenarios. This is a difficult question without more information on their molecular characteristics to allow for their simultaneous identification in living cells. However, it is worth noting that in tobacco leaf cells expressing GFP-targeted to the chloroplast, almost all vesicular structures containing GFP outside plastids are labeled by Lysotracker Red, *i.e.*, they are acidic (Figure 3), and in wild type cells almost all Lysotracker Red labeled SAVs contain high peptidase activity, which are distinctive features of SAVs. This might imply that, under certain conditions, SAVs may be quite prevalent. On the other hand, based on calculations of the amount of GFP released into the central vacuole in a line expressing Rubisco fused to GFP, Ono *et al.* [49] calculated that at least 41% of Rubisco might be degraded through an autophagic pathway involving RCBs. The independent operation of SAVs and RCBs is also suggested by the phenotype of autophagy knock out mutants, in which chloroplast protein degradation proceeds normally, or in many cases is accelerated, rather than delayed [66–68]. A careful quantitative analysis of Rubisco levels in senescing leaves of Arabidopsis shows that the time-course of Rubisco loss is very similar in wild type and two autophagy mutants, except for a slight retention of Rubisco very late during senescence (*i.e.*, when leaves have already lost more than 95% of their chlorophyll [69]. Recent studies clearly show the accumulation of degradation fragments of Rubisco and glutamine synthetase II in an *atg5* knock out line [70]. Since RCBs presumably do not operate in an *atg5* KO, this suggests the operation of an alternative proteolytic pathway. SAG12 expression increases in some *atg* mutants [71], therefore SAVs might represent the pathway alternative to autophagy bringing about chloroplast protein degradation in autophagy mutants.

Figure 5. Disappearance of the SAG12-GFP signal due to concanamycin-A treatment. Detached tobacco leaves expressing SAG12 fused to GFP were induced to senesce by incubation in darkness for 3 days. Leaf disks were then excised and incubated in 1 μm concanamycin-A for 15 hours. Leaf pieces were observed directly with an epifluorescence microscope (Olympus BX61) fitted with an FITC filter. Note that the bright fluorescent vesicles (SAVs, a few representative SAVs are indicated by arrowheads) in the control, left panel, disappeared completely in leaf cells treated with concanamycin-A (right panel).

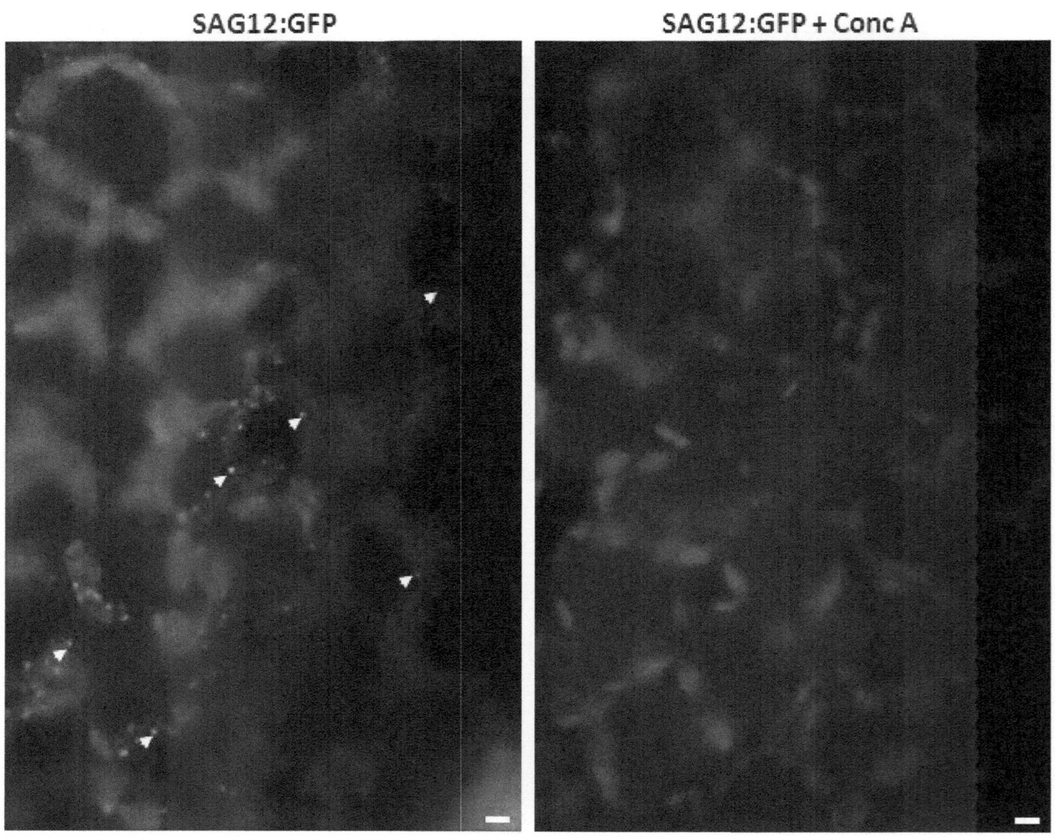

7. Future Directions

The available evidence indicates that Senescence-Associated Vacuoles are lytic compartments possibly involved in the degradation of chloroplast proteins during senescence. There are a number of questions that still need to be resolved, and these include, among others, (1) the biogenesis of SAVs, (2) the identification of SAVs proteases responsible for protein degradation, (3) an assessment of the quantitative importance of SAVs in chloroplast protein breakdown and (4) a detailed study of the interrelations between SAVs and other extraplastidial chloroplast degradation pathways. Undoubtedly, identifying molecular targets to impair SAVs biogenesis and/or function will be critical to facilitate progress in the understanding of this novel cellular structure.

Acknowledgments

Funding from CONICET (PIP 487) and ANPCYT (PICT 2012-1862, PICT 061), Argentina, is gratefully acknowledged.

Author Contributions

Cristian Carrión supplied unpublished data and revised the manuscript, Dana Martínez, María Lorenza Costa and Juan José Guiamet wrote different sections of this manuscript and revised the final version.

Conflicts of Interest

The authors declare no conflict of interest.

References

1. Noodén, L.D.; Guiamet, J.J.; John, I. Whole plant senescence. In *Plant Cell Death Processes*; Noodén, L.D., Ed.; Academic Press: San Diego, CA, USA, 2004; pp. 227–244.
2. Li, P.O.; Kim, H.J.; Nam, H.G. Leaf senescence. *Annu. Rev. Plant. Biol.* **2007**, *58*, 115–136.
3. Masclaux-Daubresse, C.; Reisdorf-Cren, M.; Orsel, M. Leaf nitrogen remobilisation for plant development and grain filling. *Plant Biol.* **2008**, *10*, 23–36.
4. Mae, T. Leaf senescence and nitrogen metabolism. In *Plant Cell Death Processes*; Noodén, L.D., Ed.; Academic Press: San Diego, CA, USA, 2004; pp. 157–168.
5. Killingbeck, K.T. Nutrient resorption. In *Plant Cell Death Processes*; Noodén, L.D., Ed.; Academic Press: San Diego, CA, USA, 2004; pp. 215–226.
6. Peoples, M.B.; Dalling, M.J. The interplay between proteolysis and amino acid metabolism during senescence and nitrogen reallocation. In *Senescence and Aging in Plants*; Noodén, L.D., Leopold, A.C., Eds.; Academic Press: San Diego, CA, USA, 1988; pp. 181–217.
7. Vergutz, L.; Manzoni, S.; Portorato, A.; Ferreira Novais, R.; Jackson, M.B. Global resorption efficiencies and concentrations of carbon and nutrients in leaves of terrestrial plants. *Ecol. Monogr.* **2012**, *82*, 205–220.
8. Martínez, D.E.; Guiamet, J.J. Senescence-related changes in the leaf apoplast. *J. Plant Growth Regul.* **2014**, *33*, 44–55.
9. Del Río, L.A.; Pastori, G.M.; Palma, J.M.; Sandalio, L.M.; Sevilla, F.; Corpas, F.J.; Jiménez, A.; López-Huerta, E.; Hernández, J.A. The activated oxygen role of peroxisomes in senescence. *Plant Physiol.* **1998**, *116*, 1195–1200.
10. Keech, O.; Pesquet, E.; Ahad, A.; Askne, A.; Nordvall, D.; Vodnala, S.M.; Tuominen, H.; Hurry, V.; Dizengremel, P.; Gardeström, P. The different fates of mitochondria and chloroplasts during dark-induced senescence in Arabidopsis leaves. *Plant Cell Environ.* **2007**, *30*, 1523–1534.

11. Guiamét, J.J.; Luquez, V.M. Leaf senescence and related processes. In *Crops: Growth, Quality, and Biotechnology*; Dris, R., Ed.; WFL Publisher: Helsinki, Finland, 2005; pp. 1238–1253.
12. Krupinska, K. Fate and activities of plastids during leaf senescence. In *The Structure and Function of Plastids*; Wise, R.R., Hoober, J.K., Eds.; Springer: Dordrecht, The Netherlands, 2007; pp. 433–449.
13. Gepstein, S. Photosynthesis. In *Senescence and Aging in Plants*; Noodén, L.D., Leopold, A.C., Eds.; Academic Press: San Diego, CA, USA, 1988; pp. 85–109.
14. Costa, M.L.; Martínez, D.E.; Gomez, F.M.; Carrión, C.; Guiamét, J.J. Chloroplast protein degradation: Involvement of senescence-associated vacuoles. In *Advances in Photosynthesis and Respiration Series*; Biswal, B., Krupinska, K., Biswal, U.C., Eds.; Springer Science + Business Media: Dordrecht, The Netherlands, 2013; pp. 417–433.
15. Meir, S.; Philosoph-Hadas, S. Metabolism of polar lipids during senescence of watercress leaves. *Plant Physiol. Biochem.* **1995**, *33*, 241–249.
16. Ougham, H.; Hörtensteiner, S.; Armstead, I.; Donninson, I.; King, I.; Thomas, H.; Mur, L. The control of chlorophyll catabolism and the status of yellowing as a biomarker of leaf senescence. *Plant Biol.* **2008**, *10*, 4–14.
17. Hörtensteiner, S.; Kräutler, B. Chlorophyll breakdown in higher plants. *Biochim. Biophys. Acta* **2011**, *1807*, 977–988.
18. Krupinska, K.; Humbeck, K. Photosynthesis and chloroplast breakdown. In *Plant Cell Death Processes*; Noodén, L.D., Ed.; Academic Press: San Diego, CA, USA, 2004; pp. 169–187.
19. Kamachi, K.; Yamaya, T.; Hayakawa, T.; Mae, T.; Ojima, K. Changes in cytosolic glutamine synthetase polypeptide and its mRNA in a leaf blade of rice plants during natural senescence. *Plant Physiol.* **1992**, *98*, 1323–1329.
20. Schmutz, D.; Wyss, H.-R.; Brunold, C. Activity of sulfate-assimilating enzymes in primary leaves of *Phaseolus vulgaris* L. during dark-induced senescence. *Z. Pflanzenphysiol.* **1983**, *110*, 211–219.
21. Hidema, J.; Makino, A.; Mae, T.; Ojima, K. Photosynthetic characteristics of rice leaves aged under different irradiances from full expansion through senescence. *Plant Physiol.* **1991**, *97*, 1287–1293.
22. Zaltsman, A.; Ori, N.; Adam, Z. Two types of FtsH subunits are required for chloroplast biogenesis and Photosystem II repair in Arabidopsis. *Plant Cell* **2005**, *17*, 2782–2790.
23. Kapri-Pardes, E.; Naveh, L.; Adam, Z. The thylakoid lumen protease DegP1 is involved in the repair of photosystem II from photoinhibition in *Arabidopsis*. *Plant Cell* **2007**, *19*, 1039–1047.
24. Sun, X.; Peng, L.; Guo, J.; Chi, W.; Ma, J.; Lu, C.; Zhang, L. Formation of Deg5 and Deg8 complexes and their involvement in the degradation of photodamaged Photosystem II reaction center D1 protein in *Arabidopsis*. *Plant Cell* **2007**, *19*, 1347–1361.
25. Kato, Y.; Miura, E.; Ido, K.; Ifuku, K.; Sakamoto, W. The variegated mutants lacking chloroplastic FtsHs are defective in D1 degradation and accumulate reactive oxygen species. *Plant Physiol.* **2009**, *151*, 1790–1801.
26. Sun, X.; Fu, T.; Chen, N.; Guo, J.; Ma, J.; Zou, M.; Lu, C.; Zhang, L. The stromal chloroplast Deg7 protease participates in the repair of photosystem II after photoinhibition in *Arabidopsis*. *Plant Physiol.* **2010**, *152*, 1263–1273.

27. Yang, D.-H.; Webster, J.; Adam, Z.; Lindahl, M.; Andersson, B. Induction of acclimative proteolysis of the light-harvesting chlorophyll a/b protein of photosystem II in response to elevated light intensities. *Plant Physiol.* **1998**, *118*, 827–834.
28. Zelisko, A.; Jackowski, G. Senescence-dependent degradation of Lhcb3 is mediated by a thylakoid membrane-bound protease. *J. Plant Physiol.* **2004**, *161*, 1157–1170.
29. Zelisko, A.; García-Lorenzo, M.; Jackowski, G.; Jansson, S.; Funk, C. AtFtsH6 is involved in the degradation of the light-harvesting complex II during high-light acclimation and senescence. *Proc. Natl. Acad. Sci. USA* **2005**, *102*, 13690–13704.
30. Guiamét, J.J.; Tyystjärvi, E.; Tyystjärvi, T.; John, I.; Kairavuo, M.; Pichersky, P.; Noodén, L.D. Photoinhibition and loss of photosystem II reaction center proteins during senescence of soybean leaves. Enhancement of photoinhibition by the "stay-green" mutation *cytG*. *Physiol. Plant* **2002**, *115*, 468–478.
31. Wagner, R.; Aigner, H.; Pruzinská, A.; Jänkänpää, H.J.; Jansson, S.; Funk, C. Fitness analysis of *Arabidopsis thaliana* mutants depleted of FtsH metalloproteases and characterization of three FtsH6 deletion mutants exposed to high light stress, senescence and chilling. *New Phytol.* **2011**, *191*, 449–458.
32. Martínez, D.E.; Costa, M.; Guiamet, J.J. Senescence associated degradation of chloroplast proteins inside and outside the organelle. *Plant Biol.* **2008**, *10*, 15–22.
33. Chiba, A.; Ishida, H.; Nishizawa, N.K.; Makino, A.; Mae, T. Exclusion of ribulose-1,5-bisphosphate carboxylase/oxygenase from chloroplasts by specific bodies in naturally senescing leaves of wheat. *Plant Cell Physiol.* **2003**, *44*, 914–921.
34. Ishida, H.; Yoshimoto, K.; Izumi, M.; Reisen, D.; Yano, Y.; Makino, A.; Ohsumi, Y.; Hanson, M.R.; Mae, T. Mobilization of rubisco and stroma-localized fluorescent proteins of chloroplast to the vacuole by an ATG gene-dependent autophagic process. *Plant Physiol.* **2008**, *148*, 142–155.
35. Martínez, D.E.; Costa, M.L.; Gomez, F.M.; Otegui, M.S.; Guiamet, J.J. "Senescence-associated vacuoles" are involved in the degradation of chloroplast proteins in tobacco leaves. *Plant J.* **2008**, *56*, 196–206.
36. Lin, W.; Wittenbach, V.A. Subcellular localization of proteases in wheat and corn mesophyll protoplasts. *Plant Physiol.* **1981**, *67*, 969–972.
37. Wittenbah, V.A.; Lin, W.; Herbert, R.H. Vacuolar localization of proteases and degradation of chloroplasts in mesophyll protoplasts from senescing primary wheat leaves. *Plant Physiol.* **1982**, *69*, 98–102.
38. Srivalli, B.; Bharti, S.; Kanna-Chopra, R. Vacuole cysteine proteases and ribulose-1,5-bisphosphate carboxylase/oxygenase degradation during monocarpic senescence in cowpea leaves. *Photosynthetica* **2001**, *39*, 87–93.
39. De Deepesh, N. *Plant Cell Vacuoles. An Introduction*; CSIRO Publishing: Collingwood, Australia, 2000; pp. 1–304.
40. Müntz, K. Protein dynamics and proteolysis in plant vacuoles. *J. Exp. Bot.* **2007**, *58*, 2391–2407.
41. Hensel, L.L.; Grbic, V.; Baumgarten, D.A.; Bleecker, A.B. Developmental and age-related processes that influence the longevity and senescence of photosynthetic tissues in *Arabidopsis*. *Plant Cell* **1993**, *5*, 553–564.

42. Drake, R.; John, I.; Farrell, A.; Cooper, W.; Schuch, W.; Grierson, D. Isolation and analysis of cDNAs encoding tomato cysteine proteases expressed during leaf senescence. *Plant Mol. Biol.* **1996**, *30*, 755–767.
43. Martínez, D.E.; Bartoli, C.G.; Grbic, V.; Guiamet, J.J. Vacuolar cysteine proteases of wheat (*Triticum aestivum* L.) are common to leaf senescence induced by different factors. *J. Exp. Bot.* **2007**, *58*, 1099–1107.
44. Minimikawa, T.; Toyooka, K.; Okamoto, T.; Hara-Nishimura, I.; Nishimura, M. Degradation of ribulose-bisphosphate carboxylase by vacuolar enzymes of senescing French bean leaves: Immunocytochemical and ultrastructural observations. *Protoplasma* **2001**, *218*, 144–153.
45. Niwa, Y.; Kato, T.; Tabata, S.; Seki, M.; Kobayashi, M.; Shinozaki, K.; Moriyasu, Y. Disposal of chloroplasts with abnormal function into the vacuole in *Arabidopsis thaliana* cotyledon cells. *Protoplasma* **2004**, *223*, 229–232.
46. Wada, S.; Ishida, H.; Izumi, M.; Yoshimoto, K.; Ohsumi, Y.; Mae, T.; Makino, A. Autophagy plays a role in chloroplast degradation during senescence in individually darkened leaves. *Plant Physiol.* **2009**, *149*, 885–893.
47. Evans, M.; Rus, A.M.; Belanger, E.M.; Kimoto, M.; Brusslan, J.A. Dismantling of *Arabidopsis thaliana* mesophyll cell chloroplasts during natural leaf senescence. *Plant Biol.* **2010**, *12*, 1–12.
48. Matile, P. The vacuole and cell senescence. *Adv. Bot. Res.***1997**, *25*, 87–112.
49. Ono, Y.; Wada, S.; Izumi, M.; Makino, A.; Ishida, H. Evidence for contribution of autophagy to Rubisco degradation during leaf senescence in *Arabidopsis thaliana. Plant Cell Environ.* **2013**, *36*, 1147–1159.
50. Swanson, S.J.; Bethke, P.C.; Jones, R.L. Barley aleurone cells contain two types of vacuoles: Characterization of lytic organelles by means of fluorescent probes. *Plant Cell* **1998**, *10*, 685–698.
51. Epimashko, S.; Meckel, T.; Fischer-Schliebs, E.; Lüttge, U.; Thiel, G. Two functional different vacuoles for static and dynamic purposes in one plant mesophyll leaf cell. *Plant J.* **2004**, *37*, 294–300.
52. Frigerio, L.; Hinz, G.; Robinson, D.G. Multiple vacuoles in plant cells: Rule or exception? *Traffic* **2008**, *9*, 1564–1570.
53. Otegui, M.; Noh, Y.-S.; Martínez, D.; Vila-Petroff, M.; Staehelin, A.; Amasino, R.; Guiamet, J.J. Senescence-associated vacuoles with intense proteolytic activity develop in senescing leaves of *Arabidopsis* and soybean. *Plant J.* **2005**, *41*, 831–844.
54. Carrión, C.A.; Costa, M.L.; Martínez, D.E.; Mohr, C.; Humbeck, K.; Guiamet, J.J. *In vivo* inhibition of cysteine proteases provides evidence for the involvement of "senescence-associated vacuoles" in chloroplast protein degradation during dark-induced senescence of tobacco leaves. *J. Exp. Bot.* **2013**, *64*, 4967–4980.
55. Peptidase substrates. Available online: http://www.lifetechnologies.com/ar/es/home/references/molecular-probes-the-handbook/enzyme-substrates/detecting-peptidases-and-proteases.html#head1 (accessed on 7 October 2014).
56. Greenbaum, D.; Medzihradsky, K.; Burlingame, A.; Bogyo, M. Epoxide electrophiles as activity-dependent cysteine protease profiling and discovery tools. *Chem. Biol.* **2000**, *7*, 569–581.
57. Van der Hoorn, R.A.L.; Leeuwenbergh, M.A.; Bogyo, M.; Joosten, M.H.A.; Peck, S.C. Activity profiling of papain-like cystein proteases in plants. *Plant Physiol.* **2004**, *135*, 1170–1178.

58. Lohman, K.N.; Gan, S.; John, L.M.; Amasino, R.M. Molecular analysis of natural leaf senescence in *Arabidopsis thaliana*. *Physiol. Plant.* **1994**, *92*, 322–328.
59. Noh, Y.-S.; Amasino, R.M. Identification of a promoter region responsible for the senescence-specific expression of SAG12. *Plant Mol. Biol.* **1999**, *41*, 181–194.
60. Grbic, V. SAG2 and SAG12 protein expression in senescing *Arabidopsis* plants. *Physiol. Plant.* **2003**, *119*, 1–7.
61. Köhler, R.H.; Cao, J.; Zipfel, W.R.; Webb, W.W.; Hanson, M.R. Exchange of protein molecules through connections between higher plant plastids. *Science* **1997**, *276*, 2039–2042.
62. Thoenen, M.; Herrmann, B.; Feller, U. Senescence in wheat leaves: Is a cysteine endopeptidase involved in the degradation of the large subunit of Rubisco? *Acta Physiol. Plant.* **2007**, *29*, 339–350.
63. Carrión, C.A. Instituto de Fisiología Vegetal (INFIVE, CONICET-UNLP), La Plata, Argentina. Unpublished data, 2013.
64. Tamura, K.; Shimada, T.; Ono, E.; Tanaka, Y.; Nagatani, A.; Higashi, S.-I.; Watanabe, M.; Nishimura, M.; Hara-Nishimura, I. Why green fluorescent fusion proteins have not been observed in the vacuoles of higher plants. *Plant J.* **2003**, *35*, 545–555.
65. Dettmer, J.; Hong-Hemesdorf, A.; Stierhof, Y.-D.; Schumacher, K. Vacuolar H^+-ATPase activity is required for endocytic and secretory trafficking in *Arabidopsis*. *Plant Cell* **2006**, *18*, 715–730.
66. Doelling, J.H.; Walker, J.M.; Friedman, E.M.; Thompson, A.R.; Vierstra, R.D. The APG8/12-activating enzyme APG7 is required for proper nutrient recycling and senescence in *Arabidopsis thaliana*. *J. Biol. Chem.* **2002**, *277*, 33105–33114.
67. Thompson, A.R.; Doelling, J.H.; Suttangkakul, A.; Vierstra, R.D. Autophagic nutrient recycling in *Arabidopsis* directed by the ATG8 and ATG12 conjugation pathways. *Plant Physiol.* **2005**, *138*, 2097–2110.
68. Liu, Y.; Bassham, D.C. Autophagy: Pathways for self-eating in plant cells. *Ann. Rev. Plant Biol.* **2012**, *63*, 215–237.
69. Lee, T.A.; Vande Wetering, S.W.; Bruslan, J.A. Stromal protein degradation is incomplete in *Arabidopsis thaliana* autophagy mutants undergoing natural senescence. *BMC Res. Notes* **2013**, *6*, 17.
70. Guiboileau, A.; Avila-Ospina, L.; Yoshimoto, K.; Soulay, F.; Azzopardi, M.; Marmagne, M.; Lothier, J.; Masclaux-Daubresse, C. Physiological and metabolic consequences of autophagy deficiency for the management of nitrogen and protein resources in *Arabidopsis* leaves depending on nitrate availability. *New Phytol.* **2013**, *199*, 683–694.
71. Yoshimoto, K.; Jikumaru, Y.; Kamiya, Y.; Kusano, M.; Consonni, C.; Panstruga, R.; Ohsuni, Y.; Shirasu, K. Autophagy negatively regulates cell death by controlling NPR1-dependent salicylic acid signaling during senescence and the innate immune response in *Arabidopsis*. *Plant Cell* **2009**, *21*, 2914–2927.

11

Nutritive Value Response of Native Warm-Season Forage Grasses to Harvest Intervals and Durations in Mixed Stands

Vitalis W. Temu [1,*], Brian J. Rude [2,†] and Brian S. Baldwin [3,†]

[1] Agricultural Research Station, Virginia State University, Petersburg, VA 23806, USA
[2] Animal and Dairy Science Department, Mississippi State University, Mississippi State, MS 39762, USA; E-Mail: BRude@ads.msstate.edu
[3] Plant and Soil Sciences Department, Mississippi State University, Mississippi State, MS 39762, USA; E-Mail. BBaldwin@pss.msstate.edu

† These authors contributed equally to this work.

* Author to whom correspondence should be addressed; E-Mail: vtemu@vsu.edu

---

Abstract: Interest in management of native warm-season grasses for multiple uses is growing in southeastern USA. Forage quality response of early-succession mixed stands of big bluestem (BB, *Andropogon gerardii*), indiangrass (IG, *Sorghastrum nutans*), and little bluestem (SG, *Schizachyrium scoparium*) to harvest intervals (30-, 40-, 60-, 90 or 120-d) and durations (one or two years) were assessed in crop-field buffers. Over three years, phased harvestings were initiated in May, on sets of randomized plots, ≥90 cm apart, in five replications (blocks) to produce one-, two-, and three-year-old stands, by the third year. Whole-plot regrowths were machine-harvested after collecting species (IG and LB) sample tillers for leafiness estimates. Species-specific leaf area (SLA) and leaf-to-stem ratio (LSR) were greater for early-season harvests and shorter intervals. In a similar pattern, whole-plot crude protein concentrations were greatest for the 30-d (74 $g \cdot kg^{-1}$ DM) and the least (40 $g \cdot kg^{-1}$ DM) for the 120-d interval. Corresponding neutral detergent fiber (NDF) values were the lowest (620 $g \cdot kg^{-1}$ DM) and highest (710 $g \cdot kg^{-1}$ DM), respectively. *In vitro* dry matter and NDF digestibility were greater for early-season harvests at shorter intervals (63 and 720 $g \cdot kg^{-1}$ DM). With strategic harvesting, similar stands may produce quality hay for beef cattle weight gain.

Keywords: native grass; warm-season; bluestem; digestibility; composition; crude protein; defoliation; forage quality; mixed stand; harvest interval

1. Introduction

In the southeastern USA, frequent summer forage shortages have aroused a growing interest in native warm-season grasses (NWSG) as alternative forage resources [1]. Favored NWSG species include; indiangrass (IG, *Sorghastrum nutans* (L.). Nash), big bluestem (BB, *Andropogon gerardii* Vitman), little bluestem (LB, *Schizachyrium scoparium* (Michx). Nash). switchgrass (*Panicum virgatum* L.) and eastern gamagrass (*Tripsacum dactyloides* L.) [2]. Most NWSGs are considered valuable summer forage resournces for ruminants [3,4] since they can grow well at elevated temperatures to produce digestible biomass [4]. However, their nutritive value changes with defoliation management [5] and information on NWSGs in mixed stands is scarce.

Usually, forage nutritive value of a plant can be predicted based on leafiness, which varies positively with crude protein (CP) concentration and digestibility [6]. For grasses, leafiness usually declines towards plant maturity, due to factors like shuttering of senescent dry leaves and translocation of leaf carbohydrates to the crown and roots. Forage nutritive value assessment can also include such attributes as leaf-to-stem ratio (LSR), based on weight and species-specific leaf area (SLA), which correlates negatively with leaf-fiber content and positively with digestibility [6]. Other indicators of forage nutritive value are types and concentration of: neutral detergent fiber (NDF)—containing hemicellulose, cellulose and lignin—acid detergent fiber (ADF)—with its high levels of cellulose and lignin—and acid detergent lignin (ADL)—lignin—which is an indigestible herbage component [7]. Together, they affect total forage dry matter (DM) intake and digestibility [8,9]. For a percentage increase in lignin, digestible DM decreases by three to four units [9]; this is true for both native and introduced grasses [10]. Unfortunately, most warm-season grasses have short-lived active vegetative growth [11,12] and are early maturing. As grasses mature, they grow reproductive tillers with more proportions of structural carbohydrates and phenolic compounds, mainly cellulose and lignin [3,10], while CP decreases [12]. At maturity, LSR also decreases while the proportion of senescent leaves increases [9,12].

Although most NWSGs get more lignified and less digestible in late-summer, they still make quality forages when cut at early stages [11,13]. Even at the same phenological stage, recovering defoliated plants usually have more nutritious leaves than their undefoliated counterparts [14]. This happens as the recovering plants preferentially allocate more reserve carbohydrates to leaf growth ahead of roots and reproductive structures [14,15]. In mixed stands, however, forage nutritive value of NWSGs is greatly influenced by how the dominant species respond to management. There is lack of information on how yield and forage quality of mixed stands dominated by BB, IG, and LB may respond to haying regimes. Knowing how yield and forage quality of these NWSGs, in mixed stands, may respond to harvesting for consecutive years will be useful to producers making informed management decisions. However, the objectives of this component of the study, focused on the effects of harvest interval and duration on forage nutritive value of early-succession mixed-stands dominated by BB, IG, and LB in terms of chemical composition, *in vitro* digestibility, and species morphological components.

2. Results and Discussion

2.1. Chemical Composition of Forage Dry Matter

To determine harvest interval effects on forage nutritive value, mean concentrations of CP, NDF, ADF, and ADL in whole-plot forage samples were compared by harvest interval, harvest date, and number of years in production. The comparison only involved data from June–September regrowth harvests after the first equalizing May harvest. The May harvest data were excluded from the analyses since fields are not usually harvested for forage this early.

2.1.1. Crude Protein Concentration

Forage nutritive value assessment by CP concentration in DM was based on the recommended minimum of 70 $g \cdot kg^{-1}$ DM for beef cattle [16]. Harvest intervals, harvest year, and number of years in production did not interact to affect CP concentration, so means were pooled across harvest years and years in production for comparison (Table 1). For the second harvest after mid-May (the only one with five entries), CP concentrations were lower (42 $g \cdot kg^{-1}$ DM) for the 120-d harvest interval ($p < 0.01$) compared with the 30-d and 40-d (74 $g \cdot kg^{-1}$ DM). A similar trend was also observed for the third and fourth harvests. Except for the 60-d harvest interval, CP concentration values decreased across the season to the lowest in the fifth harvest of the 30-d (47 $g \cdot kg^{-1}$ DM) and fourth of the 40-d (52 $g \cdot kg^{-1}$ DM) harvest intervals.

Table 1. Effects of harvest interval and harvest date on forage nutritive value based on crude protein (CP) concentration in dry matter (DM) after the first harvest (mid-May) from mixed native grass stands [†] pooled across two harvest years and durations.

Frequency	Harvested regrowth				$p > F\alpha$ [§]
	2nd	3rd	4th	5th	
	$g \cdot kg^{-1}$ DM				
120 (2) [‡]	42 E				
90 (2)	44 DE				
60 (3)	54 C	52 CD			0.54
40 (4)	69a [#] AB	64a B	52b CD		<0.01
30 (5)	74a A	67ab AB	63b B	47c CDE	<0.01

[†] Stands of indiangrass (*Sorghastrum nutans*), big bluestem (*Andropogon gerardii*) and little bluestem (*Schizachyrium scoparium*). [‡] Days between successive harvests with number of harvests per season in brackets. [§] Probability of difference between means within a row. Means followed by different upper case letters differ significantly. [#] Means, within a row, followed by different lower case letters differ. Mean differences declared significant at $\alpha = 0.05$.

The CP concentration values were between reported averages of 97 $g \cdot kg^{-1}$ DM (leaf) and 43 $g \cdot kg^{-1}$ DM (stem) for big bluestem and switchgrass at early head emergence [3]. In part, these trends were probably influenced by the fact that CP in the current study was measured on mixed whole-plot samples, and that proportions of broadleaf material in subsequent harvests decreased. The observed lower CP concentrations when days to the second harvest were longer (Table 1), portrayed a common phenomenon associated with cell wall lignification towards maturity to ensure structural support to leaves and inflorescence [15]. This would generally increase cell wall proportions while decreasing cell contents in stems and leaves. For example, in five weeks, cell contents of bermudagrass can decrease from 65% to 40% and CP concentration from 120 to 80 $g \cdot kg^{-1}$ DM, as that of cell walls increases from 35% to 65%, respectively [4]. Similarly, leaf CP concentration of big bluestem declines for prolonged regrowth periods [5].

The observed decrease in late-season CP concentrations could also be due to changes in environmental factors, mainly temperature, solar radiation, soil moisture, and nutrient availability [12]. Even at the same age, forage nutritive value usually decreases as temperature increases [17]. Lower CP concentration in late-season harvests for grasses also results from temperature- and photoperiod-induced transition into the reproductive phase, which is usually richer in stem components [18]. For the 30-d harvest interval, this transition might have coincided with the fourth harvest, August 15 (Figure 1). Comparable declines in CP concentration values from 110 $g \cdot kg^{-1}$ DM (May) to only 20 $g \cdot kg^{-1}$ DM (beyond September), have also been reported [19]. Additionally, most harvesting events in the current study were followed by heavy rains, which usually leach soil nitrogen [20] and thus reduce CP concentration in the biomass.

2.1.2. Across-Season Crude Protein Fluctuations

To identify possible combinations of harvest intervals and dates for good quality hay, CP concentrations of independent harvests, pooled across harvest-years and number of years in production were compared (Table 1). The CP values for the first (74 $g \cdot kg^{-1}$ DM) and second (67 $g \cdot kg^{-1}$ DM) harvests of the 30-d and the first (69 $g \cdot kg^{-1}$ DM) of the 40-d interval were similar and ranked higher than the third of 30-d (63 $g \cdot kg^{-1}$ DM) and second of 40-d (64 $g \cdot kg^{-1}$ DM). However, only values in the higher rank met or closely approached the minimum of 70 $g \cdot kg^{-1}$ DM for quality hay [16]. Other values (<55 $g \cdot kg^{-1}$ DM) were far below this minimum. Similarly, the second harvest of the 40-d interval with 67 $g \cdot kg^{-1}$ DM CP (nearly 70) could also make good-quality hay. This ranking showed that both timing and days between successive harvests have implications on hay quality, which is compromised by a >10-d delay, for the 30-d intervals.

2.1.3. Forage Fiber Concentration

Nutritive value assessment also involved comparing the proportions of NDF, ADF, and ADL in the DM samples (Table 2) because they negatively affect intake and digestibility [21]. Results on fiber concentration in the whole-plot forage samples are discussed with reference to their respective desirable levels to support weight gains of beef cattle.

2.1.3.1. Neutral Detergent Fiber (NDF) Concentration

Though NDF is partially digestible by ruminants, the extent varies between plant species and maturity stage [22]. Usually, grass hay with 550 $g \cdot kg^{-1}$ DM to 600 $g \cdot kg^{-1}$ DM NDF concentration is of good quality, while those with 700 to 800 g·kg^{-1} DM are considered poor in quality [23]. This is so because NDF concentration is inversely related to voluntary DM intake [20]. The ADF fraction is of less digestible fiber and is negatively correlated with intake [24] which should not exceed 300 to 350 g·kg^{-1} DM for good-quality grasses [23]. Within the 30-d harvest interval, there were year differences in relation to NDF concentrations which could be attributed to changes in weather conditions (Figure 2) and a 5-d delay of the second harvest (22 June) in 2009 (Figure 1). In 2008, NDF values only differed between the 30-d and 90-d harvest intervals and between the 60-d harvests (Table 2). In 2009, however, NDF concentrations had lesser values for the second, compared to later harvests in the season ($p < 0.03$). For the second harvest of the 30-d, 40-d, and 60-d intervals, NDF values were similar and slightly above the upper limit of 600 g·kg^{-1} DM for high-quality forage, but far below 700 g·kg^{-1} DM to be considered poor-quality ones [23]. Even as the season advanced, NDF values remained about 670 g·kg^{-1} DM, which tends to be of poor-quality hay.

Table 2. Effects of harvest interval on forage quality, based on neutral detergent fiber (NDF), acid detergent fiber (ADF), and acid detergent lignin (ADL) concentrations in the dry matter for each harvest, after the first (mid-May) from mixed native grass stands † in 2008 and 2009 harvest-years pooled over two harvest durations ‡.

Interval (days)	June–September 2008					June–September 2009				
	2nd	3rd	4th	5th	$p > F\alpha$ §	2nd	3rd	4th	5th	$p > F\alpha$
	g·kg^{-1} DM					g·kg^{-1} DM				
NDF										
120 (2)	670 ABC #	-	-	-	-	710 A	-	-	-	-
90 (2)	650 C	-	-	-	-	660 B	-	-	-	-
60 (3)	680a AB	650b	-	-	<0.02	640b BC	680a	-	-	<0.01
40 (4)	660 BC	650	630	-	0.33	630b BC	660ab	670a	-	<0.02
30 (5)	690 A	670	670	670	0.08	620b C	670a	670a	660a	<0.03
ADF										
120 (2)	360	-	-	-	-	400 A	-	-	-	-
90 (2)	360	-	-	-	-	360 B	-	-	-	-
60 (3)	370	360	-	-	0.41	340b BC	370a	-	-	<0.01
40 (4)	350	350	350	-	0.90	330c C	350b	370a	-	<0.01
30 (5)	380	360	380	370	0.13	340b BC	370a	330b	380a	<0.01
ADL										
120 (2)	54 A	-	-	-	-	52 A	-	-	-	-
90 (2)	53 A	-	-	-	-	47 AB	-	-	-	-
60 (3)	51 AB	58	-	-	0.19	37 C	43	-	-	0.14
40 (4)	37b C	41b	55a	-	<0.011	43a BC	31b	46a	-	<0.01
30 (5)	42c BC	48bc	53b	67a	<0.01	46 AB	44	44	45	0.9

† Stands of indiangrass (*Sorghastrum nutans*), big bluestem (*Andropogon gerardii*) and little bluestem (*Schizachyrium scoparium*). ‡ Combined data from plots harvested for the first or second time in the year. § Probability of difference between means within a row, in the same year. Days between successive harvests with number of harvests per season in brackets. # Means followed by different letters; uppercase within column or lowercase within row differ significantly at $\alpha = 0.05$.

Figure 1. Actual harvest days by harvest interval over the experimental period.

Year	TRT	May	June				July	August				Sep	Oct-Dec
2007			1†	2	3	4		1	2	3	4		
	120 (2)‡	15§										15	
	90 (2)	15							15				
	60 (3)	15					15					15	
	40 (4)	15				25		5				15	
	30 (5)	15		14			15		15			15	
2008													
	120 (2)	13										23	
	90 (2)	13							18				
	60 (3)	13					16					23	
	40 (4)	13				23		4				23	
	30 (5)	13		16			16		18			23	
2009													
	120 (2)	21										30	
	90 (2)	21								24			
	60 (3)	21					21					30	
	40 (4)	21				30			10			30	
	30 (5)	21			22		21			24		30	
2010													
	All	25											

Greater NDF concentrations for longer harvest intervals are usually associated with growth of reproductive tillers and increased proportions of structural carbohydrates for physical support [8,10,12]. Greater NDF concentration in late-season harvests is usually a phenological response to hot temperatures and short day lengths, which induces faster development with reduced LSR [25]. Similar findings have been reported for switchgrass with NDF concentration at maturity, averaging 640 $g \cdot kg^{-1}$ DM in May, but up to 790 $g \cdot kg^{-1}$ DM in September and beyond [19]. Measured NDF concentrations could also be influenced by changes in types and proportions of annual forbs in the stands across the season due to defoliation and weather conditions. In fact, a notable increase in short-growing annual grasses and forbs was associated with the 30-d harvest interval (data not included) although the proportions of BB, IG, and LB in the stands were not altered.

2.1.3.2. Acid Detergent Fiber (ADF) Concentration

Differences in ADF concentration between harvest intervals or harvest dates within a season were not observed in 2008 (Table 2), but were in 2009. As with NDF, ADF concentrations in the second harvest after mid-May of 2009 were greater for longer intervals, which suggests increased lignification of the cell walls at maturity. These ADL values were consistent with their corresponding decrease in CP concentrations. However, within harvest interval, and across the season, ADF concentrations for the 30-d were lesser at the second and fourth harvests, averaging 335 $g \cdot kg^{-1}$ DM, but greater at the third and fifth (375 $g \cdot kg^{-1}$ DM). At the second and fourth harvests of the 30-d interval, ADF values were actually below the upper limit of 350 g·kg^{-1} DM for good-quality hay [23]. This pattern of changes in ADF concentrations also matched the unusual weather conditions, in 2009 (Figure 2a,b), with heavy rainfall in May and September and a prolonged hot June–August dry spell. It seems that elevated temperatures, in 2009, created drought conditions towards the third and fifth harvests, which usually result with reduced plant size and less lignified tissues [25].

Figure 2. (**a**) Temporal trends in monthly rainfall totals (mm); (**b**) Trends in monthly mean temperature (°C) during the study period, 2007 to 2009, Aberdeen, MS.

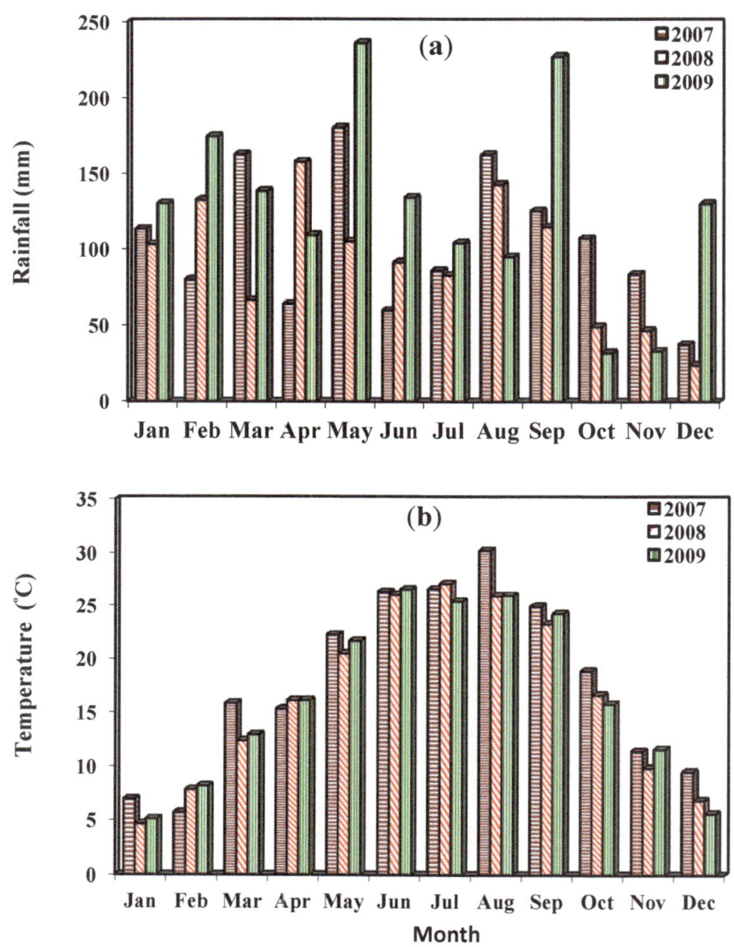

2.1.3.3. Acid Detergent Lignin (ADL) Concentration

Lignin, an anti-quality component deposited in plant cell walls at maturity [17], is thought to interfere with fiber digestion. Such interference may involve obstructing microbial enzymes [26], or forming chemical cross-linkages with fiber polysaccharides [12]. Therefore, forages with lower ADL concentrations are more desirable. In this study, ADL data showed effects due to harvest year, harvest intervals and dates across the season (Table 2). During both the 2008 and 2009 seasons, ADL concentrations for the second harvest, after mid-May, were about 40 g·kg^{-1} DM at shorter harvest intervals and increased to slightly above 50 g·kg^{-1} DM for the 120-d. However, year differences were also observed in ADL concentrations, across the season, with the greatest value (67 g·kg^{-1} DM) being for the fifth harvest in 2008, while in 2009 differences were only observed between the 40-d harvests.

The observed greater ADL concentrations at longer harvest intervals were characteristic of mature forage plants [17]. However, there were some discrepancies between values from this study and those in the literature. For example, while some values were above a reported pooled average of 47 $g \cdot kg^{-1}$ DM for big bluestem and switchgrass leaf tissues [3], most values were below this average. Still all values came below a reported 56 $g \cdot kg^{-1}$ DM for big bluestem [4]. Yet, all were mostly above the range of 34 to 43 $g \cdot kg^{-1}$ DM for mixed warm-season grasses [27]. The noted discrepancies could be explained by the fact that ADL values in the current study were of mixed stands, whose biomass was not sorted by species, or leaf and stem components. As in the case of NDF above, increased proportion of annuals in the biomass at the 30-d harvest interval, likely influenced changes in ADL concentrations in the succeeding harvest samples.

2.2. Forage in Vitro Digestibility

As another indicator of forage quality to ruminants, *in vitro* digestibility values of the DM (IVDMD) and NDF (NDFD in the harvested material were compared for effects of harvest interval and harvest dates across the season, in two consecutive years. The results (Table 3) are discussed in two subsections.

2.2.1. Harvest Interval and DM Digestibility

Usually, grasses with a minimum of 550 $g \cdot kg^{-1}$ IVDMD are considered good quality forages [19]. This is necessary information when estimating potential DMI by animals, which is usually influenced by how easily digestible the forage is. Significant year-effects on measured IVDMD were observed, so mean comparison results are presented by year (Table 3). Effects of harvest interval or harvest date on IVDMD were not observed in 2008 and values averaged 460 $g \cdot kg^{-1}$ DM. In 2009, IVDMD values for the second harvest were above 600 $g \cdot kg^{-1}$ DM for the 30-d and 40-d intervals and decreased for the longer harvest-intervals to 450 $g \cdot kg^{-1}$ DM in the 120-d.

The observed lower digestibility values for longer harvest intervals were likely a result of: (1) Most tillers being harvested at their transition into reproductive phase, usually characterized by jointed stems and lignified cell walls [25]; (2) Translocation of carbohydrates to the crown and roots, that left standing biomass richer in cell walls than cell contents; and (3) Shuttering of senescent leaves at harvest, which usually increases the proportion of stems in the biomass [8]. These results are similar to reported declines in IVDMD from 520 and 450 $g \cdot kg^{-1}$ DM, for big bluestem shoots, clipped at 30, 41, and 51 cm stages [5]. Similar declines in DM digestibility from 708 to 566 $g \cdot kg^{-1}$ DM between late vegetative and early heading stages have also been reported on mixed switchgrass and big bluestem hay [11]. On the other hand, greater digestibility values for shorter harvest intervals were also consistent with the observed decrease in fiber concentrations (NDF, ADF and ADL) discussed above. Plant biomass with less fiber is usually less lignified and therefore easily digested [13,14]. In the current study, differences in DM digestibility indicate greater availability of nutrients in the native grasses harvested at early rather than late-growth stages.

Table 3. Effect of harvest interval on *in vitro* dry matter (DM) and neutral detergent fiber (NDF) digestibility for each harvest after the first (mid-May) from mixed native grass stands † in 2008 and 2009 harvest-years pooled over two harvest durations ‡.

Interval (days)	June–September 2008					June–September 2009				
	Harvest Timing					Harvest Timing				
	2nd	3rd	4th	5th	$p > F\alpha$ §	2nd	3rd	4th	5th	$p > F\alpha$
	g·kg^{-1} DM					g·kg^{-1} DM				
DM										
120 (2)	420	-	-	-	-	450 D #	-	-	-	-
90 (2)	470	-	-	-	-	530 C	-	-	-	-
60 (3)	460	480	-	-	0.48	580a †† BC	490b	-	-	<0.01
40 (4)	490	510	470	-	0.26	650a A	570b	480c	-	<0.01
30 (5)	440	440	430	430	0.87	630a AB	540b	640a	500b	<0.01
NDF										
120 (2)	560 C	-	-	-	-	550 D	-	-	-	-
90 (2)	600 BC	-	-	-	-	650 C	-	-	-	-
60 (3)	620 AB	620	-	-	0.93	680a BC	600b	-	-	<0.01
40 (4)	670a A	650ab	610c	-	0.02	740a A	670b	590c	-	<0.01
30 (5)	620 AB	620	590	570	0.11	720a AB	650b	740a	600b	<0.01

† Stands of indiangrass (*Sorghastrum nutans*), big bluestem (*Andropogon gerardii*) and little bluestem (*Schizachyrium scoparium*). ‡ Combined data from plots harvested for the first or second time in the year. § Probability that, in the same year, means within a row differed significantly. Days between successive harvests, with number of harvests per season in brackets. # Means followed by different letters; uppercase within column or lowercase within row, are different. Mean differences declared significant at $\alpha = 0.05$.

2.2.2. Harvest Timing and DM Digestibility

To assess the influence of weather-induced phenological changes on forage quality indicators, IVDMD values were compared between harvest dates, within harvest intervals (Table 3). For the 30-d harvest interval, IVDMD values of the second and fourth harvests were similar (>600 g·kg^{-1} DM) and greater ($p < 0.01$) than those of the third and fifth harvests by about 100 units (Table 4). Within the 40-d harvest interval, IVDMD was greater for the second ones (650 g·kg^{-1} DM) than the third (570 g·kg^{-1} DM) and fourth (480 g·kg^{-1} DM) harvests ($p < 0.01$). These IVDMD values are similar to others in literature for leaves (604 g·kg^{-1}) and stems (500 g·kg^{-1}) of mixed big bluestem and switchgrass at early head emergence [3,28]. Dry matter digestibility of big bluestem hay has also been found to decline from 678 to 546 g·kg^{-1} between late July and early August [11]. Although all IVDMD values for the 30-, 40-, and 60-d harvest intervals were well above the minimum of 550 g·kg^{-1} for quality forage [16], issues of acceptability associated with NDF concentration would make the 60-d harvests undesirable.

Table 4. Effects of harvest interval and harvest duration on mean June–September leaf:stem ratio (LSR) and specific leaf area (SLA) of indiangrass (*Sorghastrum nutans*) and little bluestem (*Schizachyrium scoparium*) tillers from mixed native grass stands † at their first and second harvest year recorded at each harvest after the first (mid-May) in 2008 and 2009.

Interval (days)	LSR				SLA			
	2008		2009		2008		2009	
	Y108 ‡	Y207	Y109	Y208	Y108	Y207	Y109	Y208
					$cm^{-2}\cdot g^{-1}$			
				Indiangrass				
Control	-	-	0.4b §	0.4c	-	-	98d	98d
120 (2)	1.6	1.9	0.7b	0.8bc	109c	102b	101d	110d
90 (2)	1.8	1.3	0.8b B	1.4ab A	107c	100b	108d	110d
60 (3)	1.7	1.7	1.5a	1.4ab	112c	117b	227c	243c
40 (4)	2.1	2.1	1.4a B	1.8a A	137b	150a	403b	368b
30 (5)	1.8	2.0	1.3a B	1.9a A	152a	154a	567a A	529a B
$p > F_\alpha$ #	0.57	0.13	<0.01	<0.01	<0.01	<0.01	<0.01	<0.01
				Little bluestem				
Control	-	-	0.4b	0.4c	-	-	118d	118d
120 (2)	1.2	1.0	0.6b	0.7bc	132b	134c	119d	128d
90 (2)	1.3	2.1	0.8b B	1.2ab A	133b	136c	131d	156d
60 (3)	1.5	1.6	1.2a B	1.5a A	162a	156bc	284c	306c
40 (4)	1.7B	1.9A	1.3a	1.2ab	168a	166ab	475ab	470b
30 (5)	1.9	2.3	1.4a	1.5a	162a	194a	732a A	674a B
$p > F_\alpha$	0.14	0.21	<0.01	<0.01	<0.01	<0.01	<0.01	<0.01

† Stands of indiangrass (*Sorghastrum nutans*), big bluestem (*Andropogon gerardii*) and little bluestem (*Schizachyrium scoparium*). ‡ Y108, Y109, Y207 and Y208 are plots in their first and second harvesting year, established in 2008, 2009 and 2007, respectively. § Means of the same attribute followed by different letters; lowercase within column or uppercase within row differ. Days between successive harvests with number of harvests per season in brackets. # Probability that, means in respective columns differ significantly at $\alpha = 0.05$.

It is also likely that increased rainfall and warm temperatures in June of 2009 induced faster plant development, similar to earlier findings [25]. Such growing conditions often result with reduced LSR and increased forage fiber, characteristic of less digestible biomass. The fact that weather conditions in June were generally reproduced in August may also explain the observed patterns of forage IVDMD values. It appears that forage nutritive value in 2009, for the third harvest at the 30-d interval, was more influenced by the prolonged June–August dry spell (Figure 1). Usually, drought stresses, when not severe enough to kill plants, tend to increase their forage digestibility [20]. This is so because affected plants grow less vigorously with less lignified cell walls due to reduced demand for mechanical support [25]. However, most IVDMD values at the 30- and 40-d harvest intervals (Table 4) were above or very close to 550 $g\cdot kg^{-1}$ that is recommended for good quality forages [16,20]. Therefore, management interventions intended to boost biomass production under similar conditions should target the early-season harvests at 30- to 40-d intervals.

2.2.3. Harvest Interval on NDF Digestibility

Neutral detergent fiber digestibility values were in patterns similar to those of IVDMD (Table 3) indicating that forage quality was better for the early-season biomass at shorter harvest intervals. In 2008, the second harvest at the 40-d interval had greater NDFD (670 $g \cdot kg^{-1}$ DM) than the 90-d (600 $g \cdot kg^{-1}$ DM) and 120-d (560 $g \cdot kg^{-1}$ DM) harvests, but not different from the 30-d and 60-d (620 $g \cdot kg^{-1}$ DM). In 2009, NDFD for the first regrowth of the 40-d harvest interval (740 $g \cdot kg^{-1}$ DM) was greater than for the 60-d (680 $g \cdot kg^{-1}$ DM), 90-d (650 $g \cdot kg^{-1}$ DM) and 120-d (550 $g \cdot kg^{-1}$ DM), but not different from the 30-d harvests (720 $g \cdot kg^{-1}$ DM). The observed lower NDFD at longer harvest intervals was similar to reported decline in grass NDFD of up to 400 $g \cdot kg^{-1}$ DM due to maturity alone [29]. This suggests that fiber lignin concentrations increased as more plants in the mixed stands matured. Such declines in forage nutritive value are usually associated with developmental changes. These changes usually involve development of xylem vessels for water transport, accumulation of cellulose and complex carbohydrates, all bound together by lignin deposition [26]. Such changes would make plant cell walls less digestible.

2.2.4. Harvest Timing on NDF Digestibility

Means, for both 2008 and 2009 data, were also compared between harvest dates, within harvest intervals (Table 3). In 2008, effects of harvest date on NDFD were only significant for the 40-d harvest interval, being greater ($p < 0.03$) at the second harvest (670 $g \cdot kg^{-1}$ DM) than the fourth (610 $g \cdot kg^{-1}$ DM), but not the third (650 $g \cdot kg^{-1}$ DM). As was for IVDMD in 2009, NDFD within the 30-d harvest interval was not different between the second and fourth harvests which averaged above 700 $g \cdot kg^{-1}$ DM. These were also greater ($p < 0.01$) than 625 $g \cdot kg^{-1}$ DM for the third and fifth harvests. Within the 40-d harvest interval, NDFD was 740 $g \cdot kg^{-1}$ DM for the second harvest and greater than the third (670 $g \cdot kg^{-1}$ DM) and fourth (590 $g \cdot kg^{-1}$ DM). With mixed grasses, NDFD values have been found to decline from 800 $g \cdot kg^{-1}$ DM for early-May harvest to 440 $g \cdot kg^{-1}$ DM by late-June while lignin concentrations increased from 17 to 53 $g \cdot kg^{-1}$ DM, respectively [7]. However, most NDFD values for the 30- and 40-d harvest intervals were between 650 and 540 $g \cdot kg^{-1}$ DM, recommended for good- to medium-quality grass hay, respectively [30]. This implies that increased NDF concentration in the late-season harvests might not severely limit DMI, owing to their greater digestibility values. Decline in NDFD is usually a result of amounts and types of lignin deposited, which differ between species [26]. Additionally, forbs in the current study appeared to account for a substantial proportion of the regrowth biomass, which possibly improved the observed average digestibility.

2.3. Species Tiller Leafiness

How harvest intervals or number of consecutive harvest years might affect species forage quality was also assessed on two morphological components associated with leafiness. This was based on the fact that leafiness of plant material, expressed as LSR or SLA, is a common indicator of forage nutritive value [6]. For both IG and LB, in 2008 and 2009 harvest years, changes in LSR and SLA due to harvest dates, within harvest interval, were rare and inconsistent (Table 4). Effects of harvest interval and harvest duration were, therefore, assessed on their respective June–September averages.

2.3.1. Tiller Leaf-to-Stem Ratio

In 2008, IG showed no effect of harvest interval or harvest duration (number of years in production) on LSR, which averaged 1.8 (Table 4). Similarly, LSR for LB was not affected by harvest interval and values ranged from 1.5 to 1.8 for plants in their first and second year of production, respectively. Only the 40-d harvest interval had greater LSR (1.9) for the second-year plants. In 2009, means of LSR, within harvest interval, differed ($p < 0.01$) between plants in first (Y109) and second (Y208) year of production. For Y109 plots, harvesting IG at <90-d intervals resulted with greater LSR (1.5) for the 60-d harvest interval, almost 4-fold the 0.4 at first harvest (control). Still, IG harvested at <120-d intervals had greater LSR in the second- (Y208) rather than in first-year (Y109) plots. A similar trend was observed for LB whose LSR values for the control plants (0.4) were lower ($p < 0.01$) by more than 50% compared to plants harvested at intervals ≤90-d, which ranged from 0.8 to 1.5 (Table 4).

The absence of harvest interval effect on tiller LSR in 2008 was likely due to adequate rainfall distribution (Figure 1a) that possibly allowed compensatory growth to override the negative effects of defoliation. Having greater LSR ($p < 0.01$) for shorter harvest intervals in 2009 suggests that respective harvest events mostly coincided with vegetative stage of the regrowth, usually characterized by faster growth of leaf blades than stems and leaf sheaths [15,21]. The decreased LSR values observed for the 90-d and 120-d harvest intervals were also characteristic of assimilate translocation to the crown and shuttering of senescent dry leaves [21]. These LSR values also reflected late-season changes in temperature and photoperiod, which usually prompt grasses to transition into reproductive phase with reduced LSR [18]. Hence grasses at the reproductive phase would make poor hay even at short harvest intervals. Greater LSR values for the second- rather than first-year plants were expected because defoliation often results with thinner tillers, mostly vegetative, and with more leaves than stems [31].

2.3.2. Tiller Specific Leaf Area

The SLA of IG in the first and second year of production was affected ($p < 0.01$) by harvest interval during both 2008 and 2009 (Table 4). In 2008, SLA ranged from <110 $cm^{-2}\cdot g^{-1}$ for plants harvested at intervals ≥90-d to >135 $cm^{-2}\cdot g^{-1}$ for harvesting at <60-d intervals. While the control plants and those harvested at a >60-d interval in 2009 had SLA (104 $cm^{-2}\cdot g^{-1}$) comparable to that of 2008; SLA for plants harvested at a 30-d interval reached over 500 $cm^{-2}\cdot g^{-1}$ (Table 1). These SLA values were greater for plants in their first rather than second year of production by over 35 units. In similar trends, SLA of LB in Y108 and Y207 plots was greater ($p < 0.01$) for plants harvested at shorter intervals (Table 1). Within harvest intervals, SLA across years of production, in 2008, ranged from 133 $cm^{-2}\cdot g^{-1}$ (120-d) to 176 $cm^{-2}\cdot g^{-1}$ (30-d) with no difference between Y108 and Y207 plots (Table 4).

In 2009, SLA of LB across years of production for the control plants and those harvested at a >60-d interval averaged 128 $cm^{-2}\cdot g^{-1}$. However, values for plants harvested at shorter intervals were greater than 250 $cm^{-2}\cdot g^{-1}$ and reached well over 700 $cm^{-2}\cdot g^{-1}$ (Y109) and 650 $cm^{-2}\cdot g^{-1}$ (Y208) for the 30-d interval (Table 4). The observed greater SLA values for shorter harvest intervals are characteristic of vegetative growth phase of grasses, at which leaf blade elongation and expansion rates exceed that of stems [21]. At longer harvest intervals, plants are more likely to transition into the reproductive phase during which growth is more of stem elongation and cell wall lignification [15,21] consistent with observed increase in fiber content. These observed changes in tiller LSR and SLA values indicated that the native grasses produced better quality forage when harvested at shorter rather than longer intervals. These results on species morphological components, therefore, further stress the importance of timely harvesting for similar mixed stands dominated by IG and LB to produce quality hay.

3. Experimental Section

3.1. Study Location and Field Layout

This study was conducted at Bryan Farms, Clay County, (33°39'N; 88°34'W) MI, USA, in unfertilized conservation field buffers planted with mixed NWSGs, at their early-succession stages. Dominant soils in the study area are Griffith silty clay, classified as Fine, smectitic, thermic Aquic Hapludert with pH ranging from 5.0 to 5.6 and Okolona silty clay, classified as Fine, smectitic, thermic Oxyaquic Hapludert with pH range of 6.0 to 7.8. A seed mixture of 1.12 kg BB, 2.24 kg LB, and 1.12 kg IG per hectare of prepared seedbed was sown in 2005, and allowed to grow undisturbed for two years. Extended post-emergence herbicide (imazapic at 0.28 kg a.i ha^{-1}) {(±)-2-[4,5-dihydro-4-methyl-4-(1-methylethyl)-5-oxo-1H-imidazol-2-yl]-5-methyl-3-pyridinecarboxylic acid} was applied to control competitive weeds. In late spring of 2007, five 7.5 × 1-m parallel strips, at least 3 m apart were randomly assigned to five, four, and three harvests at 30-, 40-, and 60-d intervals, respectively, or only two harvests at 90- or 120-d interval (Figure 3), giving five harvest intervals per block. The 90-d interval mimicked a standard practice of harvesting a hay crop early in the growing season, and then stockpiling the regrowth for late-season grazing or conservation uses. In a randomized complete block design, these five harvest intervals were replicated in five blocks, three in two buffers of one crop field and two in another field, about 5 km away, on similar soils.

During the spring of 2008, other 7.5 × 1-m plots were marked next to each previous-year plot with 90-cm alleys between the first- and second-year plots for each harvest interval. Plots harvested first in 2007 were designated Y207, indicating they were in their second harvest year (Y2), but started in 2007 (07). Plots harvested first in 2008 adjacent to Y207 plots were designated Y108, indicating they were in their first harvest year (Y1), but started in 2008 (08). In 2009, a third set of five 7.5 × 1-m, plots separated by 90-cm alleys were marked on one end of each block; a total of three plots per harvest interval per block. Adding the third set of plots on the respective block ends was necessary to avoid possible negative effects of the two-year feet and machine traffic on plant growth. For each block, however, an area with relatively uniform species composition, terrain, and plant vigor, large enough to accommodate all three sets of plots, was clearly defined in the first harvest-year. With this arrangement, there were no notable differences in plant performance between third year plots and the rest, within a harvest interval. Plots started in 2009, were designated Y109 while the Y108 plots re-designated Y208 and the Y207 became Y307 (Figure 3). In spring of 2009, the Y307 plots were harvested only once, in May, to assess post-season recovery and then removed from the harvest regime. To avoid shedding, plants in the separating alleys next to harvested plots were also trimmed to the same height, using a hand-held weed eater on each harvested day.

Figure 3. Plot arrangement, in one replication, showing establishment sequence. Five first-year plots (Y1) established in mid-May from 2007 to 2009, each 7.5 m long and 1 m wide, with marked and monitored indiangrass and little bluestem plants assigned to 30-, 40-, 60-, 90-, and 120-d harvest intervals. In each year, plots are labeled Y1, Y2, or Y3 indicating plots beginning their first, second and third harvest year, respectively.

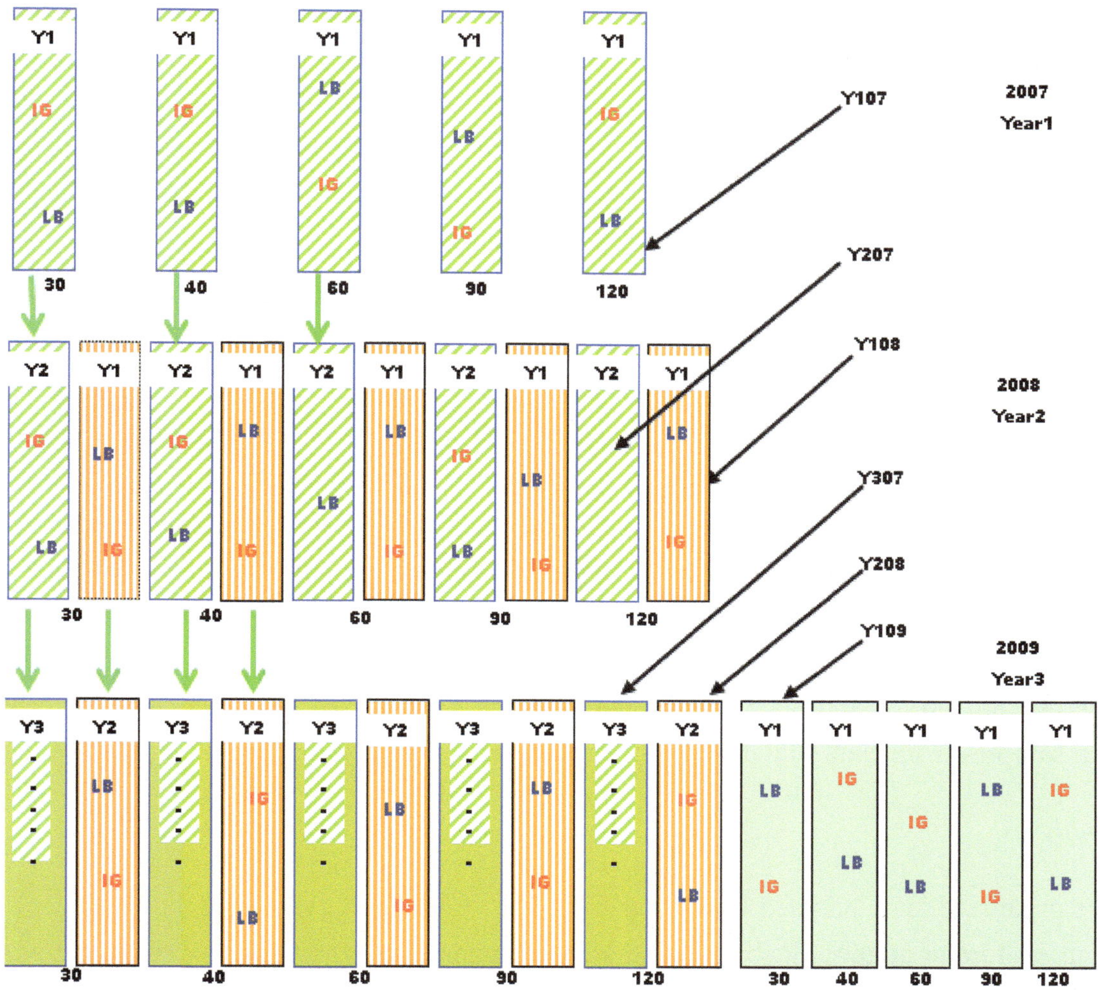

3.2. Harvesting and Forage Sampling

In mid-May of each year, all study plots received a common/equalizing harvest, after which regrowth was harvested on assigned dates throughout the summer (Figure 2). Occasionally, harvesting was hastened by one to two, or delayed for up to six days (Figure 2) to avoid major rainfall events, thus allowing optimum machine operation. Whole-plot forage was harvested by a 1.0 m wide Carter Flail Forage Harvester (Carter Manufacturing Company, Inc., Brookston, IN, USA). At harvest, fresh whole-plot forage sample was collected from each plot and later dried in a forced-air oven at 65 °C to constant weight and processed for analyses, described below.

3.3. Species Morphological Assessment

Assessment of species leafiness during the growing season was based on measurements taken on tiller leaf and stem components. A day before each plot-harvest event, three tillers of IG and LB were clipped at ground level and later separated into leaves and stems by cutting through the collar, leaving leaf sheaths as stem components. Each block had a reference plot, not-harvested, from which sample tillers were collected and used as control in assessing species morphological response to the harvest intervals. In the interest of time, BB was excluded in the species assessment. Leaf blades of each tiller were run through a portable area meter (LI-COR, Model No LI-3000 LI-COR Biosciences, Lincoln, NE, USA) to determine total leaf area (LA, cm^{-2}). The stem and leaf sections were then oven-dried separately at 60 °C to constant weights, cooled in desiccators and weighed on a microbalance AG 104 (Mettler-Toledo Inc., Columbus, OH, USA), to determine dry weights of tiller leaf (LW) and stem (SW) components. Tiller leaf-to-stem ratio (LSR) was calculated as LW/SW and specific leaf area (SLA, $cm^2 \cdot g^{-1}$) as LA/LW. For each year, LSR values, after the equalizing (mid-May) harvest, were averaged within harvest interval across harvest dates.

3.4. Forage Nutritive Value Assessment

Dried whole-plot samples were ground to pass a 1-mm sieve (Willey mill, Standard model 3, Arthur H. Thomas Co., Philadelphia, PA, USA) and stored in plastic sample bags until analyzed for their chemical composition and digestibility. Samples were analyzed for crude protein (CP) by block digestion method [32]. Concentrations of neutral detergent fiber (NDF), acid detergent fiber (ADF) and acid detergent lignin (ADL) in the DM as well as DM and NDF digestibility were determined according to ANKOM Technology method 3 [33]. Ash content was determined by combustion in a muffle furnace at 550 °C for four hours [33]. Percent organic matter content (DM basis) was calculated by subtracting ash content (DM basis) from 100. During each year, values for each harvesting event were averaged within harvest interval and recorded as mean sample DM, CP, NDF, *etc.*, concentration. Data for grab samples from each May harvest were handled separately.

3.5. Data Analyses

Data were organized and analyzed for effects of harvest interval, harvest year, and number of years in production, on forage quality attributes. The latter compared yields of first- and second-year plots assigned to a harvest interval, within a harvest year. Data were subjected to analysis of variance (ANOVA) in a randomized complete block design with harvest intervals, species, year, and harvest duration as fixed effects in five replications, using the general linear model of SAS Institute [34]. Means separation was by Fisher's protected least significant difference (LSD) and were declared different at $\alpha = 0.05$.

4. Conclusions

In this study, IG and LB had greater SLA and LSR for early-season tillers in the mixed stands harvested 30 and 40 days after the equalizing mid-May harvest. This showed that early-season regrowths of IG and LB under comparable conditions will take more than 30 days to begin transitioning into the reproductive phase and that after 40 days biomass increase may be more of stem than leaf elongation. Of the two, LB seemed to take longer to transition into the reproductive phase, a desirable forage quality feature. Crude protein concentrations of whole-plot forage for early-season (the second- and third) harvests at the 30-d and 40-d intervals were about the acceptable minimum of 70 $g \cdot kg^{-1}$ DM for good quality forage. Crude protein concentrations in whole-plot forage samples were not affected by harvest year, or number of years in production, thereby implying that forage quality of the studied native grasses, in mixed stands, may not be compromised by previous intensity of defoliation. The fact that nutritive values for the same harvest intervals change significantly after the July–August dry spell shows that drought stresses can override the forage-quality significance of harvest intervals.

There were year differences in the measured forage fiber attributes making effects of harvest intervals and date practically undetectable in the year with favorable rainfall distribution. Although measured fiber concentrations were generally indicative of good to medium quality forage—NDF, for example, ranged 600–700 $g \cdot kg^{-1}$ DM—their crude protein and digestibility values were limiting. This suggests that actual chemical composition of the fiber components were more responsible for observed disagreements between the measured parameters. *In vitro* DM and NDF digestibility was greater for shorter harvest intervals and fluctuated across the harvest season with greater values for the second harvest, early-season, and the regrowth coinciding with the dry spell.

Overall, data shows that mixed stands of BB, IG, and LB can produce quality hay when harvested early in the season at 30-d intervals before the summer dry spell and that a 10-d harvest delay may not compromise forage quality. Data also suggests that forage quality of NWSGs in mixed stands may be manipulated by in-season management since it was not affected by defoliation history. More studies on the effects of harvest interval and duration on individual species' forage nutritive value aspects are needed. Similarly, feeding studies with ruminant animals to assess effects of harvest interval and duration on forage acceptability are also needed. The effects of other management practices, including spring-burning and fertilizer application, also merit further studies.

Acknowledgments

We are grateful to USDA NRCS Agriculture Wildlife Conservation Center for funding the research, B. Bryan Farms for allowing us to do the research on their property, the Department of Plant and Soil Sciences, Mississippi State University, and the Agricultural Research Station at Virginia State University, for various logistic, technical and moral supports provided during the research and while writing the manuscript. We also appreciate the lab assistance provided by Cathy Aultman.

Author Contributions

The second and third listed authors contributed to the experimental design and data analyses.

Conflicts of Interest

The authors declare no conflict of interest.

References

1. Shea, A.B. The Return of Native Grasses to Tennessee. 1999. Available online: http://www.tn.gov/environment/conservationist/archive/grass.htm (accessed on 12 August 2009).
2. Jones, J.; Coggin, D.S.; Cummins, J.L.; Hill, J. *Restoring and Managing Native Prairies: A Handbook for Mississippi Landowners*; Wildlife Mississippi: Starkville, MS, USA, 2007.

3. Griffin, J.L.; Jung, G.A. Leaf and stem forage quality of big bluestem and switchgrass. *Agron. J.* **1983**, *75*, 723–726.
4. Mitchell, R.B.; Vogel, K.P.; Klopfenstein, T.J.; Anderson, B.E.; Masters, R.A. Grazing evaluation of big bluestems bred for improved forage yield and digestibility. *Crop Sci.* **2005**, *45*, 2288–2292.
5. Forwood, J.R.; Magai, M.M. Clipping frequency and intensity effects on big bluestem yield, quality, and persistence. *J. Range Manag.* **1992**, *45*, 554–559.
6. Al Haj Khaled, R.; Duru, M.; Decruyenaere, V.; Journey, C.; Cruz, P. Using leaf traits to rank native grasses according to their nutritive value. *Rangel. Ecol. Manag.* **2006**, *59*, 648–654.
7. Cherney, D.J.R.; Cherney, J.H.; Lucey, R.F. *In vitro* digestion kinetics and quality of perennial grasses as influenced by forage maturity. *J. Dairy Sci.* **1993**, *76*, 790–797.
8. Twidwell, E.K.; Johnson, K.D.; Cherney, J.H.; Volenec, J.J. Forage quality and digestion kinetics of switchgrass herbage and morphological components. *Crop Sci.* **1988**, *28*, 778–782.
9. Linn, J.G.; Martin, N.P. *Forage Quality Test and Interpretations*; AG-FO-2637; Minnesota Extension Service, University of Minnesota: St. Paul, MN, USA, 1999.
10. Wilson, J.R.; Hatfield, R.D. Structural and chemical changes of cell wall types during stem development: Consequences for fibre degradation by rumen microflora. *Aust. J. Agric. Res.* **1997**, *48*, 165–180.
11. Vona, L.C.; Jung, G.A.; Reid, R.L.; Sharp, W.C. Nutritive value of warm-season grass hays for beef cattle and sheep; digestibility, intake and mineral utilization. *J. Anim. Sci.* **1984**, *59*, 1582–1593.
12. Buxton, D.R.; Redfearn, D.D. Plant limitations to fiber digestion and utilization. *J. Nutr.* **1997**, *127*, 814S–818S.
13. Difante, G.S.; Nascimento Júnior, D.; Silva, S.C.; Euclides, V.P.B.; Zanine, A.M.; Adese, B. Tillering dynamics of marandu palisadegrass submitted to two cutting heights and three cutting intervals. *Rev. Bras. Zootec.* **2008**, *37*, 189–196.
14. Manske, L.L.; Ske, L.L. *General Description of Grass Growth and Development and Defoliation Resistance Mechanisms*; Range Management Report DREC 98–1022, USA; NDSU Dickinson Research Extension Center: Dickinson, ND, USA, 1998.
15. Hannaway, D.B.; Hannaway, K.J.; Sohn, P.; Griffith, S.; Wycoff, H. Grass Growth and Regrowth for Improved Management. 2000. Available online: http://www.fsl.orst.edu/forages/projects/regrowth/main.cfm?PageID=7 (accessed on 30 August 2008).
16. National Research Council. *Nutrient Requirement of Beef Cattle*, 7th revised ed.; NRC, National Academy Press: Washington, DC, USA, 1996.
17. Nordheim-Viken, H.; Jørgensen, M.; Volden, H. Effects of maturity stage, temperature and photoperiod on growth and nutritive value of timothy (*Phleum pratense* L.). *Anim. Feed Sci. Technol.* **2009**, *152*, 204–218.
18. Beaty, E.R.; Engel, J.L. Forage quality measurements and forage research: A review, critique and interpretation. *J. Range Manag.* **1980**, *33*, 49–54.
19. Sanderson, M.A.; Voigt, P.; Jones, R.M. Yield and quality of warm-season grasses in central Texas. *J. Range Manag.* **1999**, *52*, 145–150.
20. Ball, D.M.; Hoveland, C.S.; Lacefield, G.D. *Southern Forages*, 1st ed.; Potash & Phosphate Inst. and the Foundation for Agronomic Research: Atlanta, GA, USA, 1991; p. 322.

21. Briske, D.D. Developmental morphology and physiology of grasses. In *Grazing Management: An Ecological Perspective*; Heitschmidt, R.K., Sahlu, T., Eds.; Timber Press: Portland, OR, USA, 1991; pp. 85–108.
22. Van Soest, P.J.; Robertson, J.B.; Lewis, B.A. Methods for dietary fiber, neutral detergent fiber, and nonstarch polysaccharides in relation to animal nutrition. *J. Dairy Sci.* **1991**, *74*, 3583–3597.
23. Weiss, W.P.; Eastridge, M.L.; Underwood, J.F. Forages for Dairy Cattle. 2010. Available online: http://ohioline.osu.edu/as-fact/0002.html (accessed on 20 January 2011).
24. Reid, R.L.; Jung, G.A.; Thayne, W.V. Relationships between nutritive quality and fiber components of cool season and warm season forages: A retrospective study. *J. Anim. Sci.* **1988**, *66*, 1275–1291.
25. Buxton, D.R. Quality-related characteristics of forages as influenced by plant environment and agronomic factors. *Anim. Feed Sci. Technol.* **1996**, *59*, 37–49.
26. Moore, K.J.; Jung, H.J.G. Lignin and fiber digestion. *J. Range Manag.* **2001**, *54*, 420–430.
27. Mulkey, V.R.; Owens, V.N.; Lee, D.K. Management of warm-season grass mixtures for biomass production in South Dakota USA. *Bioresour. Technol.* **2008**, *99*, 609–617.
28. Hendrickson, J.R.; Moser, L.E.; Moore, K.J.; Waller, S.S. Leaf nutritive value related to tiller development in warm-season grasses. *J. Range Manag.* **1997**, *50*, 116–122.
29. Hoffman, P.C.; Lundberg, K.M.; Bauman, L.M.; Shaver, R.D. The effect of maturity on NDF digestibility. *Focus Forage* **2003**, *5*, No. 15.
30. Hoffman, P.C.; Lundberg, K.M.; Barman, L.M.; Shaver, R.D. NDF digestibility: Reference values for forages, byproducts and total mixed rations. *Focus Forage* **2006**, 5, No. 17.
31. Sbrissia, A.F.; Silva, S.C.; Matthew, C.; Carvalho, C.A.B.; Carnevalli, R.A.; Pinto, L.F.M.; Fagundes, J.L.; Pedreira, C.G.S. Tiller size/density compensation in grazed Tifton 85 bermudagrass swards. *Pesqui. Agropecu. Bras.* **2003**, *38*, 1459–1468.
32. Association of Official Analytical Chemists. *Official Methods of Analysis*; Association of Official Analytical Chemist: Washington, DC, USA, 2001.
33. ANKOM Technologies Method 3. *In Vitro* True Digestibility Using the Daisy^{II} Inclubator. ANKOM Technologies -08/05. 2005. Available online: http://www.ankom.com/media/documents/IVDMD_0805_D200.pdf (accessed on 13 September 2010).
34. SAS Institute. *SAS Version 9.2*; SAS Institute: Cary, NC, USA, 2007.

12

A Substantial Fraction of Barley (*Hordeum vulgare* L.) *Low Phytic Acid* Mutations Have Little or No Effect on Yield across Diverse Production Environments

Victor Raboy [1,†,*], Kevin Peterson [1,†,‡], Chad Jackson [2,†], Juliet M. Marshall [2], Gongshe Hu [1,†], Hirofumi Saneoka [3,†] and Phil Bregitzer [1]

[1] USDA-ARS Small Grains and Potato Research Unit, 1600 South 2700 West, Aberdeen, ID 83210, USA; E-Mails: Peterson.kevin1@mayo.edu (K.P.); Gongshe.hu@ars.usda.gov (G.H.); phil.bregitzer@ars.usda.gov (P.B.)

[2] Idaho Falls Research and Extension Center, University of Idaho, 1776 Science Center Drive, Suite 205, Idaho Falls, ID 83402, USA; E-Mails: chadj@uidaho.edu (C.J.); jmarshall@uidaho.edu (J.M.M.)

[3] Graduate School of Biosphere Science, Hiroshima University, 1-4-4 Kagamiyama, Higashi-Hiroshima 739-8528, Japan; E-Mail: saneoka@hiroshima-u.ac.jp

† These authors contributed equally to this work.

‡ Current Address: Mayo Clinic, 200 First St. S.W., Rochester, MN 55905, USA.

* Author to whom correspondence should be addressed; E-Mail: victor.raboy@ars.usda.gov

Academic Editors: Francesca Sparvoli, Roberto Pilu and Eleonora Cominelli

---

Abstract: The potential benefits of the *low phytic acid* (*lpa*) seed trait for human and animal nutrition, and for phosphorus management in non-ruminant animal production, are well documented. However, in many cases the *lpa* trait is associated with impaired seed or plant performance, resulting in reduced yield. This has given rise to the perception that the *lpa* trait is tightly correlated with reduced yield in diverse crop species. Here we report a powerful test of this correlation. We measured grain yield in lines homozygous for each of six barley (*Hordeum vulgare* L.) *lpa* mutations that greatly differ in their seed phytic acid levels. Performance comparisons were between sibling wild-type and mutant lines obtained following backcrossing, and across two years in five Idaho (USA) locations that greatly differ in crop yield potential. We found that one *lpa* mutation (*Hvlpa*1-1) had no detectable effect on yield and a second (*Hvlpa*4-1) resulted in yield losses of only 3.5%, across all locations. When comparing yields in three relatively non-stressful production environments, at least three *lpa* mutations (*Hvlpa*1-1, *Hvlpa*3-1, and *Hvlpa*4-1) typically had yields similar to or within 5% of the wild-type sibling isoline. Therefore in the case of barley, *lpa* mutations can be readily identified that when simply incorporated into a cultivar result in adequately performing lines, even with no additional breeding for performance within the *lpa* line. In conclusion, while some barley *lpa* mutations do impact field performance, a substantial fraction appears to have little or no effect on yield.

Keywords: barley; *Hordeum vulgare*; low phytic acid; grain yield

---

1. Introduction

The potential benefits of the *low phytic acid* (*lpa*) crop seed phenotype, in the contexts of human and animal nutrition and in the context of management of phosphorus (P) in animal agricultural production, are well documented [1]. However, there is a commonly encountered view that this seed trait is conditionally associated with reduced seed performance and reduced seed yield. This view often reflects sound science since in many cases the *lpa* trait is associated with impaired seed or plant performance [2]. However, many studies reporting on field performance of *lpa* genotypes present relatively limited data using lines obtained following limited or no breeding, which are often evaluated in a growth chamber or greenhouse bench, or one or a few field locations or replications over years. This limited experimental approach can also lead to the opposite but equally problematic and often not well-supported conclusion that a given *lpa* mutation has little or no impact on yield [3]. Further, comparisons are often made between lines that greatly differ in genetic background other than the *lpa* loci. Even comparisons made between a wild-type (for seed phytic acid) cultivar and an *lpa* mutation isolated following mutagenesis of that cultivar, or following most transgenic approaches [4], are less than optimal. Many genetic differences that can impact field performance, other than the *lpa* alleles, may exist between these types of lines. The strongest test of the relationship between the *lpa* trait and any other trait, whether it be yield or nutritional quality, is comparisons between sibling near-isogenic lines, obtained following sufficient backcrossing, and following selection for field performance in fixed genotypes that are either homozygous wild-type or homozygous mutant. Such sibling near-isogenic lines are genetically similar at most loci other than the *lpa* loci, so any differences observed in field performance or nutritional quality can validly be attributed to the different *lpa* locus alleles.

Here we report the results of what we believe is, to date, the most robust field evaluation of *lpa* variants of any crop. Six independently isolated barley (*Hordeum vulgare* L.) *lpa* mutations, representing lesions mapped to four different chromosomes therefore representing a minimum of four loci, were backcrossed to the cultivar (*cv.* Harrington) in which they were initially induced and isolated. For each mutation a set of near-isogenic sibling lines were then established, each set consisting of one mutant and one wild-type line. Field trials were then conducted over two years at five locations with widely divergent production conditions, providing a robust test of performance across environments.

2. Results

2.1. Barley Low Phytic Acid Mutations Selected for Isoline Development

A screen of sodium azide-induced mutations in barley for the "high seed inorganic P" phenotype associated with *lpa* mutants had identified 23 putative mutants, of which 20 were selected for initial seed P phenotype evaluation (Supplementary Methods and Supplementary Table 1), and initial visual evaluation of plant performance. These 20 were grown in a nursery in 2004 in Aberdeen, Idaho, and following grain harvest, assayed for whole grain P fractions. Of these all but one displayed the low grain phytic acid trait (Supplementary Table 1). The one exception was *Hv*-MAZ423, which appears to be a high grain total P mutant in which all grain P fractions, including inorganic P, are elevated, as compared with wild-type. This putative high grain total P mutant is the subject of separate, ongoing study. Of the remaining 19 mutants, a total of 17 have been chromosomally mapped to date (Figure 1; Supplementary Tables 2 and 3).

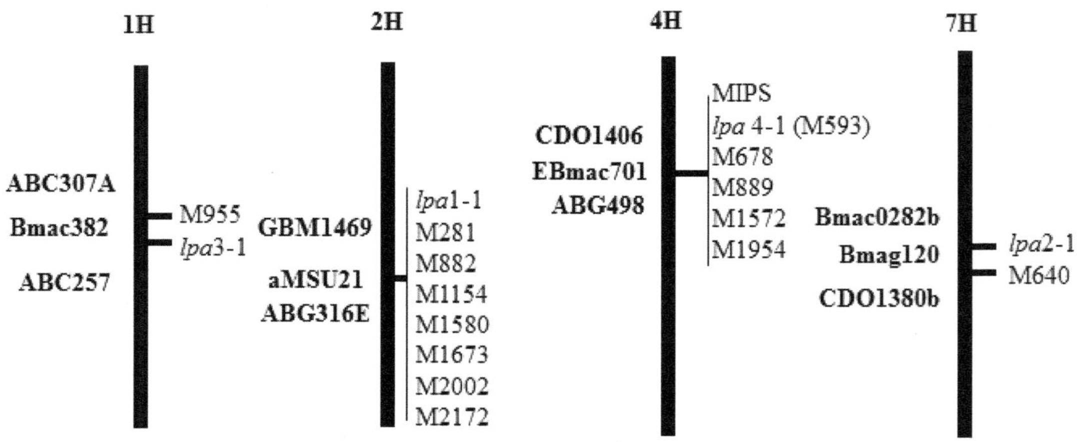

Figure 1. Chromosomal position of 17 barley *low phytic acid* mutations and the barley genome's single *myo*-Inositol-3-P_1 synthase (MIPS) gene. Map positions are approximate and only meant to be illustrative of linkage groups. As indicated, all mutations fell into four linkage groups, but allelic or non-allelic relationships within each linkage group are not yet known. The molecular markers used in mapping are on the left of each chromosome. The *lpa* mutations comprising each linkage group are on the right of each chromosome.

Previous studies focused on four mutations that appeared to represent the range of seed P phenotypes observed, and that mapped to three linkage groups: *Hvlpa*1-1 (chromosome 2H), *Hvlpa*2-1 (chromosome 7H), *Hvlpa*3-1 (chromosome 1H), and *Hvlpa*-M955 (chromosome 1H). Based on initial observation of adequate plant performance, and to add mutations mapping to new chromosomal sites, for the present study we selected two additional mutations: *Hvlpa*-M593 and *Hvlpa*-M640. Initial quantitative analyses (Supplementary Table 1) indicated that *Hvlpa*-M593 conditioned a ~53% reduction in seed phytic acid, and that *Hvlpa*-M640 conditioned a ~47% reduction, but in the case of *Hvlpa*-M640, the effect appeared to be endosperm-specific.

Segregation analyses conducted during mapping indicated that these additional two mutants displayed an inheritance pattern of recessive alleles at single gene (Supplementary Table 2). Linkage analyses (Supplementary Table 3) revealed that *Hvlpa*-M593 mapped to chromosome 4H, defining a new barley *lpa* locus. Therefore hereafter, *Hvlpa*-M593 will be referred to as *Hvlpa*4-1. *Hvlpa*-M640 mapped to chromosome 7H, linked to *lpa*2-1. However, the allelic relationship if any between *lpa*2-1 and *lpa*M640 on chromosome 7H, or between *lpa*3-1 and *lpa*-M955 on chromosome 1H, is not known at present and is the subject of ongoing study. Thus these six mutations represent a total of four linkage groups and a minimum of four loci.

2.2. Confirmation of Isoline Seed Phytic Acid Phenotype

The seed inorganic P test was used for routine genotyping (homozygous wild-type *versus* homozygous mutant) of *lpa* isolines used in this study (Supplementary Figure 1 and Experimental Section). However, to provide a more in-depth confirmation of the seed P and phytic acid P phenotype of grain produced by these lines when grown at the locations used here, a quantitative analysis was conducted with grain samples from test plots at two 2011 locations, Aberdeen and Soda Springs (Table 1). These two locations were selected for this analysis because they represent the extremes in yield potential over all locations in this study (see below). This analysis confirmed that when grown in these specific locations, the predicted *lpa* phenotype of each mutant line (based on previous analyses such as given in Supplementary Table 1) was observed. Apparent reductions in grain phytic acid P, as compared with levels observed in grain produced by sibling wild-type lines, ranged from a low of ~22% in *Hvlpa*2-1, to the moderate reductions of ~35% in *Hvlpa*4-1, ~40% in *Hvlpa*-M640 and ~48% in *Hvlpa*1-1, to the more substantial reductions of ~57% in *Hvlpa*3-1 and ~80% in *Hvlpa*-M955. These are referred to as "apparent reductions" since the method used here probably overestimates phytic acid by 5% to 10%, an inflation resulting from reading components of the "extraction matrix" as phytic acid P when using this method. The best illustration of this is the data for *Hvlpa*-M955. Previous analyses of the grain phytic acid levels in this mutant using more accurate methods consistently indicated that the reduction in phytic acid P in this mutant, as compared with wild-type, is >90%. This analysis also confirmed that with one exception, *Hvlpa*1-1, the *lpa* mutations did not greatly impact grain total P: the reductions in grain phytic acid P were largely matched by molar-equivalent (in terms of P) increases in grain inorganic P. In the case of *Hvlpa*1-1, this analysis also confirmed that homozygosity for the mutation causes a reduction in whole-grain total P of 15% to 20% (Table 1).

2.3. Yield Evaluations

To provide the most valid statistical approach to test for differences between sibling near-isogenic lines, the Analysis of Variance of the grain yield data presented in Table 2 was conducted using only the data for the six isoline pairs. The greatest variation observed was due to the main effects of Year and Location and the interaction of Year × Location. The large main effect of Year is clearly seen in the greater overall productivity in 2011 as compared with 2010; overall mean grain yield in 2011 (5.47 ± 0.28 Mg·ha^{-1}) was ~20% greater than that observed in 2010 (4.52 ± 0.22 Mg·ha^{-1}) (Tables 3 and 4). The large main effect of Location is also clear in the data, and critical to the main objective of this study. Clear and large differences in overall mean grain yield per location, a measure of location yield potential, were observed in both years and the location ranking were similar in both years (from highest overall yield to lowest): Aberdeen > Idaho Falls > Rupert > Ashton > Soda Springs. The interaction effect of Year × Location was largely due to the very large difference in overall yield at one location, Ashton, where the yield in 2010 (2.43 ± 0.21 Mg·ha^{-1}) was only 40% of that observed in 2011 (5.96 ± 0.32 Mg·ha^{-1}). The large differences in productivity observed across the two years and five locations provides a powerful test of the impact on yield of an *lpa* mutation, the primary objective of this study.

Table 1. Seed total phosphorus (Total P), phytic acid P and inorganic P (P_i) in pairs of sibling barley near-isogenic and cultivar controls grown in Aberdeen and Soda Springs, ID, in 2011.

Isoline Pair	Genotype	Idaho Field Location					
		Aberdeen			Soda Springs		
		Total P	Phytic Acid P	Inorganic P	Total P	Phytic Acid P	Inorganic P
		mg·g^{-1}					
1	WT	2.96	2.86	0.29	3.10	2.62	0.26
1	*Hvlpa*1-1	2.52	1.51	1.05	2.44	1.35	0.83
2	WT	2.86	2.47	0.24	3.09	2.75	0.25
2	*Hvlpa*3-1	2.76	1.10	1.45	2.88	1.15	1.66
3	WT	2.78	2.32	0.22	3.12	2.52	0.24
3	*Hvlpa*-M955	2.91	0.53	2.12	3.33	0.51	2.21
4	WT	2.75	2.38	0.23	2.93	2.54	0.25
4	*Hvlpa*2-1	2.93	1.83	0.67	3.18	1.99	0.90
5	WT	3.10	2.49	0.39	2.98	2.59	0.27
5	*Hvlpa*4-1	2.91	1.53	1.18	3.31	1.76	1.34
6	WT	2.86	2.44	0.26	3.15	2.69	0.26
6	*Hvlpa*-M640	2.98	1.61	0.80	3.36	1.51	1.32
		Cultivar Controls					
Harrington		2.73	2.37	0.26	3.07	2.59	0.30
Baronesse		2.89	2.41	0.26	3.12	2.65	0.28
Conrad		3.01	2.52	0.19	3.49	3.10	0.31
Std. Dev. of Mean		0.12	0.14	0.07	0.15	0.14	0.05
LSD $p = 0.05$		0.34	0.41	0.20	0.43	0.39	0.15

Table 2. Analysis of variance of yield of twelve isolines grown in five locations in Idaho in two years.

Source of Variation	*df*	Mean Square	*F* Value	Prob. > F
Model	132	10.20	28.82	<0.0001
Year	1	97.24	274.70	<0.0001
Location	4	224.79	635.05	<0.0001
Year × Loc	4	61.43	173.54	<0.0001
Rep (Loc)	15	0.65	1.85	0.03
Isoline	11	3.70	10.45	<0.0001
Year × Line	11	0.22	0.62	0.82
Loc × Line	44	0.73	2.06	0.0002
Year × Loc × Line	42	0.47	1.31	0.11
Residual Error	330	0.35		

Table 3. Yield of barley *lpa* near-isogenic lines and three cultivar controls when grown in five locations in Idaho, 2010. Data represent means of four replicates at each location.

Isoline Pair	Genotype	Idaho Location					Line Mean
		Aberdeen	Ashton	Idaho Falls	Rupert	Soda Springs	
		Grain Yield (Mg·ha^{-1})					
1	WT	6.04	2.52	6.15	6.20	2.13	4.61
1	*Hvlpa*1-1	6.61	2.78	6.57	5.47	2.00	4.69
2	WT	5.83	2.37	5.84	6.20	2.22	4.49
2	*Hvlpa*3-1	6.17	2.03	5.71	5.15	1.88	4.19
3	WT	5.72	2.59	5.98	5.45	2.80	4.60
3	*Hvlpa*-M955	5.27	1.88	5.71	4.74	1.47	3.81
4	WT	6.48	2.34	6.15	5.89	2.41	4.91
4	*Hvlpa*2-1	6.51	2.59	5.98	5.23	1.81	3.86
5	WT	6.14	2.32	6.45	5.71	2.52	4.63
5	*Hvlpa*4-1	6.06	2.30	6.74	5.45	1.96	4.50
6	WT	6.59	2.30	6.77	5.89	2.10	4.73
6	*Hvlpa*-M640	6.30	2.40	6.45	4.44	1.61	3.58
		Cultivar Controls					
Harrington		6.38	2.42	5.54	5.54	2.22	4.42
Baronesse		7.17	3.57	6.81	7.45	2.59	5.52
Conrad		7.06	2.37	6.08	6.91	2.05	4.89
Location Mean ± SE		6.29 ± 0.34	2.43 ± 0.21	6.19 ± 0.29	5.78 ± 0.22	2.10 ± 0.19	4.52 ± 0.22
LSD $p = 0.05$		0.98	0.59	0.83	0.63	0.55	0.64

Table 4. Yield of barley *lpa* near-isogenic lines and three cultivar controls when grown in five locations in Idaho, 2011. Data represent means of four replicates at each location.

Isoline Pair	Genotype	Idaho Location					Line Mean
		Aberdeen	Ashton	Idaho Falls	Rupert	Soda Springs	
		Grain Yield (Mg·ha^{-1})					
1	WT	6.51	5.79	6.10	5.22	3.05	5.33
1	*Hvlpa*1-1	7.35	6.69	5.57	5.57	3.25	5.68
2	WT	6.80	6.15	4.93	4.98	3.55	5.37
2	*Hvlpa*3-1	6.98	5.32	4.96	5.01	2.78	5.01
3	WT	6.40	6.34	5.57	4.84	3.59	5.35
3	*Hvlpa*-M955	5.07	4.83	4.86	4.61	2.47	4.37
4	WT	7.03	6.32	5.52	5.10	3.42	5.48
4	*Hvlpa*2-1	7.51	4.57	6.27	5.03	2.78	5.23
5	WT	7.80	6.37	5.96	5.25	3.44	5.76
5	*Hvlpa*4-1	6.95	5.84	5.27	5.88	2.99	5.51
6	WT	7.29	6.01	5.40	5.48	3.20	5.48
6	*Hvlpa*-M640	6.98	5.03	5.93	4.95	2.47	5.21
		Cultivar Controls					
Harrington		6.96	6.27	5.69	5.30	3.17	5.44
Baronesse		8.61	7.13	6.86	6.08	4.05	6.55
Conrad		8.00	6.71	6.64	6.74	3.03	6.22
Location Mean ± SE		7.08 ± 0.37	5.96 ± 0.32	5.69 ± 0.29	5.34 ± 0.36	3.16 ± 0.21	5.47 ± 0.28
LSD $p = 0.05$		1.05	0.93	0.83	1.02	0.59	0.79

A simple, statistically-valid approach to evaluating the impact of these barley *lpa* mutations on yield, and one that also reflects a good genetics design, is to conduct pair-wise "*T*-tests", using the "LSD $p = 0.05$" value, to compare yields of sibling backcross lines that are either homozygous wild-type or homozygous mutant. Using this approach, if the difference in yield between wild-type and mutant sibling isolines is greater than the "LSD $p = 0.05$" value, one can conclude that the yield difference is statistically significant at the $p = 0.05$ level and is attributable to the allelic variant at the given *lpa* locus.

While variation due to isoline was of a lesser magnitude as compared with the main effects of Year and Location, it still was highly statistically significant. It is first important to note that no statistically-significant differences in yield were observed between the six wild-type isolines obtained after backcrossing and the original cultivar Harrington included as a control (Tables 3 and 4). Yet differences are clear between the six wild-type isolines and Harrington on the one hand, and the cultivars Baronesse and Conrad on the other. This supports the experimental model that any observed differences in yield between a given wild-type and sibling *lpa* line are most likely attributable to the *lpa* allelic difference and not due to background genomic effects [5].

Of importance to this study are the two interactions of Year × Line and Location × Line. The Year × Line interaction was not statistically significant, indicating that isoline performances ranked similarly in each year. In contrast, the statistically significant Location × Line interaction indicates that isoline performance was differentially impacted by environment. This can be readily observed in the yield data when viewed over both years (Tables 3 and 4): while the ranking of lines by yield might not have greatly differed between years, the magnitude of these differences in line yield varied. As a result, the large Year main effect adds power to the evaluation of the magnitude of differences in line yields between locations.

Inspection of the yield data meaned over locations within each year indicates that one line, *Hvlpa*1-1, displayed grain yield equal to or greater than that of its sibling wild-type isoline in both years (Tables 3 and 4). *Hvlpa*1-1 yielded as much or more than its sibling wild-type line in seven out of ten location/years, only yielding less in statistically-significant terms in one location/year (Rupert in 2010). In fact the results obtained for the Rupert 2010 location/year are anomalous. Despite being a presumed "non-stressful" location, all *lpa* isolines except for *lpa*4-1 yielded less than their wild-type siblings in that location/year, including the *lpa*1-1 isoline. We have no explanation for this anomalous result other than that it might be due to an unknown or unrecognized stress (or stresses) impacting the *lpa* lines more than the wild-type lines, or that it might indicate an example of unpredictability for *lpa* line yield, itself potentially important.

Despite this caveat, *Hvlpa*1-1 was similarly productive, as compared with its sibling wild-type isoline, in nearly all locations/years, including both stressful (Soda Springs in both years and Ashton in 2010) and non-stressful (Aberdeen, Idaho Falls, Rupert/2011) environments. A second mutation, *Hvlpa*4-1, yielded within 4% of its sibling wild-type isoline across all locations and years. Further, when yield comparisons are made between *lpa* and sibling wild-type lines when grown in the three least stressful, most productive environments (Aberdeen, Idaho Falls and Rupert), for three of the six *lpa* lines (*Hvlpa*1-1, *Hvlpa*3-1 and *Hvlpa*4-1), yields were consistently within 5% of the given sibling wild-type line.

Where the impact of the *lpa* trait on grain yield was most pronounced was in the most stressful, least productive environments, which in this experiment was Soda Springs, a "Dry-Land" production site where no irrigation was provided. Statistically-significant impacts on yield were observed for eight of the ten possible isoline/year pair-wise comparisons between mutant and wild-type at Soda Springs. The exceptions were *Hvlpa*1-1 in both years, and for *Hvlpa*3-1 in 2010 and *Hvlpa*4-1 in 2011.

In broad terms there does appear to be a correlation between the magnitude of grain phytic acid reduction conditioned by a given mutation and impact on yield. The consistently poorest performer was *lpa*-M955, which as a homozygote conditions a reduction in grain phytic acid that is >80%, and results in a ~20% reduction in grain yield, as compared with its sibling wild-type line. In the most stressful, poorest yield-potential location, Soda Springs, homozygosity for *lpa*-M955 results in a 30% to 50% reduction in yield. However, the correlation between the extent of grain phytic acid reduction conditioned by a mutant and its yield breaks down when one considers mutations, which condition more moderate reductions in grain phytic acid. The three best performers, *Hvlpa*1-1, *Hvlpa*3-1 and *Hvlpa*4-1 condition reductions in grain phytic acid ranging from ~40% to 60% (depending on environment), as compared with sibling wild-type lines, whereas the two mutations with intermediate performance, *Hvlpa*2-1 and *Hvlpa*-M640, condition reductions in the range of ~30%. Also, the best performer, *Hvlpa*1-1, is known to impact grain phytic acid accumulation in a tissue-specific manner, perturbing accumulation only in the endosperm. Therefore while the magnitude of grain phytic acid reduction conditioned by a mutation may be an important determinant of yield impact, gene function and tissue-specificity also play an important role.

3. Discussion

A previous study evaluated field performance of less advanced versions of four of these pairs of barley *lpa* near-isogenic lines, and was conducted at four sites in Idaho (USA) that include three used in the present study [5]. *Hvlpa*1-1, *Hvlpa*2-1 or *Hvlpa*3-1, conditioning seed phytic acid reductions of 30% to 60%, had yields similar to their wild-type sibling lines when grown in two relatively non-stressful production environments that utilized irrigation. *Hv*-M955, with greater than 90% reduction in seed phytic acid, yielded 13% less than its wild-type isoline in these environments. In the two non-irrigated, more stressful production environments included in the initial study, only *Hvlpa*1-1 displayed yields statistically similar to wild-type, but its yield was nevertheless 11% less than that recorded for its wild-type sibling. Furthermore, *Hvlpa*2-1 and *Hvlpa*3-1 each had 25% reduction in seed yield and *Hv*-M955 had nearly 35% reduction in seed yield. Thus in this initial study three out of four barley *lpa* isolines performed well in non-stressful, "high-yield potential" environments, but yield in more stressful environments was clearly reduced, and a trend for a yield reduction was observed even for the best performing *Hvlpa* line. This initial study was conducted over two years at two sites but only in one year at two additional sites.

The good field performance of *Hvlpa*1-1 lines observed in initial studies led to the breeding and registration of the first *lpa* cultivar of any crop species [6]. *Hvlpa*1-1 probably encodes a putative member of the sulfate transporter gene family [7]. Perturbation of this gene in barley also results in a second desirable crop seed phenotype, reduced seed total P [8]. Therefore this gene might prove to be an excellent target for engineering the ideal seed P phenotype, low-phytic acid and low total P, in other crops species.

In the present study performance comparisons were made using more advanced versions of the initial four sets of near-isogenic lines, obtained following one additional backcross, and an additional two sets of near-isogenic lines developed using *Hvlpa*-M640 and *Hvlpa*4-1. Furthermore, all six sets of isolines were evaluated across two years in five Idaho (USA) locations that greatly differ in crop yield potential. Confirming the results of the initial study [5], *Hvlpa*1-1 performed well in all locations, and had no detectable negative impact on yield.

A second mutation, *Hvlpa*4-1, resulted in yield losses of only 3.5%. Furthermore, when comparing yields in three relatively non-stressful production environments, at least three *lpa* mutations (*Hvlpa*1-1, *Hvlpa*3-1, and *Hvlpa*4-1) typically had yields similar to or within 5% of the wild-type sibling isoline. Thus while some *lpa* mutations do greatly or consistently impact field performance, a substantial fraction of barley *lpa* mutations appear to have little or no effect on yield. Therefore in the case of barley, *lpa* mutations can be readily identified that when simply incorporated into a cultivar result in adequately performing lines, at least in many production environments, even with no additional breeding for performance within the *lpa* line. Nevertheless, *lpa* line yield in more stressful environments remains problematic, as does consistency of performance for any given line in any environment. Therefore additional breeding via recurrent selection for field performance would still be needed to produce lines that are consistently high-yielding across diverse production environments (see below).

*Hvlpa*1-1 differs from the other three genotypes evaluated in the initial study [5] in that its recessive alleles result in a localized, tissue-specific block in phytic acid accumulation. Here we report that *Hvlpa*M640 is phenotypically similar to barley *lpa*1-1 in that it also appears to be endosperm-specific. However, the *Hvlpa*-M640 isoline did not perform as consistently well as did the *Hvlpa*1-1 isoline. This may indicate that it is not solely the endosperm-specificity of *Hvlpa*1-1 that results in its consistent performance, but rather that there are gene/function-specific differences impacting yield. Nevertheless, in contrast to the results of the present study yield trials indicate that a low-phytate barley cultivar produced using *Hv*-M640 has yields comparable with conventional cultivars in a wide variety of production environments [9]. Thus conventional genetics and breeding approaches can achieve at least one of the advantages provided by genetic engineering; genotypes in which blocks in the phytic acid pathways are restricted to specific tissues of the seed, such as the embryo or aleurone layer, thus minimizing the negative impact on whole plant performance and yield.

The only statistically valid experimental approach to attributing a quantitative difference in field performance, or in any trait, to a specific allelic variant of an *lpa* gene, is via comparisons using wild-type and mutant sibling near-isogenic lines. This is true regardless of the method used in generating the novel allelic variants that confer the *lpa* trait; whether the *lpa* trait is conferred by recessive alleles obtained following chemical or radiation mutagenesis, or obtained using any method that involves transformation or "genome modification", methods that require tissue culture, regeneration or the use of selectable markers such as those conferring herbicide tolerance. A good example of this is "genome modification" [4], a powerful and highly valuable method that targets specific endogenous gene sequences and thus in theory avoids non-linked genomic change such as that resulting from chemical mutagenesis, or from non-homologous insertion/mutation resulting from earlier genetic modification methods. Genome modification has been used to target one of the maize (*Zea mays* L.) genome's *IPK1* genes, which encodes an inositol-1,3,4,5,6-pentakisphosphate 2-kinase [4]. However, since tissue culture and regeneration can greatly impact genome expression and epigenetics in offspring [10], as can expression of reporter genes that confer herbicide resistance [11], assessing field performance using *lpa* progeny obtained via this method still first requires sufficient backcrossing, certainly a minimum of four to six backcrosses to selected recurrent parents, followed by isolation and evaluation of near-isogenic lines.

Even the use of near-isogenic lines has a potentially serious limitation. If there was no selection for performance during or after the incorporation of an allele from a divergent donor line into an elite line or cultivar, the resulting *lpa* or wild-type sibling lines might differ in yield due to inheritance of undesirable allelic variants at loci not subject to selection, or via "linkage-drag" of undesirable loci linked to the *lpa* locus in the donor parent [12]. This is particularly important because elite field performance reflects inheritance of favorable alleles, both genetic and epigenetic, at multiple loci throughout the genome. Thus simply incorporating an *lpa* allele via backcrossing, regardless of its source or method of induction, without some form of accompanying selection for performance, might result in lines with reduced performance that is not linked to the *lpa* trait or the *lpa* allele. Clearly, recurrent selection for field performance, within an *lpa* genotype over several generations, is required to obtain a line with favorable alleles genome-wide, both genetic and epigenetic, that retains the desired *lpa* phenotype and that also display elite performance. Despite this caveat, the use of near-isogenic lines, even without additional breeding, remains the most statistically appropriate approach to attributing a difference in yield or field performance to inheritance of an allelic difference at an *lpa* locus.

Few studies of field performance of *lpa* types have in fact made comparisons between truly near-isogenic sibling lines that, that to the best approximation available with current science, essentially differ only in the allelic variants of a single *lpa* locus. The first study reporting evaluation of the impact on crop performance or yield of an *lpa* type was Ertl *et al.* [12]. In this study, the maize *lpa*1-1 allele (*Zmlpa*1-1), originally isolated in a "Flint corn-like" genetic background referred to as "Early ACR" [13], was backrossed to the BC_3 (four crosses to the recurrent parent inbred line), into a sufficient number of diverse elite maize "Dent corn" inbred lines to produce a total of 14 pairs of near-isognenic hybrids, each pair consisting of a wild-type and *Zmlpa*1-1 version. The performance evaluation of these pairs of maize hybrids consisted of studies conducted in one year, at three locations with two replications, in central Iowa, USA. In six of the 14 near-isogenic pairs no statistically-significant effect of the *Zmlpa*1-1 allele on yield was observed. In eight of the 14 near-isogenic pairs a yield loss due to the *Zmlpa*1-1 was observed, and overall the yield loss was 6%, as compared with wild-type. Thus this first report of performance evaluation of low-phytate versions of a crop reported results similar to those reported here. However, while ground-breaking, this first study had inherent limitations. The production environments and practices were similar with no clearly stressful or low-productivity environments, and replication was minimal. More importantly the Flint and Dent corn genetic backgrounds used greatly differ, and only four crosses to the recurrent parent (the Dent corn inbred lines) were completed. Therefore the resulting pairs of lines can only be considered "early generation" near-isogenic lines, with residual genomic contribution of the very divergent donor parent of approximately 6.25%. In the present study the *lpa* mutations were isolated in the same genetic background that they were subsequently backcrossed to, cv. Harrington. Thus in this case even "early-generation" backcross lines, such as the *Hvlpa*4-1 and *Hvlpa*-M640 lines, are more genetically similar to their wild-type sibling lines than those evaluated in the Ertl *et al.* study [12].

A third series of studies that evaluated field performance of *lpa* types using near-isogenic lines addressed issues of germination and field emergence in *lpa* soybean [*Glycine max* (L.) Merr.]

types [14–16]. Germination and emergence is reduced in many soybean *lpa* genotypes, and the effect is highly pronounced when seed is produced in tropical *versus* temperate environments, a phenomenon referred to as the "seed source effect" [14]. When low-phytate soybean lines were produced via incorporation of recessive alleles of the soybean MIPS-encoding gene (the LR33 mutations) [17], or the *pha1* and *pha2* alleles first isolated as the M153 mutant [18], and seed was produced in a tropical environment, subsequent field emergence was greatly reduced when compared to that of seeds of the same genotype when tested seed was produced in a more moderate season in Puerto Rico, or in a temperate region [14,16]. While the extreme effects on germination and emergence that maturation in a tropical environment produces has not been overcome via subsequent breeding, evaluation of BC_4-derived near-isogenic lines indicated that selection could identify lines with improved performance [16].

The studies with soybean and barley *lpa* genotypes illustrate the potential of classical genetics and breeding approaches for overcoming agronomic problems associated with the *lpa* trait. Selection within lines segregating for a given *lpa* allele can identify lines with enhanced agronomic performance [15]. Identification of genes that, if perturbed, have little impact on agronomic performance, such as barley *lpa*1 and M640 [6,9], or soybean *lpa*-ZC-2 [19], can lead to low-phytate cultivars with superior agronomic performance.

4. Experimental Section

4.1. Isoline Pair Development and Cultivar Controls

Each of six *lpa* mutations were crossed and subsequently backcrossed to varying degrees to the cultivar Harrington in which they were originally isolated. BC_xF_2 progenies were obtained following self-pollination of a given BC_xF_1, and sibling homozygous wild-type and homozygous mutant plants were identified. These were then used for increases to provide seed for the present studies. Seed used in the 2010 yield trial were: BC_5F_4 (*Hvlpa*1-1 and its wild-type sibling line); BC_4F_4 (*Hvlpa*2-1, *Hvlpa*3-1, and *Hvlpa*-M955 and their wild-type sibling lines); F_4 (*Hvlpa*4-1 and *Hvlpa*-M640). Seed used in the 2011 yield trial were advanced further generations as follows: BC_5F_5 (*Hvlpa*1-1 and its wild-type sibling line); BC4F5 (Hvlpa2-1, Hvlpa3-1, and Hvlpa-M955 and their wild-type sibling lines); BC1F3 (Hvlpa4-1 and *Hvlpa*-M640). Since each mutation was first isolated in the genetic background represented by the cultivar Harrington, then crossed or backcrossed to the same cultivar, following which sibling plants either homozygous wild-type or homozygous mutant were increased, each pair of sibling lines can be considered near-isogenic. For confirming the *lpa* genotype and phenotype of a given plant or line a microtitre plate assay for seed inorganic P was used (Supplemental Figure 1): homozygous wild-type lines typically have relatively low levels of inorganic P (<0.5 mg·g^{-1}), whereas homozygous *lpa* lines typically have greatly elevated seed inorganic P (>1.0 mg·g^{-1} in most cases). This "high seed inorganic P" phenotype provides a simple assay visual inspection of which is sufficient to genotype lines and test for homogeneity (Supplemental Figure 1).

In addition to the six pairs of near-isogenic lines, three barley cultivars were included in the field trials as controls. These were: (1) Harrington, a "2-rowed malting barley" that represents the genetic background in which these mutations were isolated; (2) Baronesse, a "two-rowed feed barley" that is

4. Shukla, V.K.; Donyon, Y.; Miller, J.C.; DeKelver, R.C.; Moehle, E.A.; Worden, S.E.; Mitchell, J.C.; Arnold, N.L.; Gopolan, S.; Meng, X.; *et al.* Precise genome modification in the crop species *Zea mays* using zinc-finger nucleases. *Nature* **2009**, *459*, 437–443.
5. Bregitzer, P.; Raboy, V. Effects of four independent Low-Phytate mutations on barley agronomic performance. *Crop Sci.* **2006**, *46*, 1318–1322.
6. Bregitzer, P.; Raboy, V.; Obert, D.E.; Windes, J.M.; Whitmore, J.C. Registration of "Herald" Barley. *Crop Sci.* **2007**, *47*, 441–442.
7. Ye, H.; Zhang, X.-Q.; Broughton, S.; Westcott, S.; Wu, D.; Lance, R.; Li, C. A nonsense mutation in a putative sulphate transporter gene results in low phytic acid in barley. *Funct. Integr. Genomics* **2011**, *11*, 103–110.
8. Ockenden, I.; Dorsch, J.A.; Reid, M.M.; Lin, L.; Grant, L.K.; Raboy, V.; Lott, J.N.A. Characterization of the storage of phosphorus, inositol phosphate and cations in grain tissues of four barley (*Hordeum vulgare* L.) *low phytic acid* genotypes. *Plant Sci.* **2004**, *167*, 1131–1142.
9. Bregitzer, P.; Raboy, V.; Obert, D.E. Registration of "Clearwater" low-phytate hulless spring barley. *J. Plant Regist.* **2008**, *2*, 1–4.
10. Stelpflug, S.C.; Eichten, S.R.; Hermanson, P.J.; Springer, N.M.; Kaeppler, S.M. Consistent and heritable alterations of DNA methylation are induced by tissue culture in maize. *Genetics* **2014**, *198*, 209–218.
11. Brown, R.H.; Raboy, V.; Bregitzer, P. Unintended consequences: High phosphinothricin acetyltransferase activity related to reduced fitness in barley. *In Vitro Cell. Dev. Biol.* **2013**, *49*, 240–247.
12. Ertl, D.; Young, K.A.; Raboy, V. Plant genetic approaches to phosphorus management in agricultural production. *J. Environ. Qual.* **1998**, *27*, 299–304.
13. Raboy, V.; Gerbasi, P.F.; Young, K.A.; Stoneberg, S.D.; Pickett, S.G.; Bauman, A.T.; Murthy, P.P.N.; Sheridan, W.F.; Ertl, D.S. Origin and seed phenotype of maize *low phytic acid* 1-1 and *low phytic acid* 2-1. *Plant Physiol.* **2000**, *124*, 355–368.
14. Meis, S.J.; Fehr, W.R.; Schnebly, S.R. Seed source effect on field emergence of soybean lines with reduced phytate and raffinose saccharides. *Crop Sci.* **2003**, *43*, 1336–1339.
15. Oltmans, S.E.; Fehr, W.R.; Welke, G.A.; Raboy, V.; Peterson, K.L. Agronomic and seed traits of soybean lines with low-phytate phosphorus. *Crop Sci.* **2005**, *45*, 593–598.
16. Anderson, B.P.; Fehr, W.R. Seed source affects field emergence of low-phytate soybean lines. *Crop Sci.* **2008**, *48*, 929–932.
17. Hitz, W.; Carlson, T.J.; Kerr, P.S.; Sebastian, S.A. Biochemical and molecular characterization of a mutation that confers a decreased raffinosaccharide and phytic acid phenotype on soybean seeds. *Plant Physiol.* **2002**, *128*, 650–660.
18. Wilcox, J.R.; Premachandra, G.S.; Young, K.A.; Raboy, V. Isolation of high seed inorganic P, low-phytate soybean mutants. *Crop Sci.* **2000**, *40*, 1601–1605.
19. Yuan, F.J.; Zhao, H.J.; Ren, X.L.; Zhu, S.L.; Fu, X.J.; Shu, Q.Y. Generation and characterization of two novel low phytate mutations in soybean (*Glycine max* L. Merr.). *Theor. Appl. Genet.* **2007**, *115*, 945–957.

20. Dorsch, A.J.; Cook, A.; Young, A.K.; Anderson, N.J.; Bauman, T.A.; Volkmann, J.C.; Murthy, P.N.P.; Raboy, V. Seed phosphorus and inositol phosphate phenotype of barley low phytic acid genotypes. *Phytochemistry* **2003**, *62*, 691–706.
21. Haug, W.; Lantzsch, H.-J. Sensitive method for the rapid determination of phytate in cereals and cereal products. *J. Sci. Food Agric.* **1983**, *34*, 1423–1426.

13

Accumulation of Phosphorus-Containing Compounds in Developing Seeds of Low-Phytate Pea (*Pisum sativum* L.) Mutants

Arun S.K. Shunmugam [1], Cheryl Bock [2], Gene C. Arganosa [1], Fawzy Georges [2], Gordon R. Gray [1] and Thomas D. Warkentin [1,*]

[1] Crop Development Centre, Department of Plant Sciences, University of Saskatchewan, 51 Campus Drive, Saskatoon, SK S7N 5A8, Canada; E-Mails: s.ask@usask.ca (A.S.K.S.); gene.arganosa@usask.ca (G.C.A.); gr.gray@usask.ca (G.R.G.)

[2] National Research Council Canada, 110 Gymnasium Place, Saskatoon, SK S7N 0W9, Canada; E-Mails: Cheryl.bock@nrc-cnrc.gc.ca (C.B.); fawzy.georges@usask.ca (F.G.)

* Author to whom correspondence should be addressed; E-Mail: tom.warkentin@usask.ca

Academic Editor: Francesca Sparvoli

Abstract: Low phytic acid (*lpa*) crops are low in phytic acid and high in inorganic phosphorus (P_i). In this study, two *lpa* pea genotypes, 1-150-81, 1-2347-144, and their progenitor CDC Bronco were grown in field trials for two years. The *lpa* genotypes were lower in IP_6 and higher in Pi when compared to CDC Bronco. The total P concentration was similar in *lpa* genotypes and CDC Bronco throughout the seed development. The action of *myo*-inositol phosphate synthase (MIPS) (EC 5.5.1.4) is the first and rate-limiting step in the phytic acid biosynthesis pathway. Aiming at understanding the genetic basis of the *lpa* mutation in the pea, a 1530 bp open reading frame of *MIPS* was amplified from CDC Bronco and the *lpa* genotypes. Sequencing results showed no difference in coding sequence in *MIPS* between CDC Bronco and *lpa* genotypes. Transcription levels of MIPS were relatively lower at 49 days after flowering (DAF) than at 14 DAF for CDC Bronco and *lpa* lines. This study elucidated the rate and accumulation of phosphorus compounds in *lpa* genotypes. The data also demonstrated that mutation in *MIPS* was not responsible for the *lpa* trait in these pea lines.

Keywords: D-*myo*-inositol phosphate synthase (MIPS); phosphorus; phytate biosynthesis; phytic acid

1. Introduction

Phytic acid (*myo*-inositol-1,2,3,4,5,6-hexa*kis*phosphate; IP_6) is the major storage form of phosphorus (P) in most plant seeds [1]. Phytate is present within subcellular protein inclusions in all seeds. Particularly, in cereal grains, phytates are essentially localized in the germ and aleurone tissues whereas in dicotyledons, phytate is distributed throughout the cotyledon [2,3]. Endogenous phytase enzymes break down phytate during seed germination and release its phosphorus, *myo*-inositol, and mineral contents for use by the growing seedling [4]. IP_6 also accumulates in other plant tissues and organs that accumulate nutrient stores for subsequent redistribution, such as pollen, roots, and tubers [5].

Applied interest in seed IP_6 primarily concerns its roles in human health and animal nutrition. It is a strong chelator of mineral elements such as iron, zinc, calcium, and potassium, forming mixed salts that are largely excreted by humans and other non-ruminant animals such as poultry, swine, and fish [6,7]. Excretion of seed-derived IP_6 can contribute to dietary iron and zinc deficiencies, a major public health problem in the developing world. In addition, undigested phytate excreted by non-ruminant animals represents an important source of phosphorus pollution in the environment [8]. Due to nutritional and environmental concerns, the development of cultivars with a low-phytate trait has become an attractive breeding objective in many crop species. Chemically induced, non-lethal recessive mutants that decrease seed phytic acid content have been isolated and genetically mapped in maize (*Zea mays* L.) [9,10], barley (*Hordeum vulgare* L.) [11,12], and soybean (*Glycine max* L. Merr) [13]. Recently, Warkentin *et al.* [14] developed and characterized two low-phytic acid mutants of field pea (*Pisum sativum* L.). The low phytic acid (*lpa*) mutations have the potential to alleviate the environmental and nutritional problems associated with phytic acid in animal feeds [15]. Moreover, *lpa* crops may also offer improved nutrition for human populations that depend upon grains and legumes as staple foods. In addition, these *lpa* mutants provide a valuable system to study seed phytic acid synthesis.

In plants, the six-carbon cyclitol *myo*-inositol gives rise to compounds with roles in such diverse functions as signal transduction, membrane biogenesis, stress tolerance, and the generation of seed storage compounds including IP_6 [16,17]. The *de novo* synthesis of *myo*-inositol involves the conversion of glucose 6-phosphate to *myo*-inositol-1-phosphate (IP_1) that is subsequently dephosphorylated to release free *myo*-inositol. The former reaction is catalyzed by *myo*-inositol-1-phosphate synthase (MIPS; EC 5.5.1.4), the first and rate-limiting enzyme of the pathway [17–19]. This makes *MIPS* an attractive target for manipulation to produce low-phytate crops. Previous studies have demonstrated that 50% to 95% reductions in phytic acid can be obtained when this enzyme is targeted through mutagenesis or by genetic engineering methodologies [20–26]. In soybean, mutations in *MIPS* coding sequences conferred a decreased phytic acid phenotype, and an effective reduction in phytate content (90% to 95%) has been observed when one of the four *MIPS* in soybean, *GmMIPS1* was silenced through a RNA interference (RNAi) approach [24,26]. An *lpa* phenotype was also produced by manipulating the *MIPS* gene through an antisense approach in *Oryza sativa* L. [22].

The objective of this study was to investigate the accumulation of phytic acid and other phosphorus compounds in developing seeds of normal and low-phytate genotypes of field pea (*Pisum sativum* L.). We also examined *MIPS* gene expression and analyzed the sequence at the nucleotide and protein levels to ascertain if variation in *MIPS* coding sequences was responsible for the *lpa* trait. This will help us to understand the nature of the low phytate mutation(s) and develop markers for the low phytate trait furthering the development of low-phytate cultivars.

2. Results

2.1. Agronomic Characteristics of Low-Phytate Pea Genotypes

The agronomic characteristics of the *lpa* pea genotypes were similar to their normal phytate progenitor CDC Bronco except for 1000 seed weight and grain yield at maturity (Figure 1, Table 1). CDC Bronco and genotypes 1-150-81 and 1-2347-144 did not differ in percent emergence, days to flowering, plant height, mycosphaerella blight score, lodging and days to maturity at all four site years (Table 1). CDC Bronco was higher in mean 1000 seed weight (219 g) than 1-150-81 and 1-2347-144 with 207 and 205 g, respectively. CDC Bronco had the highest grain yield (2.83 $t{\cdot}ha^{-1}$), significantly greater than genotypes 1-150-81 and 1-2347-144 with 2.36 and 2.33 $t{\cdot}ha^{-1}$, respectively.

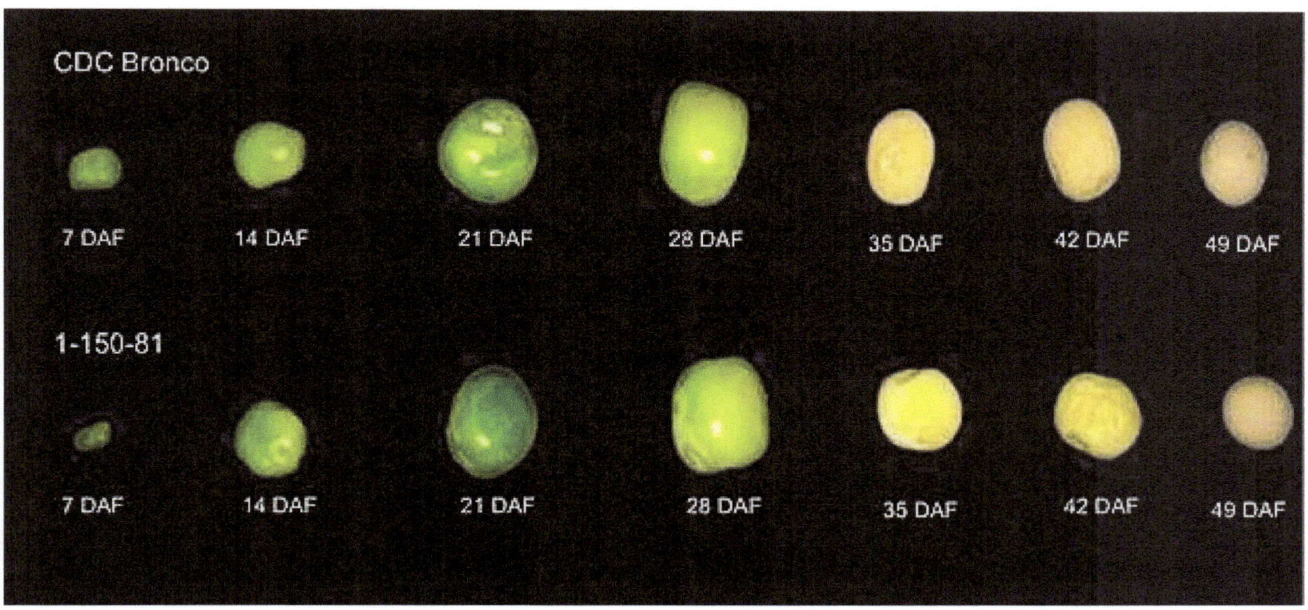

Figure 1. Developing seeds of pea (*Pisum sativum* L.). Indicated are CDC Bronco and low-phytate genotype 1-150-81. Representative photographs are shown from the Rosthern site in 2010. DAF stands for "days after flowering".

2.2. Accumulation of Phosphorus and Phosphorus-Containing Compounds during Seed Development

In seed coat extracts of CDC Bronco, 1-150-81 and 1-2347-144, there was no detectable amount of IP_6 or other inositol polyphosphates during seed development from 14 to 49 DAF (data not shown [27]). In addition, no lower inositol phosphate intermediates (IP_1, IP_3, IP_4, IP_5) other than phytic acid were detected in cotyledons of CDC Bronco, 1-150-81, and 1-2347-144 in any of the developmental stages analyzed (data not shown [27]).

Table 1. Agronomic traits of pea (*Pisum sativum* L.) cultivar CDC Bronco and low-phytate genotypes 1-150-81 and 1-2347-144 assessed at Saskatoon and Rosthern, Saskatchewan in 2010 and 2011. Values represent means ± SE, $n = 16$. Different letters within a column indicate a significant difference at $p < 0.05$ based on LSD test.

Genotype	Emergence Count (%) [d]	Days to Flower	Plant Height (cm) [e]	Mycosphaerella Blight Score (0–9 Scale) [f]	Lodging Score (1–9 Scale) [g]	Days to Mature	Grain Yield ($t \cdot ha^{-1}$) [h]	1000 Seed Weight (g)
CDC Bronco	56 ± 3.0 [a]	58 ± 1.0 [a]	79 ± 1.0 [a]	5.3 ± 0.1 [a]	5.4 ± 0.2 [a]	100 ± 1.0 [a]	2.83 ± 0.1 [a]	219 ± 2.2 [a]
1-150-81	55 ± 3.0 [a]	59 ± 1.0 [a]	78 ± 2.0 [a]	5.2 ± 0.1 [a]	5.3 ± 0.1 [a]	102 ± 1.0 [a]	2.36 ± 0.2 [b]	207 ± 1.6 [b]
1-2347-144	53 ± 3.0 [a]	59 ± 1.0 [a]	76 ± 2.0 [a]	5.3 ± 0.1 [a]	5.3 ± 0.1 [a]	101 ± 1.0 [a]	2.33 ± 0.2 [b]	205 ± 2.3 [b]

[d] Based on seedlings in a 1 m^2 section of each plot counted 5 weeks after planting; [e] Measured when the pod set was completed; [f] Assessed base on 0–9 scale, where 0 = no disease, 9 = completely blighted; [g] Assessed based on 1–9 scale, where 1 = erect, 9 = completely lodged; [h] Residual harvest weighed after sampling developing seeds (120 pods from each plot).

The PsMIPS protein sequence of CDC Bronco was used in a phylogenetic analysis with 14 other MIPS sequences obtained from the National Center for Biotechnology Information (NCBI) database for a variety of plants. The phylogenetic tree presented in Figure 5 shows the evolutionary divergence among the plant MIPS sequences analyzed. The analysis confirmed PsMIPS protein sequences of CDC Bronco and the two *lpa* genotypes, 1-150-81 and 1-2347-144, to be 100% identical and they clustered together. The monocots, *Z. mays*, *O. sativa*, *T. aestivum*, *and A. sativa* clearly cluster together in one branch. Two distinct sub-branches were obtained for dicots. *P. sativum*, *M. truncatula*, and *C. arietinum* clustered together in a branch, while *G. max* and *P. vulgaris* formed the other branch. MIPS from *R. communis*, *S. tuberosum*, *A. deliciosa*, and *G. hirsutum* were placed in a separate branch. *A. thaliana* and *B. napus* aligned together as an individual branch.

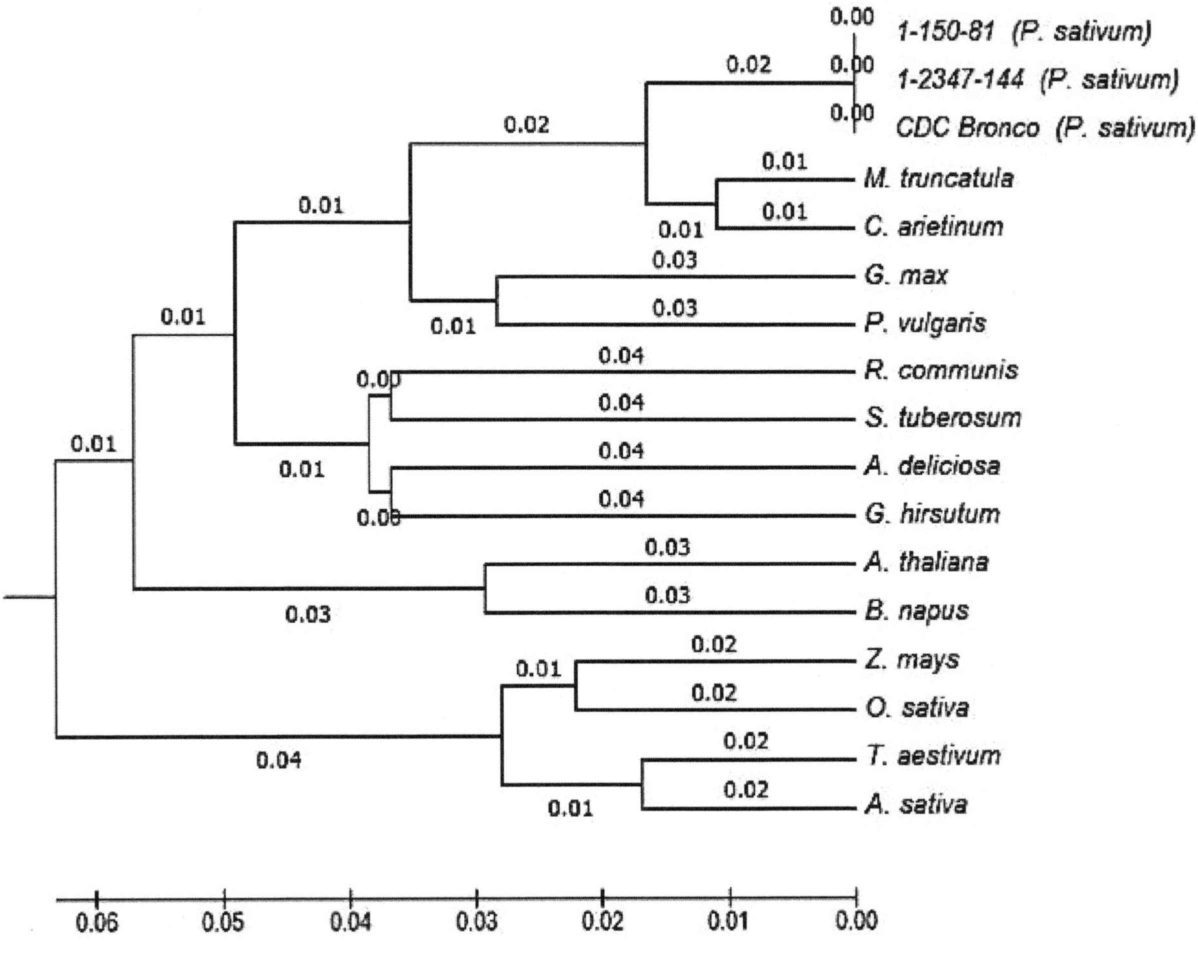

Figure 5. Phylogenetic analysis of MIPS proteins from different plant species including *Pisum sativum* L., CDC Bronco, and low-phytate genotypes 1-150-81 and 1-2347-144. GenBank Accession Numbers include: *Medicago truncatula*, XP_003601987.1; *Glycine max*, ABC55420.1; *Cicer arietinum*, NP_001266035.1; *Phaseolus vulgaris*, XP_007159720.1; *Ricinus communis*, ACU30131.1; *Zea mays*, ACG33827.1; *Oryza sativa*, BAA25729.1; *Triticum aestivum*, AEQ61648.1; *Arabidopsis thaliana*, NP_179812.1; *Actinidia deliciosa*, AFV31635.1; *Brassica napus* ACJ65004.1; *Solanum tuberosum*, XP_006366474.1; *Gossypium hirsutum*, ACJ11714.1; *Avena sativa*, BAB40956.2. The scale bar represents 0.05 substitutions per amino acid site, reflected in the lengths of the branches. Bootstrap values from 1000 iteration analyses are shown in italics.

2.4. *PsMIPS* Gene Expression

An examination of the *PsMIPS* transcript profile in developing seeds by sqPCR revealed decreased expression levels for CDC Bronco, 1-150-81, and 1-2347-144 at 49 DAF when compared to 14 DAF (Figure 6). In comparison to CDC Bronco, *PsMIPS* expression at 14 DAF was somewhat decreased in 1-150-81 and elevated in 1-2347-144, while levels were essentially the same at 49 DAF (Figure 6).

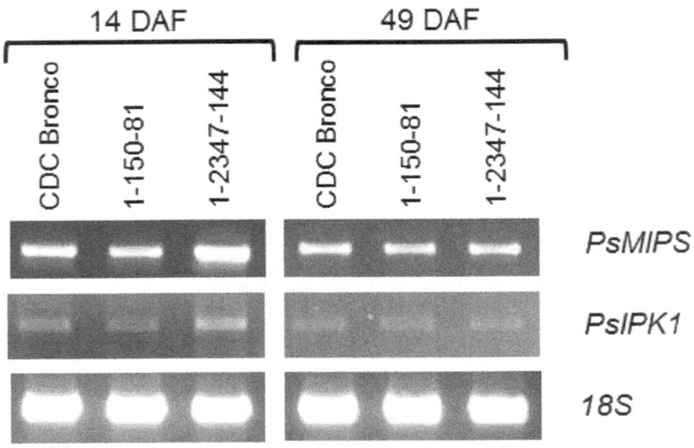

Figure 6. Transcript abundance of *PsMIPS* in developing seeds of pea (*Pisum sativum* L.) for CDC Bronco and low-phytate genotypes (1-150-81 and 1-2347-144) at 14 DAF and 49 DAF as indicated. A fragment of *18S* rRNA used as a loading control is also shown. Results are representative from a minimum of three independent experiments. DAF, days after flowering.

3. Discussion

3.1. Agronomic Traits in Low-Phytate Pea

Genes, alleles, and environmental conditions affect the agronomic performance of all crops. It has been demonstrated that *lpa* crops can be generated without major compromises in plant performance [30]. The agronomic performance of 1-150-81 and 1-2347-144 did not significantly differ from CDC Bronco when assessed for traits such as percent emergence, plant height, lodging score, days to flowering, and days to maturity. In soybean *lpa* genotypes, a reduced seedling emergence was observed [31]. The difference in temperature during seed filling was proposed as the reason for the reduced emergence in the low phytate genotypes. Warkentin *et al.* [14] showed that 1-150-81 and 1-2347-144 were slightly different from CDC Bronco in days to flowering and days to maturity. They reported that the 1-150-81 and 1-2347-144 flowered 3 days later and matured 2 to 3 days later than CDC Bronco. This was not observed in the present study. However, the grain yield at maturity was significantly different between 1-150-81, 1-2347-144 and CDC Bronco, with 1-150-81 and 1-2347-144 showing 16% and 17% lower grain yields than CDC Bronco, respectively. This is similar to differences in grain yield observed by Warkentin *et al.* [14]. Since the phytic acid biosynthesis pathway is active in most tissues of a plant, *lpa* mutations can also affect the vegetative processes apart from phytate accumulation [4]. Lower seed weight might be attributed to a reduction in starch accumulation resulting from a defective inositol phosphate synthesis pathway [32]. By targeting a specific gene or its expression to appropriate seed tissues, low-phytate crops can be produced restoring their seed weight and yield [4]. Embryo-specific silencing of expression of an ATP binding cassette (ABC) transporter in maize produced seeds with low-phytic acid with no adverse effect on seed weight [33]. When the myo-inositol methyltransferase (IMT) gene was transferred to *Brassica napus* through a transgenic approach, a 19% to 35% reduction in phytate was achieved without affecting the seed parameters [34]. In addition, rice low-phytate mutants produced through RNAi mediated seed-specific silencing of the inositol pentakisphosphate 2-kinase (IPK1) gene, had no undesirable agronomic characters [35]. Bregitzer and Raboy [36] showed that under irrigated and non-stressful production environments, the barley lpa1-1, lpa2-1, and lpa3-1 produced similar yields as that of the wild type barley. When the low phytate barley cultivar resulting from M640 was evaluated in distinct environments, it had a significantly higher yield when compared with commercial cultivars [37]. These studies provide evidence that low-phytate mutants can be produced in crop varieties without any compromise in agronomic traits.

3.2. Redistribution of Phosphorus in Low-Phytate Pea during Seed Development

A large fraction of nutrient P taken up by crop plants is ultimately packaged into seed phytic acid, and this single small molecule represents a major pool in the flux of P in the world's agricultural ecology. As phytic acid represents a significant portion of total seed P, the accumulation of phytate and inositol phosphates has been studied in different plant species. Grain crops typically contain about 10 $mg{\cdot}g^{-1}$ phytic acid on a seed dry weight basis, representing about 65 to 85% of seed total P [38,39]. Total P concentration typically ranges from 3 to 4 $mg{\cdot}g^{-1}$ in seed produced by grain crops, with phytic acid-P (P_{PhA}) ranging from 2 to 3 $mg{\cdot}g^{-1}$.

In this study, the concentration of IP_6 and its lower isomeric forms (IP_1, IP_3, IP_4 and IP_5) were analyzed separately in seed coats and cotyledons. There was no traceable amount of inositol phosphates in seed coats. The accumulation of IP_6 was continuous and linear throughout the stages of seed development and the highest concentration was observed at 49 DAF. The other *myo*-inositol polyphosphates (IP_3, IP_4 and IP_5) and IP_1 were not present in traceable amounts. In soybean wild-types [40] and maize *lpa1-1*, *lpa2-1* and wild types [10], phytate was reported to accumulate gradually during seed development. P_i concentration decreased during seed development in wild type, and total P levels remained relatively consistent. The maize *lpa* genotypes had little to no increase in phytate, and P_i concentration was high and did not decrease during development. In barley, the final levels and partitions of P forms are well documented, however it is unclear at what point during seed development the expression of the *lpa* genotype is initiated and how seed P accumulates in developing seeds of different barley *lpa* genotypes [41]. Israel *et al.* [42] compared the changes in seed phytic acid concentrations as well as *myo*-inositol phosphates during seed development between *lpa* and normal phytate genotypes and found that IP_3 content was relatively low in all genotypes and decreased during seed maturation. The other inositol phosphates such as IP_4 and IP_5 were not detected. Larson *et al.* [43] reported that a reduction in seed phytate results in a molar equivalent increase in P_i in rice. In the present study, P_i concentration in the two *lpa* genotypes was 72% to 84% higher than in CDC Bronco. Between the two *lpa* genotypes, 1-2347-144 had 11% more P_i than 1-150-81. Throughout the developing stages, P_i concentration was higher at 14 DAF and decreased as IP_6 accumulation began. Thus, there appears to be an inverse relationship between P_i accumulation and phytic acid content. The total P concentration showed no significant difference between CDC Bronco, 1-150-81, and 1-2347-144 except at 21 DAF where CDC Bronco and 1-2347-144 had 8% and 6% more total P, respectively, than 1-150-81. This demonstrates that despite variations in IP_6 and P_i levels, the total P content in the seeds is unaffected. The significantly higher Pi concentration when compared to CDC Bronco in *lpa* pea genotypes shows that the *lpa* genotypes tend to balance the total P levels within the seed to provide adequate amounts of P required for P-related mechanisms in the seed. Unlike the normal genetic and environmental effects that result in quantitative variation in seed total P, *lpa* mutants show large effects on the partitioning of P into phytic acid P and P_i.

3.3. Myo-Inositol-3-Phosphate Synthase (MIPS) Is Identical in Normal and Low-Phytate Pea Genotypes

Genetic mapping and comparison of the position of *lpa* mutation with *MIPS* loci in rice have been reported by Larson *et al.* [43]. They mapped the rice *MIPS* gene on to a locus on chromosome 3, which was orthologous to the *MIPS* gene near maize *lpa1* on chromosome 1S. Previously, *MIPS* gene expression proximal to the site of phytic acid synthesis during grain development in rice was demonstrated by Yoshida *et al.* [44]. Hitz *et al.* [26] confirmed a mutation in *MIPS* responsible for LR33 *lpa* mutation in soybean. Furthermore, higher *MIPS* expression was found in wheat genotypes with high phytic acid levels compared to an *lpa* genotype [45]. These findings have contributed to the interest in *MIPS* as a target for manipulation to produce low phytate crops. *lpa* genotypes of *Arabidopsis*, potato, rice, soybean, and canola have been generated by down-regulation (antisense, RNA interference, or cosuppression) or mutations in *MIPS* gene [24,46–48]. Bioinformatic and sqPCR analyses in this study provide evidence that the isolated MIPS gene was specific to phytic acid accumulation in seeds of both the low and normal phytate genotypes. However, the current study has shown that there are no mutations in the *PsMIPS* nucleotide sequences in the *lpa* pea genotypes.

Our sqPCR analysis revealed that at 14 DAF the expression of *PsMIPS* was higher in 1-2347-144 than CDC Bronco and 1-150-81. However, at 49 DAF, its expression was similar in both *lpa* genotypes and CDC Bronco. This further confirms that the reduction in phytate levels caused by *lpa* mutations is not directly controlled by *MIPS* expression. Moreover, mutations affecting *MIPS* are often associated with lower seed yield, seed viability, increased susceptibility to pathogens, and undesired morphology [46,47,49] and, although the seed weight in *lpa* pea genotypes was lower compared to their normal progenitor, the phytate reduction was not detrimental. Therefore, we suggest that the mutation could have occurred in other genes involved in the early stages of the phytate biosynthetic pathway, such as myo-inositol kinases and 2-phosphoglycerate kinase. For example, Shi *et al.* [50] found that the maize *lpa*2 mutant with 30% less phytic acid and three-fold more P_i was caused by a mutation in an inositol phosphate kinase gene. Stevenson-Paulik *et al.* [51] generated phytate-free seeds in *Arabidopsis* through disruption of inositol polyphosphate kinases. To gain further insights into the mutation causing the low phytate phenotype in pea, it is necessary to analyze the other enzymes related to phytate biosynthesis. The other possibility is that the mutation could have affected the transport of phytic acid to the vacuole [25]. Mutations in phytic acid ATP-binding cassette transporter reduce the phytic acid content significantly and may also result in absence of lower inositol phosphates [25]. This may possibly explain the fact that there was no lower inositol phosphates detected in this study.

4. Experimental Section

4.1. Plant Material and Growth Conditions

Seeds of two *lpa* field pea (*Pisum sativum* L.) genotypes (1-150-81 and 1-2347-144; [14], and their progenitor, a normal phytate genotype (CDC Bronco) [52] were obtained from the Crop Development Centre at the University of Saskatchewan. A four-replicate randomized complete block field trial was conducted at two locations in Saskatchewan (Rosthern and Saskatoon) in 2010 and 2011. Field trials were managed using standard techniques for field pea production in Saskatchewan that have been described earlier [14]. Seeding in both years was conducted between May 14 and May 18. Flowers were tagged at the time of flowering and developing seeds were taken 7 days after flowering (DAF) and every 7 d thereafter until maturity on day 49. Pea pods were collected in Ziploc® bags, transported to the laboratory on ice packs and stored at −80 °C until use. Final harvesting in both years was conducted between August 27 and September 17. Field plots were evaluated for several phenotypic parameters during the growing season in each year. These included percent emergence, days to flower, days to maturity, plant height, mycosphaerella blight score, lodging, grain yield and 1000 seed weight. The details of these determinations have been described previously [14].

For molecular studies, seeds of the aforementioned genotypes were sown in 15 cm plastic pots filled with Sunshine® Mix #3/LG3 (Sun Gro Horticulture Canada Ltd., Seba Beach, AB, Canada) and plants grown in a controlled environment chamber (PGR15; Conviron, Winnipeg, MB, Canada) set at 23/18 °C (day/night) temperatures with 16 h day length. The chamber was illuminated with fluorescent lights (T5/HO/835; Sylvania) to provide a photosynthetic photon flux density (PPFD) of 400 μmol photons $m^{-2} \cdot s^{-1}$. Plants were regularly provided with a water-soluble fertilizer (Plant-Prod® 20-20-20 Classic; Plant Products Co. Ltd., Brampton, ON, USA) and irrigated with ddH_2O as required. The experiment consisted of four biological replicates for each genotype. Each replicate consisted of four pots with two plants per pot. Developing seeds were collected at the same time intervals as described above and frozen in liquid nitrogen.

4.2. Extraction and Detection of Inositol Phosphates

Seed coats were dissected from the cotyledons and each component was freeze dried (FreeZone 6 Liter Console Freeze Dry System; Labconco, Kansas City, MO, USA) and stored at −80 °C until use. Seed coat or cotyledons were ground using glass beads to a 0.5 mm diameter using a custom designed mill powered with an inverter drive (SM-PLUS Sub-Micro; Leeson Corporation, Grafton, WI, USA). Samples were stored at −20 °C until extraction. Inositol phosphates were extracted from the samples (100 mg) using the extraction method described in [53] with minor modifications [54]. Extracts were filtered using 0.45 μm Acrodisc® syringe filters (25 mm; Pall Corporation, Port Washington, NY, USA) and used immediately for high-performance liquid chromatography (HPLC) analysis.

Detection of inositol phosphates was performed using anion-exchange high performance liquid chromatography (HPLC) on a Dionex ICS 3000 BioLC® system (Dionex, Sunnyvale, CA, USA) and a protocol similar to those reported previously [53,54]. The system consisted of an AS50 Autosampler with a 100 μL injection loop, an AS50 Thermal Compartment (set at 30 °C), a GP50 Gradient Pump and an ED50 Electrochemical Detector coupled with an Anion Self-Regenerating Suppressor (ASRS 300, 4-mm) running in external water mode and a current of 297 mA. Separation was achieved using an OmniPac PAX-100 analytical anion exchange column (Dionex; 8.5 μm, 4 × 250 mm) preceded by an OmniPac PAX-100 guard column (Dionex; 8.5 μm, 4 × 50 mm). Inositol phosphates were separated with a multi-step gradient using water purified by a Milli-Q Water System (Millipore, Milford, MA, USA) to a resistance of ≥18 mΩ. Mobile phases included were water (A), 200 mM NaOH (B), and water/isopropanol (50:50, v/v) (C). The total run time was 80 min, which included an equilibration to starting conditions. A column flow rate of 1.0 $mL \cdot min^{-1}$ was maintained for the mobile phase flow, with a linear gradient profile consisting of solvent A with the following proportions (v/v) of solvent B or C: 0–13 min, B = 6% and C = 12%; 13–30 min, B = 30% and C = 2%; 30–43 min, B = 56% and C = 2%; 43–55 min, B = 56% and C = 2%; 55–65 min, B = 61% and C = 8%; 65–80 min, B = 6% and C = 2%. Chromeleon software (Dionex) was used to plot chromatograms and analyze the data. A standard solution was prepared for each of IP_6 and its lower isomeric forms IP_1, IP_3, IP_4, IP_5 (#P8810, #I1267, #I7012, #I5514, #I9261, respectively; all from Sigma-Aldrich, St. Louis, MO, USA). This was used to establish retention times for these compounds, which were subsequently used to determine peak identity in the samples. Inositol phosphate quantification was afforded using external standard curves with R^2 values of 0.99 or greater for each compound. Concentrations for each of the inositol phosphate standards ranged from 1.25 to 25.0 ppm. Phytic acid phosphorus (P_{PhA}) was calculated as the number of moles of phytic acid/3.56 as described by [54], assuming all P_{PhA} comes from IP_6.

4.3. Analysis of Phosphorus Levels

Inorganic phosphorus (P_i) levels were determined as described by [14]. Ground cotyledon samples (50 mg) were extracted overnight at 4 °C in 1 mL of 0.4 M HCl followed by vigorous mixing. A 10 μL aliquot of the extract was aliquoted into a microtiter plate with 90 μL of ddH_2O and 100 μL of freshly prepared Chen's reagent. Chen's reagent contains 6 N H_2SO_4, 2.5% (w/v) $(NH_4)_2MoO_4$ (ammonium molybdate), 10% (w/v) ascorbic acid, and ddH_2O (1:1:1:2) [54]. The mixtures were incubated for two hours at room temperature before reading the A_{655} with a microplate absorbance spectrophotometer (xMark™; Bio-Rad Laboratories, Hercules, CA, USA) against a water blank. Standard curves of K_2HPO_4 were constructed ranging from 10 to 50 ppm with R^2 values of 0.95 or greater. Sample values were interpolated from these curves and expressed on a dry weight (DW) basis.

Total P in cotyledons was assayed by the wet ashing method [10]. Ground samples (50 mg) were incubated with 1 mL of concentrated (18.4 M) H_2SO_4 overnight at room temperature. Two hundred microliters of 30% (v/v) H_2O_2 were added and the samples were incubated in a heating block between 220 and 250 °C for 30 min. Samples were removed and allowed to cool at room temperature for 15 min. This cycle was repeated until the sample became clear. The volume of the samples was adjusted to 6.25 mL with ddH_2O and total extractable P was determined spectrophotometrically using the method of [55] as described above.

4.4. Reverse Transcriptase-Polymerase Chain Reaction (RT-PCR) and Sequencing

Total RNA was isolated from 75 mg of seed from 14 DAF and 49 DAF samples with an RNeasy® Plant Mini Kit (Qiagen, Valencia, CA, USA) using the RLC buffer according to the manufacturer's instructions. The RNA was eluted in 30 μL of RNase-free water. The samples were quantified (A_{260}) and purity (A_{260}:A_{280}) determined using a Nanodrop 8000 spectrophotometer (Thermo Fisher Scientific, Ottawa, ON, USA). Gel electrophoresis on denaturing 1.2% (w/v) agarose gels containing formaldehyde was used to examine the quality of the RNA. This was assessed by the sharpness of the rRNA bands and 2:1 ratio of 28S rRNA to 18S rRNA. Gels were run in 1× 3-(N-morpholino) propanesulfonic acid (MOPS) buffer and stained with ethidium bromide (Sambrook and Russell 2001). The isolated RNA was stored at −80 °C until further use. cDNA was synthesized from 100 ng of total RNA using the QuantiTect® Reverse Transcription Kit (Qiagen, Valencia, CA, USA) as described by the supplier.

The annotated Cool Season Food Legume Genome Database (http://www.coolseasonfoodlegume.org/) [56] was searched for *MIPS* and a contig in *Pisum sativum* identified (Pisum_sativum_v2_Contig5216) which contained the entire coding region of the *PsMIPS* gene, confirmed by translation and alignment with alfalfa (*Medicago sativa*) *MIPS* (GenBank Accession Number EF408869.1). Gene specific primers for *PsMIPS* were designed from this contig using the Primer-BLAST tool (http://www.ncbi.nlm.nih.gov/tools/primer-blast/) [57] at the National Center for Biotechnology Information (NCBI) to amplify a 1602-bp fragment encompassing the 1530-bp open reading frame (ORF). cDNA was amplified by polymerase chain reaction (PCR) using an iCycler (Bio-Rad Laboratories) and a thermostable DNA polymerase (Q5 High-Fidelity; New England Biolabs, Whitby, ON, USA). Forward (5'-ATGTTCATAGAGAGTTTCAAGGTTGAGAGT-3') and reverse (5'-GCTTGTGTTGGATTGGCTCCAGA-3') primers were used at a final concentration of 0.5 μM each, and 2 μL of the cDNA reaction was used as a template in the 25 μL PCR reaction. The following cycling conditions were used: cDNA denaturation at 98 °C for 30 s, then 35 cycles of denaturation at 98 °C for 10 s, annealing at 55 °C for 30 s, extension at 72 °C for 1 min, followed by a final extension at 72 °C for 2 min. PCR products were visualized on a 1.2% (w/v) agarose gel run in 1× Tris-Acetate-EDTA (TAE) and stained with ethidium bromide [58]. The PCR product corresponding to 1602-bp was excised from the gel and extracted using the QIAPrep Gel Extraction Kit (Qiagen) according to the manufacturer's instructions. Extractions from multiple PCR reactions were pooled and directly sequenced using the BigDye® Terminator v3.1 Cycle Sequencing Kit (Applied Biosystems, Foster City, CA, USA) at the National Research Council of Canada (NRC; Saskatoon, SK, Canada).

4.5. Semi-Quantitative Reverse Transcriptase-Polymerase Chain Reaction (sqRT-PCR)

Transcript levels of *PsMIPS* were examined using sqRT-PCR and performed using the Verso 1-Step RT-PCR ReddyMix™ Kit (Thermo Fisher Scientific, Ottawa, ON, USA) as recommended by the supplier. Targets were amplified using specific primers to *PsMIPS* or pea *18S* small subunit nuclear rRNA. Primers (forward primer, 5'-CATTGGAGGGCAAGTCTGGT-3'; reverse primer, 5'-CCAGCGGAGTCCTAAAAGCA-3') for pea *18S* (GenBank accession number U43011.1) generated a 510-bp amplicon that was used as a reference gene [59]. RNA was isolated as described above and 100 ng was used as template in the 25 μL reactions. The following cycling conditions were used: cDNA synthesis at 50 °C for 15 min, enzyme inactivation at 95 °C for 2 min, then 40 cycles of denaturation at 95 °C for 20 s, annealing at 55 °C for 30 s, extension at 72 °C for 1 min, followed by a final extension at 72 °C for 5 min. PCR products were visualized on a 1.2% (w/v) agarose gel run in 1× TAE and stained with ethidium bromide [58].

4.6. Sequence Analysis

Sequencing results were assembled using the BioEdit sequence alignment editor (v7.2.5, http://www.mbio.ncsu.edu/BioEdit/bioedit.html) [60]. Deduced amino acid sequences were obtained using the Translate Tool on the ExPASy SIB Bioinformatics Resource Portal (http://web.expasy.org/translate/). MIPS nucleotide and protein sequences were obtained from GenBank (www.ncbi.nih.nlm.gov) at NCBI and aligned using BLAST (http://blast.ncbi.nlm.nih.gov/Blast.cgi) [61] and CLUSTALW2.1 at the European Bioinformatics Institute (http://www.ebi.ac.uk/Tools/msa/clustalw2/) [62]. Domain analysis of MIPS amino acid sequences was performed using InterProScan (http://www.ebi.ac.uk/interpro/interproscan.html) [63]. Phylogeny analysis was performed using the MEGA6 program (http://www.megasoftware.net/) [64] and available amino acid sequences (NCBI protein sequence database; http://www.ncbi.nlm.nih.gov/nuccore).

4.7. Statistics

All statistical analyses were performed using SAS® 9.3 (SAS Institute Inc., Cary, NC, USA). The experimental data presented are mean values from two locations over two years. Levene's test was conducted to analyze the homogeneity of variance and the results represent means ± standard error (SE), based on four replications. Significant differences were determined by a one-way analysis of variance (ANOVA), $p < 0.05$. Differences among the means were analyzed by a least significant difference (LSD) post-hoc test at $p < 0.05$.

5. Conclusions

In conclusion, this study describes the accumulation patterns of phosphorus compounds in developing seeds of two low-phytate pea genotypes in comparison to their progenitor CDC Bronco. Based on the presented evidence, the possibility of a *MIPS* mutation being responsible for the low phytate trait in these pea genotypes is excluded. Interestingly, no accumulation of lower inositol polyphosphates was observed. Once the causative mutation is identified in these low-phytate pea genotypes, it can then be mapped and used in marker-assisted breeding to select low-phytate genotypes.

Acknowledgments

Financial support for this research was provided by the Saskatchewan Pulse Growers, the Natural Science and Engineering Research Council (NSERC) of Canada and the University of Saskatchewan. Work in GRG's lab was supported by a NSERC Discovery grant. The technical support of the pulse crop breeding crew at the Crop Development Centre is gratefully acknowledged.

Author Contributions

ASKS performed the fieldwork, phenotypic assays, molecular biology experiments, and drafted the manuscript. CB co-ordinated the HPLC assays and contributed to data interpretation. GCA co-ordinated the calorimetric assays. FG co-conceptualized the study and assisted with the data interpretation. GRG co-ordinated the molecular biology experiments, interpreted sequence data, and helped to draft the manuscript. TDW conceptualized and co-ordinated the study. All authors read and approved the final manuscript.

Appendix

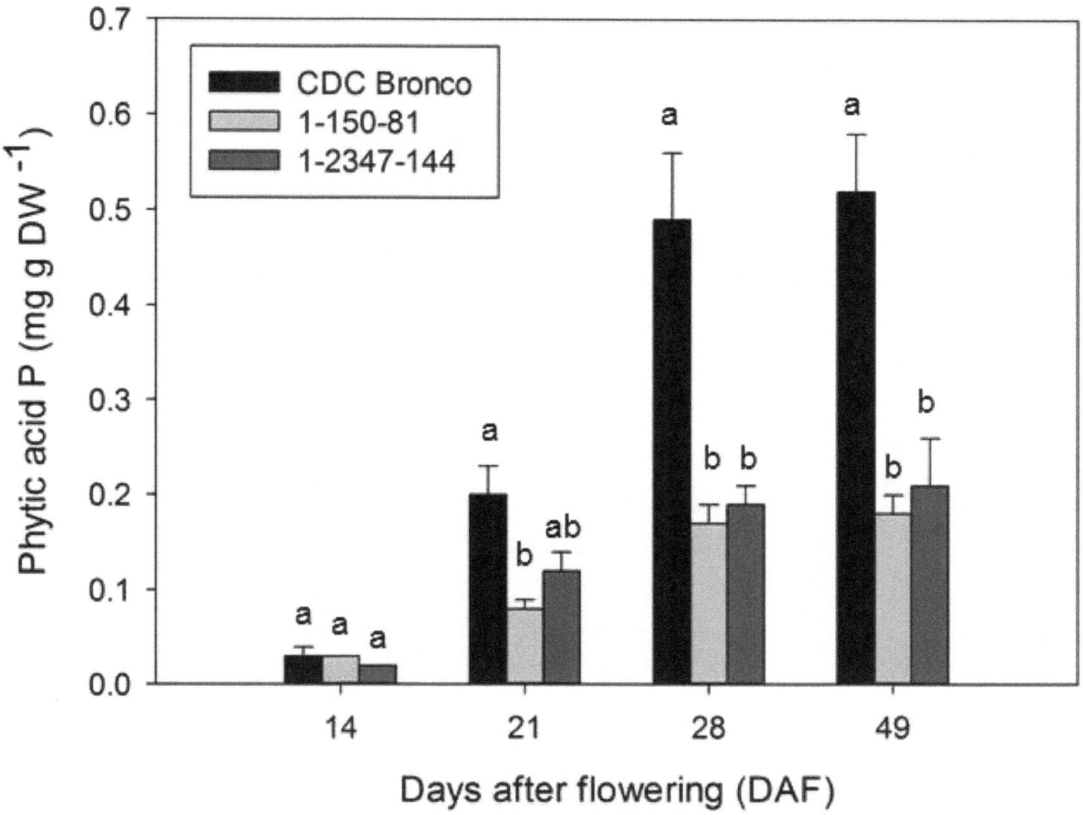

Figure A1. Content of phytic acid P (P_{PhA}) in developing seeds of pea (*Pisum sativum* L.) for CDC Bronco (black bars) and low-phytate genotypes 1-150-81 (light grey bars) and 1-2347-144 (dark grey bars). Values represent means ± SE, $n = 4$. Different letters associated with bars within each DAF are significantly different at $p < 0.05$. DAF, days after flowering; DW, dry weight; P, phosphorus P_{PhA}, phytic acid phosphorus.

```
CDC Bronco        1 CCTTTTCTCTTTGTGATTTCCATATTCACCAAAATGTTCATAGAGAGTTTCAAGGTTGAG
1-150-81          1 --TTTTCTCTTTGTGATTTCCATATTCACCAAAATGTTCATAGAGAGTTTCAAGGTTGAG
1-2347-144        1 -CTTTTCTCTTTGTGATTTCCATATTCACCAAAATGTTCATAGAGAGTTTCAAGGTTGAG
M. truncatula     1 ---------------------------------ATGTTTATCGAGAATTTCAAAGTCGAG
G. max            1 -----------------------GTGAAAAATAATGTTCATCGAGAATTTTAAGGTTGAG

CDC Bronco       61 AGTCCTAACGTGAAGTACACAGACACAGAGATTCAGTCTGTGTACAGTTACGAAACAACT
1-150-81         59 AGTCCTAACGTGAAGTACACAGACACAGAGATTCAGTCTGTGTACAGTTACGAAACAACT
1-2347-144       60 AGTCCTAACGTGAAGTACACAGACACAGAGATTCAGTCTGTGTACAGTTACGAAACAACT
M. truncatula    28 AGTCCCAATGTTAAGTACACTGAAACTGAGATTCAATCCGTTTACAATTACGAAACCACT
G. max           38 TGTCCTAATGTGAAGTACACCGAGACTGAGATTCAGTCCGTGTACAACTACGAAACCACC

CDC Bronco      121 GAACTTGTTCATCAGAACAGAAATGACACTTATCAATGGATTGTTAACCCTAAAACTGTG
1-150-81        119 GAACTTGTTCATCAGAACAGAAATGACACTTATCAATGGATTGTTAACCCTAAAACTGTG
1-2347-144      120 GAACTTGTTCATCAGAACAGAAATGACACTTATCAATGGATTGTTAACCCTAAAACTGTG
M. truncatula    88 GAACTTGTTCATGAAAATCGTAATGGCACTTATCAGTGGATTGTTAAACCTAAAACTGTT
G. max           98 GAACTTGTTCACGAGAACAGGAATGGCACCTATCAGTGGATTGTCAAACCCAAATCTGTC

CDC Bronco      181 AAATATGAATTTAAAACCGAAACTCATGTTCCTAAATTGGGGGTAATGCTTGTGGGATGG
1-150-81        179 AAATATGAATTTAAAACCGAAACTCATGTTCCTAAATTGGGGGTAATGCTTGTGGGATGG
1-2347-144      180 AAATATGAATTTAAAACCGAAACTCATGTTCCTAAATTGGGGGTAATGCTTGTGGGATGG
M. truncatula   148 AAATATGAATTTAAAACCGATATTCATGTCCCTAAATTGGGGGTAATGCTTGTGGGATGG
G. max          158 AAATACGAATTTAAAACCAACATCCATGTTCCTAAATTAGGGGTAATGCTTGTGGGTTGG

CDC Bronco      241 GGTGGAAACAACGGTTCAACCCTTACCGGTGGTGTTATTGCTAATCGAGAGGGTATTTCA
1-150-81        239 GGTGGAAACAACGGTTCAACCCTTACCGGTGGTGTTATTGCTAATCGAGAGGGTATTTCA
1-2347-144      240 GGTGGAAACAACGGTTCAACCCTTACCGGTGGTGTTATTGCTAATCGAGAGGGTATTTCA
M. truncatula   208 GGTGGAAACAACGGTTCAACCCTTACCGGTGGTGTTATTGCTAACAGAGAGGGAATTTCA
G. max          218 GGTGGAAACAACGGCTCAACCCTCACCGGTGGTGTTATTGCTAACCGAGAGGGCATTTCA

CDC Bronco      301 TGGGCAACGAAAGATAATATTCAACAAGCGAATTACTTTGGTTCTCTCACTCAAGCTTCA
1-150-81        299 TGGGCAACGAAAGATAATATTCAACAAGCGAATTACTTTGGTTCTCTCACTCAAGCTTCA
1-2347-144      300 TGGGCAACGAAAGATAATATTCAACAAGCGAATTACTTTGGTTCTCTCACTCAAGCTTCA
M. truncatula   268 TGGGCCACGAAAGATAAGATTCAACAAGCCAATTACTTTGGATCCTTGACTCAAGCTTCA
G. max          278 TGGGCTACAAAGGACAAGATTCAACAAGCCAATTACTTTGGCTCCCTCACCCAAGCCTCA

CDC Bronco      361 GCTATTCGTGTTGGATCTTTCCAAGGAGAGGAAATTCATGCTCCCTTCAAGAGCTTGTTA
1-150-81        359 GCTATTCGTGTTGGATCTTTCCAAGGAGAGGAAATTCATGCTCCCTTCAAGAGCTTGTTA
1-2347-144      360 GCTATTCGTGTTGGATCTTTCCAAGGAGAGGAAATTCATGCTCCCTTCAAGAGCTTGTTA
M. truncatula   328 GCTATTCGAGTTGGATCTTTTCAAGGAGAGGAAATTCATGCTCCTTTCAAGAGCCTCCTT
G. max          338 GCTATCCGAGTTGGGTCCTTCCAGGGAGAGGAAATCTATGCCCCATTCAAGAGCCTGCTT
```

Figure A2. *Cont.*

```
CDC Bronco      421 CCAATGGTCAACCCTGATGACATTGTTTTTGGTGGATGGGATATCAGTAACATGAACCTT
1-150-81        419 CCAATGGTCAACCCTGATGACATTGTTTTTGGTGGATGGGATATCAGTAACATGAACCTT
1-2347-144      420 CCAATGGTCAACCCTGATGACATTGTTTTTGGTGGATGGGATATCAGTAACATGAACCTT
M. truncatula   388 CCAATGGTCAACCCCGATGACATTGTTTTTGGTGGATGGGATATCAGTGACATGAACCTT
G. max          398 CCAATGGTTAACCCTGACGACATTGTGTTTGGGGGATGGGATATCAGCAACATGAACCTG

CDC Bronco      481 GCTGATGCCATGGCTAGGGCCAGGGTTTTCGACATTGATTTGCAAAAGCAATTGAGGCCT
1-150-81        479 GCTGATGCCATGGCTAGGGCCAGGGTTTTCGACATTGATTTGCAAAAGCAATTGAGGCCT
1-2347-144      480 GCTGATGCCATGGCTAGGGCCAGGGTTTTCGACATTGATTTGCAAAAGCAATTGAGGCCT
M. truncatula   448 GCTGATGCCATGGCTAGGGCCAGGGTTTTTGACATTGATTTGCAAAAGCAATTGAGGCCT
G. max          458 GCTGATGCCATGGCCAGGGCAAAGGTGTTTGACATCGATTTGCAGAAGCAGTTGAGGCCT

CDC Bronco      541 TATATGGAATCCATGGTTCCACTCCCCGGCATCTATGACCCGGATTTCATTGCTGCCAAT
1-150-81        539 TATATGGAATCCATGGTTCCACTCCCCGGCATCTATGACCCGGATTTCATTGCTGCCAAT
1-2347-144      540 TATATGGAATCCATGGTTCCACTCCCCGGCATCTATGACCCGGATTTCATTGCTGCCAAT
M. truncatula   508 TATATGGAATCCATGGTTCCACTTCCCGGTATCTATGACCCGGATTTCATTGCTGCTAAT
G. max          518 TACATGGAATCCATGCTTCCACTCCCCGGAATCTATGACCCGGATTTCATTGCTGCCAAC

CDC Bronco      601 CAAGGTGAACGTGCAAATAATGTTATTAAGGGTACAAAGAGAGAACAAATTAACCAAATC
1-150-81        599 CAAGGTGAACGTGCAAATAATGTTATTAAGGGTACAAAGAGAGAACAAATTAACCAAATC
1-2347-144      600 CAAGGTGAACGTGCAAATAATGTTATTAAGGGTACAAAGAGAGAACAAATTAACCAAATC
M. truncatula   568 CAAGGAGAACGTGCGAATAACGTTATCAAGGGTACAAAGAGAGAACAAATCAACCAAATC
G. max          578 CAAGAGGAGCGTGCCAACAACGTCATCAAGGGCACAAAGCAAGAGCAAGTTCAACAAATC

CDC Bronco      661 ATCAAAGACATTCGGGAATTTAAGGAAGCAAACAAAGTAGACAGGGTTGTTGTTCTCTGG
1-150-81        659 ATCAAAGACATTCGGGAATTTAAGGAAGCAAACAAAGTAGACAGGGTTGTTGTTCTCTGG
1-2347-144      660 ATCAAAGACATTCGGGAATTTAAGGAAGCAAACAAAGTAGACAGGGTTGTTGTTCTCTGG
M. truncatula   628 ATCAAAGACATTAAGGAATTTAAGGAAGCAAACAAAGTTGACAGGGTTGTTGTACTCTGG
G. max          638 ATCAAAGACATCAAGGCGTTTAAGGAAGCCACCAAAGTGGACAAGGTGGTTGTACTGTGG

CDC Bronco      721 ACTGCCAACACAGAGAGGTACAGTAATTTAGTTGTGGGACTCAATGACACCACAGAGAAC
1-150-81        719 ACTGCCAACACAGAGAGGTACAGTAATTTAGTTGTGGGACTCAATGACACCACAGAGAAC
1-2347-144      720 ACTGCCAACACAGAGAGGTACAGTAATTTAGTTGTGGGACTCAATGACACCACAGAGAAC
M. truncatula   688 ACTGCCAACACAGAGAGGTACAGTAACTTAGTTGTAGGACTCAATGACACCATGGAGAAC
G. max          698 ACTGCCAACACAGAGAGGTACAGTAATTTGGTTGTGGGCCTTAATGACACCATGGAGAAT

CDC Bronco      781 CTTTTTGCTGCAGTGGACAGAAATGAGTCTGAGATTTCTCCTTCCACCCTGTTTGGCATT
1-150-81        779 CTTTTTGCTGCAGTGGACAGAAATGAGTCTGAGATTTCTCCTTCCACCCTGTTTGGCATT
1-2347-144      780 CTTTTTGCTGCAGTGGACAGAAATGAGTCTGAGATTTCTCCTTCCACCCTGTTTGGCATT
M. truncatula   748 CTTTTTGCTGCTGTGGACAGAAATGAGTCTGAGATTTCACCTTCCACCCTGTTTGCCATT
G. max          758 CTCTTGGCTGCTGTGGACAGAAATGAGGCTGAGATTTCTCCTTCCACCTTGTATGCCATT
```

Figure A2. *Cont.*

```
CDC Bronco      841 GCTTGTGTTATGGAAAATGTTCCTTTCATCAATGGAAGCCCTCAGAACACTTTTGTTCCA
1-150-81        839 GCTTGTGTTATGGAAAATGTTCCTTTCATCAATGGAAGCCCTCAGAACACTTTTGTTCCA
1-2347-144      840 GCTTGTGTTATGGAAAATGTTCCTTTCATCAATGGAAGCCCTCAGAACACTTTTGTTCCA
M. truncatula   808 GCTTGTGTTATGGAGAATGTTCCTTTCATCAATGGAAGTCCTCAGAACACTTTTGTTCCA
G. max          818 GCTTGTGTTATGGAAAATGTTCCTTTCATTAATGGAAGCCCTCAGAACACTTTTGTACCA

CDC Bronco      901 GGGCTTATTGATCTTGCCATCAAGAACAACACCTTGATTGGTGGCGATGACTTCAAGAGT
1-150-81        899 GGGCTTATTGATCTTGCCATCAAGAACAACACCTTGATTGGTGGCGATGACTTCAAGAGT
1-2347-144      900 GGGCTTATTGATCTTGCCATCAAGAACAACACCTTGATTGGTGGCGATGACTTCAAGAGT
M. truncatula   868 GGGCTTATTGATCTTGCCATCAAGAACAACTGTTTGATTGGTGGTGATGATTTCAAAAGT
G. max          878 GGGCTGATTGATCTTGCCATCGCGAGGAACACTTTGATTGGTGGAGATGACTTCAAGAGT

CDC Bronco      961 GGTCAGACCAAAATGAAATCTGTTTTGGTTGATTTCCTTGTTGGAGCTGGTATCAAGCCA
1-150-81        959 GGTCAGACCAAAATGAAATCTGTTTTGGTTGATTTCCTTGTTGGAGCTGGTATCAAGCCA
1-2347-144      960 GGTCAGACCAAAATGAAATCTGTTTTGGTTGATTTCCTTGTTGGAGCTGGTATCAAGCCA
M. truncatula   928 GGTCAGACCAAAATGAAATCTGTTTTGGTGGATTTCCTTGTAGGAGCTGGTATCAAGCCA
G. max          938 GGTCAGACCAAAATGAAATCTGTGTTGGTTGATTTCCTTGTGGGGGCTGGTATCAAGCCA

CDC Bronco     1021 ACGTCGATAGTGAGTTACAATCATCTTGGAAACAATGATGGTATGAACCTCTCAGCACCA
1-150-81       1019 ACGTCGATAGTGAGTTACAATCATCTTGGAAACAATGATGGTATGAACCTCTCAGCACCA
1-2347-144     1020 ACGTCGATAGTGAGTTACAATCATCTTGGAAACAATGATGGTATGAACCTCTCAGCACCA
M. truncatula   988 ACATCAATAGTGAGTTACAATCATCTTGGAAACAATGATGGTATGAACCTCTCAGCCCCA
G. max          998 ACATCTATAGTCAGTTACAACCATCTGGGAAACAATGATGGTATGAATCTTTCGGCTCCA

CDC Bronco     1081 CAAACCTTCCGCTCCAAGGAAATCTCCAAGAGCAACGTTGTTGACGATATGGTCAACAGC
1-150-81       1079 CAAACCTTCCGCTCCAAGGAAATCTCCAAGAGCAACGTTGTTGACGATATGGTCAACAGC
1-2347-144     1080 CAAACCTTCCGCTCCAAGGAAATCTCCAAGAGCAACGTTGTTGACGATATGGTCAACAGC
M. truncatula  1048 CAAACCTTCCGCTCCAAGGAAATTTCCAAGAGCAACGTTGTTGACGATATGGTCAACAGC
G. max         1058 CAAACTTTCCGTTCCAAGGAAATCTCCAAGAGCAACGTTGTTGATGATATGGTCAACAGC

CDC Bronco     1141 AACGCTATCCTCTATGCGCCTGGTGAACATCCTGACCATGTTGTAGTCATTAAGTATGTG
1-150-81       1139 AACGCTATCCTCTATGCGCCTGGTGAACATCCTGACCATGTTGTAGTCATTAAGTATGTG
1-2347-144     1140 AACGCTATCCTCTATGCGCCTGGTGAACATCCTGACCATGTTGTAGTCATTAAGTATGTG
M. truncatula  1108 AATGCCATCCTCTATGCACCTGGCGAACATCCTGATCATGTTGTAGTCATTAAGTATGTG
G. max         1118 AATGCCATCCTCTATGAGCCTGGTGAACATCCAGACCATGTTGTTGTTATTAAGTATGTG

CDC Bronco     1201 CCATACGTCGGAGACAGCAAGAGAGCCATGGACGAGTATACTTCGGAAATATTCATGGGT
1-150-81       1199 CCATACGTCGGAGACAGCAAGAGAGCCATGGACGAGTATACTTCGGAAATATTCATGGGT
1-2347-144     1200 CCATACGTCGGAGACAGCAAGAGAGCCATGGACGAGTATACTTCGGAAATATTCATGGGT
M. truncatula  1168 CCTTATGTTGGTGACAGCAAGAGAGCTATGGATGAGTACACTTCGGAAATTTTCATGGGT
G. max         1178 CCTTACGTAGGGGACAGCAAGAGAGCCATGGATGAGTACACTTCAGAGATATTCATGGGT
```

Figure A2. *Cont.*

```
CDC Bronco    1261 GGAAAGAACACTATTGTGTTGCACAACACATGTGAGGATTCCCTCTTGGCTGCCCCTATT
1-150-81      1259 GGAAAGAACACTATTGTGTTGCACAACACATGTGAGGATTCCCTCTTGGCTGCCCCTATT
1-2347-144    1260 GGAAAGAACACTATTGTGTTGCACAACACATGTGAGGATTCCCTCTTGGCTGCCCCTATT
M. truncatula 1228 GGGAAGAACACTATTGTGTTGCACAACACATGTGAGGACTCCCTTTTGGCTGCTCCTATT
G. max        1238 GGAAAGAGCACCATTGTTTTGCACAACACATGCGAGGATTCCCTCTTAGCTGCTCCTATT

CDC Bronco    1321 ATCTTGGACTTGGTTCTTCTTGCTGAGCTTAGTACTAGAATTCAGTTCAAATCTGAAGCT
1-150-81      1319 ATCTTGGACTTGGTTCTTCTTGCTGAGCTTAGTACTAGAATTCAGTTCAAATCTGAAGCT
1-2347-144    1320 ATCTTGGACTTGGTTCTTCTTGCTGAGCTTAGTACTAGAATTCAGTTCAAATCTGAAGCT
M. truncatula 1288 ATCTTGGACTTGGTTCTTCTTGCTGAGCTTAGCACTAGAATTCAGTTTAAATCTGAAGCT
G. max        1298 ATCTTGGACTTGGTCCTTCTTGCTGAGCTCAGCACTAGAATCGAGTTTAAAGCTGAAAAT

CDC Bronco    1381 GAGAACAAGTTTCACACATTCCACCCTGTTGCTACCATCCTCAGTTATCTGACCAAGGCT
1-150-81      1379 GAGAACAAGTTTCACACATTCCACCCTGTTGCTACCATCCTCAGTTATCTGACCAAGGCT
1-2347-144    1380 GAGAACAAGTTTCACACATTCCACCCTGTTGCTACCATCCTCAGTTATCTGACCAAGGCT
M. truncatula 1348 GAGAACAAGTTCCACACCTTCCACCCTGTTGCTACCATCCTCAGTTATCTGACCAAGGCT
G. max        1358 GAGGGAAAATTCCACTCATTCCACCCAGTTGCTACCATCCTCAGCTACCTCACCAAGGCT

CDC Bronco    1441 CCTCTGGTTCCACCAGGTACACCAGTGGTGAATGCATTGTCCAAGCAGCGAGCGATGCTG
1-150-81      1439 CCTCTGGTTCCACCAGGTACACCAGTGGTGAATGCATTGTCCAAGCAGCGAGCGATGCTG
1-2347-144    1440 CCTCTGGTTCCACCAGGTACACCAGTGGTGAATGCATTGTCCAAGCAGCGAGCGATGCTG
M. truncatula 1408 CCTCTGGTTCCACCGGGCACACCAGTGGTGAATGCATTGTCAAAGCAGCGAGCAATGCTT
G. max        1418 CCTCTGGTTCCACCGGGTACACCAGTGGTGAATGCATTGTCAAAGCAGCGTGCAATGCTG

CDC Bronco    1501 GAAAACATCATGAGAGCTTGTGTTGGATTGGCTCCAGAAAACAACATGATCCTTGAGTAC
1-150-81      1499 GAAAACATCATGAGAGCTTGTGTTGGATTGGCTCCAGAAAACAACATGATCCTTGAGTAC
1-2347-144    1500 GAAAACATCATGAGAGCTTGTGTTGGATTGGCTCCAGAAAACAACATGATCCTTGAGTAC
M. truncatula 1468 GAAAACATCATGAGAGCTTGTGTTGGATTGGCTCCAGAGAACAACATGATCCTCGAGTAC
G. max        1478 GAAAACATAATGAGGGCTTGTGTTGGATTGGCCCCAGAGAATAACATGATTCTCGAGTAC

CDC Bronco    1561 AAGTGAAGCAGGGGATAGAA----------------------------------------
1-150-81      1559 AAGTGAAGCAGGGGATAGAA----------------------------------------
1-2347-144    1560 AAGTGAAGCAGGGGATAGAA----------------------------------------
M. truncatula 1528 AAGTGA------------------------------------------------------
G. max        1538 AAGTGAAGCATGGGACCGAAGAATAATATAGTTGGGGTAGCCTAGCTGAATGTTTTATGT

CDC Bronco    1581 --------------------------------------------------TCATTAGTGA
1-150-81      1579 --------------------------------------------------TCATTAGTGA
1-2347-144    1580 --------------------------------------------------TCATTAGTGA
M. truncatula      ------------------------------------------------------------
G. max        1598 TAATAATATGTTTGCTTATAATTTTGCAAGTGTAATTGAATGCATCAGCTTCATTAATGC

CDC Bronco    1591 TTAAT-------------------------------------------------------
1-150-81      1589 TTAATCC-----------------------------------------------------
1-2347-144    1590 TTA---------------------------------------------------------
M. truncatula      ------------------------------------------------------------
G. max        1658 TTTAGAGCGGGGCATATTCTGTTTACTAGGAACATGAATGAATGTAGTATAATTTTGTGT
```

Figure A2. *Cont.*

```
CDC Bronco           ----------------------
1-150-81             ----------------------
1-2347-144           ----------------------
M. truncatula        ----------------------
G. max          1718 AAAAAAAAAAAAAAAAAAAAAA
```

Figure A2. Nucleotide sequence alignment of *PsMIPS* for CDC Bronco and low-phytate genotypes 1-150-81 and 1-2347-144. Also included for comparison are nucleotide sequences for *MIPS* obtained from GenBank for *Medicago truncatula* (XM_003601939.1) and *Glycine max* (AY038802.1). Nucleotide differences are indicated in black boxes.

Conflicts of Interest

The authors declare no conflict of interest.

References

1. Cosgrove, D.J. *Inositol Phosphates: Their Chemistry, Biochemistry, and Physiology*; Elsevier: New York, NY, USA, 1980.
2. O'Dell, B.L.; de Boland, A.R.; Koirtyohann, S.R. Distribution of phytate and nutritionally important elements among the morphological components of cereal grains. *J. Agric. Food. Chem.* **1972**, *20*, 718–721.
3. Lott, J.N.A. Accumulation of seed reserves of phosphorus and other minerals. In *Seed Physiology*; Murray, D.R., Eds.; Academic Press: Sydney, Australia, 1984; Volume 1, pp. 154–196.
4. Raboy, V. Seeds for a better future: "Low phytate" grains help to overcome malnutrition and reduce pollution. *Trends Plant Sci*. **2001**, *6*, 458–462.
5. Reddy, N.R. Occurrence, distribution, content and dietary intake of phytate. In *Food Phytates*; Reddy, N.R., Sathe, S.K., Eds.; CRC Press: New York, NY, USA, 2002; pp. 25–51.
6. Erdman, J.W., Jr. Bioavailability of trace minerals from cereals and legumes. *Cereal Chem.* **1980**, *58*, 21–26.
7. Sharpley, A.N.; Charpa, S.C.; Wedepohl, R.; Sims, J.Y.; Daniel, T.C.; Reddy, K.R. Managing agricultural phosphorus for protection of surface waters: Issues and Options. *J. Environ. Qual.* **1994**, *23*, 437–451.
8. Raboy, V.; Young, K.A.; Dorsch, J.A.; Cook, A. Genetics and breeding of seed phosphorus and phytic acid. *J. Plant Physiol*. **2001**, *158*, 489–497.
9. Raboy, V.; Gerbasi, P. Genetics of myo-inositol phosphate synthesis and accumulation. In *Myo Inositol Phosphates, Phosphoinositides and Signal Transduction*; Biswas, B.B., Ed.; Plenum Publ. Co.: New York, NY, USA, 1996; pp. 257–285.
10. Raboy, V.; Gerbasi, P.F.; Young, K.A.; Stoneberg, S.D.; Pickett, S.G.; Bauman, A.T.; Murthy, P.N.; Sheridan, W.F.; Ertl, D.S. Origin and seed phenotype of maize *low phytic acid 1–1* and *low phytic acid 2–1*. *Plant Physiol*. **2000**, *124*, 355–368.
11. Larson, S.R.; Young, K.A.; Cook, A.; Blake, T.K.; Raboy, V. Linkage mapping of two mutations that reduce the phytic acid content of barley grain. *Theor. Appl. Genet*. **1998**, *97*, 141–146.

12. Rasmussen, S.K.; Hatzack, F. Identification of two low-phytate barley *(Hordeum vulgare* L.) grain mutants by TLC and genetic analysis. *Heriditas* **1998**, *129*, 107–112.
13. Wilcox, J.; Premachandra, G.; Young, K.; Raboy, V. Isolation of high seed inorganic P, low-phytate soybean mutants. *Crop Sci.* **2000**, *40*, 1601–1605.
14. Warkentin, T.D.; Delgerjav, T.; Arganosa, G.; Rehman, A.U.; Bett, K.E.; Anbessa, Y.K.; Rossnagel, B.; Raboy, V. Development and characterization of low-phytate pea. *Crop Sci.* **2012**, *52*, 74–78.
15. Ertl, D.S.; Young, K.A.; Raboy, V. Plant genetic approaches to phosphorus management in agricultural production. *J. Environ. Qual.* **1998**, *27*, 299–304.
16. Loewus, F.A.; Murthy, P.N. Myo-inositol metabolism in plants. *Plant Sci.* **2000**, *150*, 1–19.
17. Stevenson, J.M.; Perera, I.Y.; Heilmann, I.; Persson, S.; Boss, W.F. Inositol signaling and plant growth. *Trends Plant Sci.* **2000**, *5*, 252–258.
18. Loewus, F.A.; Loewus, M.W. Myo-inositol: Its biosynthesis and metabolism. *Annu. Rev. Plant Physiol.* **1983**, *34*, 137–161.
19. Loewus, F.A. Inositol biosynthesis. In *Inositol Metabolism in Plants*; Morre, D.J., Boss, W.F., Loewus, F.A., Eds.; Wiley-Liss: New York, NY, USA, 1990; pp. 13–19.
20. Pilu, R.; Panzeri, D.; Gavazzi, G.; Rasmussen, S.K.; Consonni, G.; Nielsen, E. Phenotypic, genetic and molecular characterization of a maize low phytic acid mutant (*lpa*241). *Theor. Appl. Genet.* **2003**, *107*, 980–987.
21. Kuwano, M.; Ohyama, A.; Tanaka, Y.; Mimura, T.; Takaiwa, F.; Yoshida, K.T. Molecular breeding for transgenic rice with low-phytic-acid phenotype through manipulating myo-inositol 3-phosphate synthase gene. *Mol. Breed.* **2006**, *18*, 263–272.
22. Kuwano, M.; Mimura, T.; Takaiwa, F. Generation of stable "low phytic acid" transgenic rice through antisense repression of the 1d-myo-inositol 3-phosphate synthase gene (RINO1) using the 18-kDa oleosin promoter. *Plant Biotech. J.* **2009**, *7*, 96–105.
23. Georges, F.; Hussain, A.A.K.; Keller, W.A. Method for Reducing Phytate in Canola Meal Using Genetic Manipulation Involving Myo-Inositol 1-Phosphate Synthase Gene. Patent WO 00/73473 A1, 2006.
24. Nunes, A.C.S.; Vianna, G.R.; Cuneo, F.; Amaya-Farfán, J.; de Capdeville, G.; Rech, E.L.; Aragão, F.J.L. RNAi-mediated silencing of the *myo*-inositol-1-phosphate synthase gene (*GmMIPS1*) in transgenic soybean inhibited seed development and reduced phytate content. *Planta* **2006**, *224*, 125–132.
25. Panzeri, D.; Cassani, E.; Doria, E.; Tagliabue, G.; Forti, L.; Campion, B.; Bollini, R.; Brearley, C.A.; Pilu, R.; Nielsen, E.; *et al.* A defective ABC transporter of the MRP family, responsible for the bean *lpa*1 mutation, affects the regulation of the phytic acid pathway, reduces seed myo-inositol and alters ABA sensitivity. *New Phytol.* **2011**, *191*, 70–83.
26. Hitz, D.W.; Thomas, J.C.; Phil, S.K.; Scott, A.S. Biochemical and molecular characterization of a mutation that confers a decreased raffinosaccharide and phytic acid phenotype on soybean seeds. *Plant Physiol.* **2002**, *128*, 650–660.
27. Shunmugam, A.S.K.; Bock, C.; Arganosa, G.C.; Gray, R.G.; Georges, F.; Warkentin, T.D. Department of Plant Sciences, University of Saskatchewan: Saskatoon, Saskatchewan, Canada; Unpublished raw data, 2014.

28. Majumder, A.L.; Johnson, M.D.; Henry, S.A. 1L-myo-inositol-1-phosphate synthase. *Biochim. Biophys. Acta* **1997**, *1348*, 245–256.
29. Majumder, A.L.; Chatterjee, A.; Dastida, K.G.; Majee, M. Diversification and evolution of L-*myo*-inositol 1-phosphate synthase. *FEBS Lett.* **2003**, *553*, 3–10.
30. Gillman, J.D.; Pantalone, V.R.; Bilyeu, K. The low phytic acid phenotype in soybean line CX1834 is due to mutations in two homologs of the maize low phytic acid gene. *Plant Genome* **2009**, *2*, 179–190.
31. Anderson, B.P.; Fehr, W.R. Seed source affects field emergence of low-phytate soybean lines. *Crop Sci.* **2008**, *48*, 929–932.
32. Bentsink, K.; Yuan, Y.; Koorneef, M.; Vreugdenhil, D. The genetics of phytate and phosphorus accumulation in seeds and leaves of *Arabidopsis thaliana*, using natural variation. *Theor. Appl. Genet.* **2003**, *106*, 1234–1243.
33. Shi, J.R.; Wang, H.Y.; Schellin, K.; Li, B.L.; Faller, M.; Stoop, J.M.; Meeley, R.B.; Ertl, D.S.; Ranch, J.P.; Glassman, K. Embryo-specific silencing of a transporter reduces phytic acid content of maize and soybean seeds. *Nat. Biotechnol.* **2007**, *25*, 930–937.
34. Dong, J.; Yan, W.; Bock, C.; Nokhrina, K.; Keller, W.; Georges, F. Perturbing the metabolic dynamics of *myo*-inositol in developing *Brassica napus* seeds through *in vivo* methylation impacts its utilization as phytate precursor and affects downstream metabolic pathways. *BMC Plant Biol.* **2013**, *13*, e84.
35. Ali, N.; Paul, S.; Gayen, D.; Sarkar, S.N.; Datta, K.; Datta, S.K. Development of low phytate rice by RNAi mediated seed-specific silencing of inositol 1,3,4,5,6-pentakisphosphate 2-kinase gene (*IPK1*). *PLOS ONE* **2013**, doi:10.1371/journal.pone.0068161.
36. Bregitzer, P.; Raboy, V. Effects of four independent low-phytate mutations on barley agronomic performance. *Crop Sci.* **2006**, *46*, 1318–1322.
37. Bregitzer, P.; Raboy, V.; Obert, D.E.; Windes, J.; Whitmore, J.C. Registration of "Clearwater" low-phytate hulless spring barley. *J. Plant Registr.* **2008**, *2*, 1–4.
38. Raboy, V.; Dikinson, D.B.; Neuffer, M.G. A survey of maize kernel mutants for variation in phytic acid. *Maydica* **1990**, *35*, 383–390.
39. Raboy, V. Low Phytic acid Mutants and Selection Thereof. Patent 5,689,054, 1997.
40. Raboy, V.; Dickinson, D.B. The timing and rate of phytic acid accumulation in developing soybean seeds. *Plant Physiol.* **1987**, *85*, 0841–0844.
41. Bowen, D.E.; Guttieri, M.J.; Peterson, K.; Peterson, K.; Raboy, V.; Souza, E.J. Phosphorus fractions in developing seeds of four low phytate barley (*Hordeum vulgare* L.) genotypes. *Crop Sci.* **2006**, *46*, 2468–2473.
42. Israel, D.W.; Talierico, E.; Kwanyuen, P.; Burton, J.W.; Dean, L. Inositol metabolism in developing seeds of low and normal phytic acid soybean lines. *Crop Sci.* **2011**, *51*, 282–289.
43. Larson, S.R.; Rutger, J.N.; Young, K.A.; Raboy, V. Isolation and genetic mapping of a non-lethal rice (*Oryza sativa* L.) *low phytic acid 1* mutation. *Crop Sci.* **2000**, *40*, 1397–1405.
44. Yoshida, K.T.; Wada, T.; Koyama, H.; Mizobuchi-Fukuoka, R.; Naito, S. Temporal and spatial patterns of accumulation of the transcript of *myo*-inositol-1-phosphate synthase and phytin-containing particles during seed development in rice. *Plant Physiol.* **1999**, *119*, 65–72.

45. Ma, D.; Zuo, Y.; Sun, D.; Wang, C.; Guo, T. Characterization of the *TaMIPS* gene from winter wheat (*Triticum aestivum* L.) and changes in its expression pattern with phytic acid accumulation in seeds during grain filling. *J. Cereal Sci.* **2013**, *57*, 437–443.
46. Keller, R.; Brearley, C.A.; Trethewey, R.N.; Muller-Rober, B. Reduced inositol content and altered morphology in transgenic potato plants inhibited for 1D-*myo*-inositol 3-phosphate synthase. *Plant J.* **1998**, *16*, 403–410.
47. Murphy, A.M.; Otto, B.; Brearley, C.A.; Carr, J.P.; Hanke, D.E. A role for inositol hexakisphosphate in the maintenance of basal resistance to plant pathogens. *Plant J.* **2008**, *56*, 638–652.
48. Xu, X.H.; Zhao, H.J.; Liu, Q.L.; Frank, T.; Engel, K.H.; An, G.H.; Shu, Q.Y. Mutations of the multi-drug resistance-associated protein ABC transporter gene 5 result in reduction of phytic acid in rice seeds. *Theor. Appl. Genet.* **2009**, *119*, 75–83.
49. Yuan, F.J.; Zhao, H.J.; Ren, X.L.; Zhu, S.L.; Fu, X.J.; Shu, Q.Y. Generation and characterization of two novel low phytate mutations in soybean (*Glycine max* L. Merr.). *Theor. Appl. Genet.* **2007**, *115*, 945–957.
50. Shi, J.; Wang, H.; Wu, Y.; Hazebroek, J.; Meeley, R.B.; Ertl, D.S. The maize low-phytic acid mutant *lpa*2 is caused by mutation in an inositol phosphate kinase gene. *Plant Physiol.* **2003**, *131*, 507–515.
51. Stevenson-Paulik, J.; Bastidas, R.J.; Chiou, S.T.; Frye, R.A.; York, J.D. Generation of phytate-free seeds in Arabidopsis through disruption of inositol polyphosphate kinases. *Proc. Natl. Acad. Sci. USA* **2005**, *102*, 12612–12617.
52. Warkentin, T.D.; Vandenberg, A.; Banniza, S.; Slinkard, A. CDC Bronco field pea. *Can. J. Plant Sci.* **2005**, *85*, 649–650.
53. Talamond, P.; Doulbeau, S.; Rochette, I.; Guyot, J.P. Anion exchange high-performance liquid chromatography with conductivity detection for the analysis of phytic acid in food. *J. Chromatogr. A* **2000**, *871*, 7–12.
54. Thavarajah, P.; Thavarajah, D.; Vandenberg, A. Low phytic acid lentils (*Lens culinaris* L.): A potential solution for increased micronutrient bioavailability. *J. Agric. Food Chem.* **2009**, *57*, 9044–9049.
55. Chen, P.S.; Toribara, T.Y.; Warner, H. Microdetermination of phosphorus. *Anal. Chem.* **1956**, *28*, 1756–1758.
56. Main, D.; Cheng, C-H.; Ficklin, S.P.; Jung, S.; Zheng, P.; Coyne, C.J.; McGee, R.; Mockaitis, K. In *The Cool Season Food Legume Database: An Integrated Resource for Basic, Translational and Applied Research*, Proceedings of the International Plant and Animal Genome Conference, San Diego, CA, USA, 12–16 January 2013.
57. Ye, J.; Coulouris, G.; Zaretskaya, I.; Cutcutache, I.; Rozen, S.; Madden, T. Primer-BLAST: A tool to design target-specific primers for polymerase chain reaction. *BMC Bioinform.* **2012**, *13*, e134.
58. Sambrook, J.; Russell, D.W. *Molecular Cloning, A Laboratory Manual*, 3rd ed.; Cold Spring Harbour Laboratory Press: Cold Spring Harbour, New York, NY, USA, 2001; Volume 1–3.
59. Ozga, J.A.; Yu, J.; Reinecke, D.M. Pollination-, development-, and auxin-specific regulation of gibberellin 3β-hydroxylase gene expression in pea fruit and seeds. *Plant Physiol.* **2003**, *131*, 1137–1146.
60. Hall, T.A. BioEdit: A user-friendly biological sequence alignment editor and analysis program for Windows 95/98/NT. *Nucl. Acids Symp. Ser.* **1999**, *41*, 95–98.

61. Altschul, S.F.; Madden, T.L.; Schaffer, A.A.; Zhang, J.; Zhang, Z.; Miller, W.; Lipman, D.J. Gapped Blast and Psi-Blast: A new generation of protein database search programs. *Nucleic Acids Res.* **1997**, *25*, 3389–3402.
62. Larkin, M.A.; Blackshields, G.; Brown, N.P.; Chenna, R.; McGettigan, P.A.; McWilliam, H.; Valentin, F.; Wallace, I.M.; Wilm, A.; Lopez, R.; *et al.* Clustal W and Clustal X version 2.0. *Bioinformatics* **2007**, *23*, 2947–2948.
63. Quevillon, E.; Silventoinen, V.; Pillai, S.; Harte, N.; Mulder, N.; Apweiler, R.; Lopez, R. InterProScan: Protein domains identifier. *Nucleic Acids Res.* **2005**, *33*, 116–120.
64. Tamura, K.; Stecher, G.; Peterson, D.; Filipski, A.; Kumar, S. MEGA6: Molecular Evolutionary Genetics Analysis version 6.0. *Mol. Biol. Evol.* **2013**, *30*, 2725–2729.

14

Plant Polygalacturonases Involved in Cell Elongation and Separation—The Same but Different?

Yashodar Babu and Martin Bayer *

Department of Cell Biology, Max Planck Institute for Developmental Biology, Spemannstrasse 35, Tuebingen 72076, Germany; E-Mail: yashodar.babu@tuebingen.mpg.de

* Author to whom correspondence should be addressed; E-Mail: martin.bayer@tuebingen.mpg.de

External Editor: Philip J. Harris

Abstract: Plant cells are surrounded by the primary cell wall, a rigid framework that needs to be modified in order to allow cell growth. Recent data suggest that in addition to the cellulose-hemicellulose network, the pectin matrix plays a critical role in determining the elasticity of the primary cell wall. Polygalacturonases are key homogalacturonan-hydrolyzing enzymes that function in a wide range of developmental processes. In this review, we present recent progress in understanding the role of polygalacturonases during cell elongation and separation. In discussing the specificities and possible redundancies of polygalacturonases, we focus particularly on newly discovered Arabidopsis mutants that have measurable loss-of-function phenotypes. However, data from other species are included when necessary.

Keywords: polygalacturonase; pectin; cell wall; cell elongation; genetic redundancy

---

1. Introduction

In plants, cells are surrounded by a rigid cell wall and are, therefore, fixed in their relative position. As a consequence, the overall shape of the plant body is created by a tightly controlled interplay of cell division and anisotropic cell expansion. The driving force for cell expansion is turgor pressure and it requires controlled relaxation of the cell wall while maintaining cellular integrity [1]. The high osmotic pressure inside the cell, ranging from 0.3 to 1.2 MPa in the case of *Arabidopsis thaliana*, makes this a non-trivial task [2]. Secondary cell-wall material can fortify cell walls after cell elongation has ceased but it is the primary cell wall of meristematic and elongating cells that needs to be modified to control the shape of the plant body [3].

The typical primary cell wall is a complex matrix composed of polysaccharides, a variety of proteins, and phenolics with cellulose, hemicelluloses, and pectin almost equally representing the main polysaccharide content [4]. A major constituent of this composite network is cellulose, a linear β-1,4-glucan, which co-crystalizes with other cellulose polymers into microfibrils, long, rigid fibers that can wind around the cell many times [1,5]. In seed plants (spermatophytes), the most abundant hemicelluloses in primary cell walls are usually xyloglucans (XG). However, the commelinid monocots, which include the grass family, are an exception, as in this group glucuronorarabinoxylans generally predominate [6]. Hemicelluloses are thought to cross-link cellulose microfibrils, thereby creating a rigid network while at the same time enforcing certain spacing between microfibrils to adjust primary cell wall flexibility [1,6].

In this view, the cellulose-xyloglucan network is the main structural component with load-bearing properties while pectin acts as filler matrix similar to composite materials of fiber-reinforced polymers. The extensibility of this network is modified by wall-loosening expansins, which are thought to act on the non-covalent interactions between cellulose microfibrils, as well as XG *endo*-transglycosylases [1].

This traditional model has recently been challenged by data that suggest that only a minor portion of XG is bound directly to cellulose [7] and that pectin displays load-bearing properties in the absence of XG. Therefore the pectin matrix seems to play a more important role in determining the extensibility of the primary cell wall than previously recognized [7–9].

Pectin consists of complex high molecular weight polysaccharides that can form hydrated gels. It can be classified into domains of homogalacturonan (HG), rhamnogalacturonan I (RG I) and RG II, and xylogalacturonan depending on the backbone and the degree of branching. The most abundant pectin form is homogalacturonan, a linear polymer of α-1,4-linked D-galacturonic acid [10]. It is synthesized in the *cis*-Golgi by galacturonosyltransferases (GAUTs) and usually highly methyl-esterified by pectin methyltransferases (PMTs) and to lesser degree acetylated by pectin acetyltransferases (PATs) [11]. The degree of methylesterification and acetylation plays a critical role in determining the stiffness of the cell wall since negative charges on the HG backbone can cross-link HG polymers via calcium ions [12]. Pectin methylesterases (PMEs) can hydrolyze methylester bonds and therefore have the potential to increase the degree of HG cross-linking. Interestingly, it was shown that ectopic, ubiquitous expression of PME inhibitors leads to stiffening of cell walls while ectopic expression of PMEs leads to softening [13,14]. This might be explained by competition between different HG sites for binding of calcium ions where newly de-esterified sites lead to the disruption of existing, load-bearing cross-links [9]. On the other hand, the de-esterification might also allow hydrolyzing enzymes like polygalacturonases (PGs) to act on the HG chain leading to softening of the cell wall [12]. Since the amount, structure and modification of pectin influences the physical properties of the cell wall, growth control most likely requires fine-tuning of all these variables. A more detailed description of pectin biosynthesis and structure as well as its modifications during cell growth can be found in the following comprehensive reviews [11,12,15–17].

2. Polygalacturonases

Polygalacturonases belong to the glycosyl hydrolase family 28 and are key HG hydrolyzing enzymes that have been implicated with a wide range of plant developmental processes such as cell elongation, organ abscission, fruit ripening, microspore release, and pollen tube growth [18]. Plant PG genes belong to large gene families and their expansion and diversification can be attributed to whole genome and segmental duplications in association with gene loss, as well as intron gain and (more predominantly) intron loss events [19,20]. Phylogenetic analysis of PG gene structure in Arabidopsis, rice, and other plant species reveals five distinct clades of PG genes which can be further divided into subclades indicating the occurrence of at least four ancestral PG genes before the divergence of monocots and dicots. While sequences from different clades are relatively divergent, they are rather conserved within a clade and tandem-duplicated genes generally fall into the same subclade [19–22]. Depending on their mode of action, *endo*- and *exo*-polygalacturonases can be distinguished [23]. *Endo*-PGs hydrolyze the HG polymer at random sites but require at least four consecutive GalA residues of the HG chain to be de-methylesterified [24,25]. Thus, the methylation pattern of the HG chains directly influences possible *endo*-PG-mediated HG cleavage. *Endo*-PG activity might lead to complete hydrolysis of pectin polymers and has therefore the potential to cause rapid cell elongation or even cell separation [12,26].

Exo-PGs on the other hand attack the free ends of de-methylesterified HG polymers and thereby reduce the overall polymer length. It has been speculated that the resulting modification of the pectin matrix might be subtler than the random cleavage by *endo*-PGs and might, therefore, be used to fine-tune the extensibility of the primary cell wall [27].

3. Polygalacturonases Involved in Fruit Ripening and Cell Separation

Early on, polygalacturonases were isolated from ripening fruits which implied a role in pectin degradation for tissue softening [28]. In a pioneer work in tomato, down-regulation of polygalacturonase expression by an anti-sense construct lead to decreased de-polymerization of solubilized pectins and increased storage-life of ripe fruits although there was no measureable effect on fruit softening [29,30]. Silencing of PG expression in apple and strawberry on the other hand increased the firmness of the ripe fruit significantly but did not change other ripening parameters. It could be shown that in these transgenic plants, ionically and covalently bound pectin exhibits a lower degree of de-polymerization. Microscopic analysis of transgenic fruits revealed smaller intra-cellular spaces and more cellular adhesion [31–33]. This is in agreement with earlier reports demonstrating that pectin degradation by PGs also plays a central role in cell separation in abscission events and dehiscence zones [34–36]. In apples, constitutive expression of the fruit-specific MdPG1 gene resulted in a range of novel developmental phenotypes including premature leaf shedding due to reduced cell adhesion in abscission zones, malformed leaves and malfunctioning stomata. As a consequence of the strong constitutive expression of MdPG1 in transgenic apple trees, a decrease of the average molecular weight of pectin chains could be demonstrated [37]. In rice, over-expression of *OsBURP16*, the non-catalytic PG1β-subunit of the polygalacturonase PG1, reduced cell adhesion in leaves and pectin content [38]. Together these data support the involvement of PGs in cell separation events in plants.

Many polygalacturonase genes are highly expressed in reproductive tissue [21], and for some, their involvement in cell separation events could be demonstrated by loss-of-function phenotypes: The *endo*-PG QUARTET3 (QRT3) functions in degrading the pollen mother cell wall during microsporogenesis and thus enables the release of unicellular microspores [39]. ARABIDOPSIS DEHISCENCE ZONE POLYGALACTURONASE1 (ADPG1), ADPG2, and QRT2 were reported to act in a redundant manner in anther dehiscence while ADPG2 and QRT2 function partially redundant in floral organ abscission [40].

4. Polygalacturonases Involved in Cell Expansion

The previously mentioned examples of polygalacturonases involved in fruit ripening and cell separation might lead to the conclusion that the primary role of *endo*-PGs is the more or less complete breakdown of the pectin matrix in terminal developmental situations like tissue softening, abscission, or dehiscence. This is clearly not the whole story, since, recently, the *endo*-PG POLYGALACTURONASE INVOLVED IN EXPANSION1 (PGX1) was shown to be involved in hypocotyl elongation and floral patterning [41]. Furthermore, *in vivo* assays for pectin-degrading enzymes suggested that during cotyledon expansion in cotton *endo*-PG and *exo*-PG activity could both be detected. Interestingly, their appearance differed temporally during cell elongation with high *endo*-PG activity at an early phase, followed by an increase of *exo*-PG activity during a later phase when *endo*-PG activity decreased [26]. This implies a scenario where different pectin hydrolyzing enzymes might act on the same substrate but in a consecutive manner.

The pectinase LeXPG1 was isolated from tomato seed protein extracts and gene expression was detected in the embryonic root, the developing vasculature as well as the embryo-surrounding endosperm. Based on this expression pattern, it was suggested that LeXPG1 might play a role in cell elongation as well as tissue softening in the embryo and the endosperm, respectively. Interestingly, LeXPG1 displays calcium-dependent *exo*-PG activity [42]. Furthermore, *exo*-polygalacturonase activity had been observed during abscission events in citrus explants before [43]. *Endo*-PGs and *exo*-PGs seem to be both involved in cell elongation as well as cell separation events. Therefore, their enzymatic activity does not directly correlate with one or the other process but rather a combination of both activities seems to be necessary in both events. Hence, it is tempting to speculate if cell elongation and cell separation in principal rely on identical pectin modifications.

The Arabidopsis genome contains 69 PG genes [35,44]. Several of these arose from tandem duplications and belong to the same phylogenetic subclade [21]. Intuitively, one would speculate that these tandem-duplicated genes might act in a redundant manner if the expression pattern of these genes did not diverge. Only for a handful of PG genes is there direct evidence for their function during development, based on loss-of-function phenotypes. For many other Arabidopsis PG genes, the lack of an obvious loss-of-function phenotype might indeed be a result of genetic redundancy [18].

Recent findings might shed some further light on this situation: Cell elongation defects were reported in embryos of the Arabidopsis *nimna* (*nma*) mutant [45]. *NMA* codes for a putative *exo*-polygalacturonase and is preferentially expressed in reproductive tissue. Cells of *nma* mutant embryos fail to elongate as early as the zygote stage and severe cell elongation defects can be further observed in the suspensor while cells of the embryo-proper seem to recover from their defects at later stages of embryo development (Figure 1) [45]. This might indicate that other polygalacturonases can take over NMA function in the embryo-proper.

Figure 1. Embryonic phenotype of *nma* mutants. Suspensor cells in *nma-1*$^{-/-}$ embryos show reduced cell elongation compared to wild-type. The embryo proper is false-colored in green, suspensor cells are false-colored in yellow. Scale bar: 20 µm.

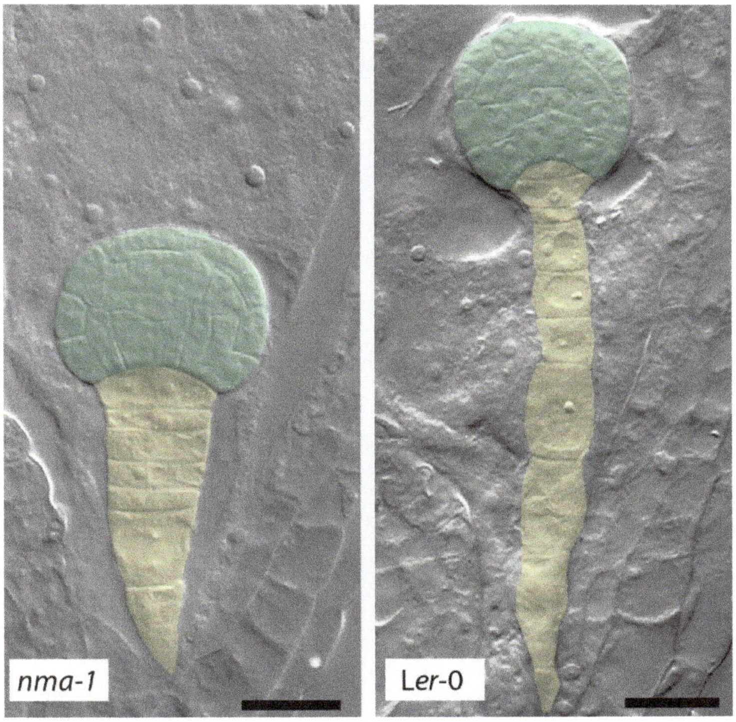

Peptides from five PGs including NMA were found to be present in cell wall fractions of 5 day-old etiolated hypocotyls of *Arabidopsis* [46]. While *NMA* is obviously present in elongating hypocotyls, the *nma* mutation does not seem to have any measurable effect on hypocotyl length [45]. It appears that other PGs are able to compensate for the loss of NMA activity in this case. For the embryonic suspensor, the situation seems to be different: While there are several closely related PG genes expressed in the suspensor of globular stage embryos according to published microarray data (Figure 2) [47], the strong cell-elongation defects observed in *nma* mutants indicate that none of these can fulfill NMA function [45]. The reason for this might be different temporal expression, sub-cellular localization, enzyme activity, or substrate specificity.

A similar situation was observed for the closely related *ADPG1*, *ADPG2*, and *QRT2* genes (Figure 2) [40]. Loss of all three genes causes an impaired pod shatter phenotype and compromises anther dehiscence. While the *pADPG1::ADPG1* transgene was able to fully complement the pod shatter defects of the triple mutant, *QRT2* and the closely related PG gene *At1g48100* failed to do so when expressed under the *ADPG1* promoter [40]. Again, this would argue for a distinct function of these proteins in the cell separation process possibly caused by different enzymatic activity or substrate preference.

Expression analysis of three closely related tomato PG genes (*TAPGl*, *TAPG2*, and *TAPG4*) indicates temporal regulation during leaf and flower abscission. The temporal expression pattern of these genes suggests that they might act consecutively to fulfill a stepwise modification of the pectin matrix [48].

Figure 2. Phylogeny and expression values of *NMA*- and *QRT2*-related polygalacturonase genes. (**A**) Phylogenetic tree of NMA- and QRT2-related polygalacturonases. Phylogeny was created with protein sequences of two sub-clades A1a and A14 [21] using neighbour-joining with bootstrap values of 100 [49]. Some of the PGs mentioned in this review (*Medicago sativa* PG3, *Solanum lycopersicum* XPG1 and a bacterial *exo*-PG from *Yersinia enterocolitica*) were also included as outgroups. Scale bar represents amino acid substitutions per site; (**B**) Mean expression values of selected Arabidopsis polygalacturonase genes closely related to *NMA* in the embryo and suspensor based on publicly available microarray data [47].

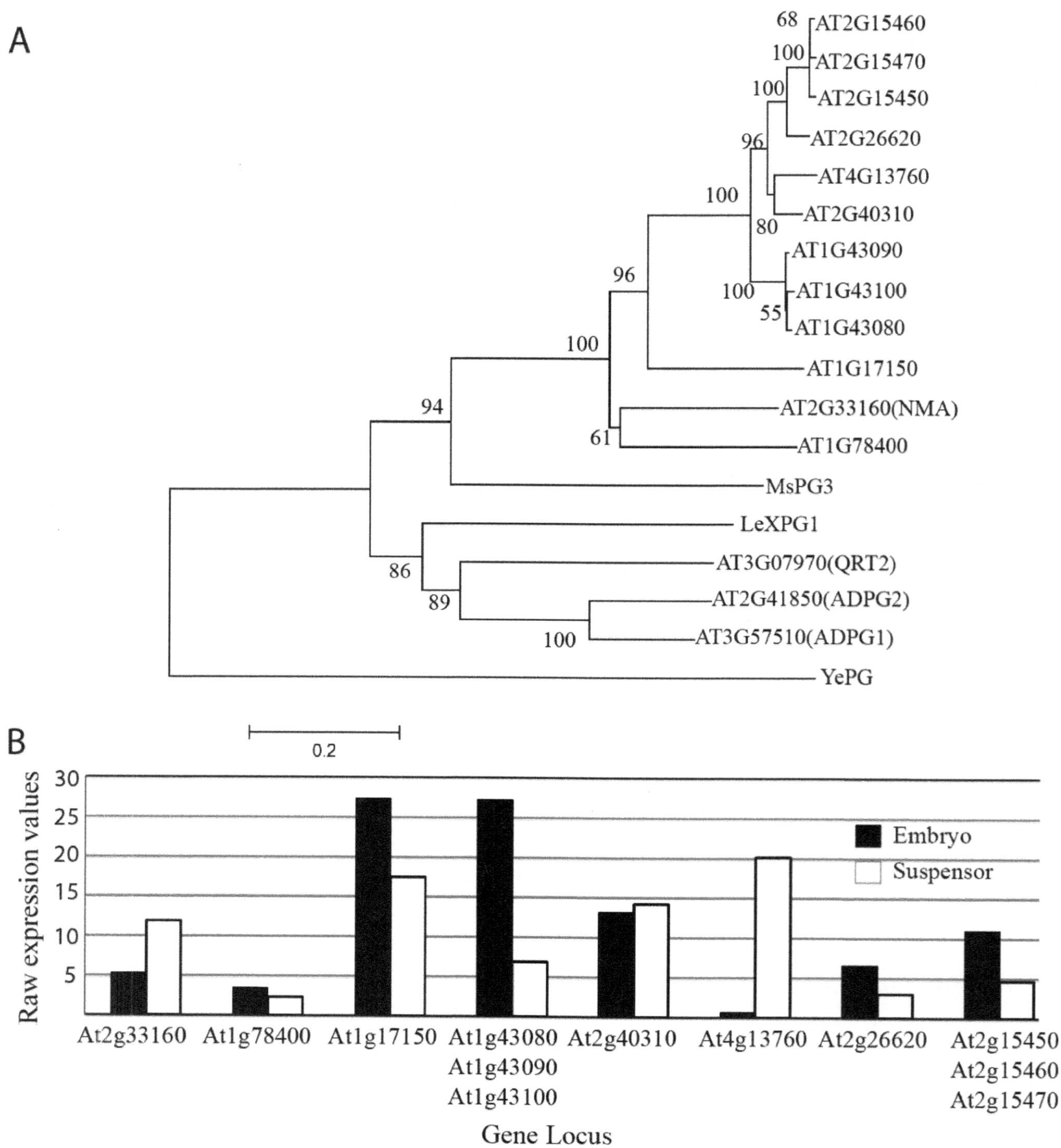

5. Perspectives

Pectin plays a central role in determining the physicochemical properties of the primary cell wall. Modifications of the pectin matrix are thus elementary for cell elongation by determining the extensibility of the cell wall. The degree of methyl-esterification is one important aspect that seems to be tightly regulated, but recent data emphasizes the importance of pectin-hydrolyzing polygalacturonases in cell elongation processes [41,45]. The temporally regulated activity of *endo-* as well as *exo*-PGs seems essential for both cell elongation and cell separation processes. To what degree the vast number of PG genes in the Arabidopsis genome reflects genetic redundancy or displays the need for a high number of specific enzymes in these cell wall processes is still unclear. With recently described Arabidopsis mutants like *nma* or *pgx1*, which show obvious and quantifiable loss-of-function phenotypes, this question can now be addressed. Promoter-swap and complementation experiments are powerful tools to support the biochemical analysis of these proteins. Complementation experiments with *endo-* and *exo*-PGs will allow a better understanding to what degree these enzymes are functionally redundant or are involved in separate non-exchangeable steps in pectin modification during cell elongation processes.

The active site in PG proteins is well conserved but substrate recognition motifs are not well understood [23]. *In vivo* complementation assays along with biochemical studies guided by protein-structure data might help unraveling specific modes of substrate recognition.

Technical advances have greatly helped our understanding of the primary cell wall composition [5,50–52]. Studying the effect of well-characterized PGs might indirectly give further insight in cell wall composition and the structure and modifications of the pectin matrix.

Furthermore, with the recent advances in genome-editing tools, like the CRISPR/Cas9 system and their application in plant biology, the study of many tandem-duplicated PG genes is now technically possible [53,54].

Understanding the substrate-specificity and the nature of the pectin modification carried out by specific polygalacturonases will not only increase our understanding of plant cell wall biology during cell elongation but will also be valuable for their use in commercial products and technical processes like biofuel production.

Acknowledgments

We apologize to our colleagues whose publications could not be included in this short review. We thank Cameron Lee, Thomas Musielak, and Daniel Slane for critical reading of the manuscript. Research in our lab is supported by the German Science Foundation (DFG, SFB1101) and the Max Planck Society.

Author Contributions

Yashodar Babu and Martin Bayer prepared the figures and wrote the manuscript.

Conflicts of Interest

The authors declare no conflict of interest.

References

1. Cosgrove, D.J. Growth of the plant cell wall. *Nat. Rev. Mol. Cell Biol.* **2005**, *6*, 850–861.
2. Forouzesh, E.; Goel, A.; Mackenzie, S.A.; Turner, J.A. *In vivo* extraction of Arabidopsis cell turgor pressure using nanoindentation in conjunction with finite element modeling. *Plant J.* **2013**, *73*, 509–520.
3. Zhong, R.; Ye, Z.H. Regulation of cell wall biosynthesis. *Curr. Opin. Plant Biol.* **2007**, *10*, 564–572.
4. Mcneil, M.; Darvill, A.G.; Fry, S.C.; Albersheim, P. Structure and function of the primary-cell walls of plants. *Annu. Rev. Biochem.* **1984**, *53*, 625–663.
5. Nishiyama, Y.; Langan, P.; Chanzy, H. Crystal structure and hydrogen-bonding system in cellulose Ibeta from synchrotron X-ray and neutron fiber diffraction. *J. Am. Chem. Soc.* **2002**, *124*, 9074–9082.
6. Scheller, H.V.; Ulvskov, P. Hemicelluloses. *Annu. Rev. Plant Biol.* **2010**, *61*, 263–289.
7. Dick-Perez, M.; Zhang, Y.; Hayes, J.; Salazar, A.; Zabotina, O.A.; Hong, M. Structure and interactions of plant cell-wall polysaccharides by two- and three-dimensional magic-angle-spinning solid-state NMR. *Biochemistry* **2011**, *50*, 989–1000.
8. Park, Y.B.; Cosgrove, D.J. Changes in cell wall biomechanical properties in the xyloglucan-deficient xxt1/xxt2 mutant of Arabidopsis. *Plant Physiol.* **2012**, *158*, 465–475.
9. Peaucelle, A.; Braybrook, S.; Hofte, H. Cell wall mechanics and growth control in plants: The role of pectins revisited. *Front. Plant Sci.* **2012**, *3*, 121.
10. Ridley, B.L.; O'Neill, M.A.; Mohnen, D. Pectins: Structure, biosynthesis, and oligogalacturonide-related signaling. *Phytochemistry* **2001**, *57*, 929–967.
11. Atmodjo, M.A.; Hao, Z.; Mohnen, D. Evolving views of pectin biosynthesis. *Annu. Rev. Plant Biol.* **2013**, *64*, 747–779.
12. Senechal, F.; Wattier, C.; Rusterucci, C.; Pelloux, J. Homogalacturonan-modifying enzymes: Structure, expression, and roles in plants. *J. Exp. Bot.* **2014**, *65*, 5125–5160.
13. Peaucelle, A.; Louvet, R.; Johansen, J.N.; Hofte, H.; Laufs, P.; Pelloux, J.; Mouille, G. Arabidopsis phyllotaxis is controlled by the methyl-esterification status of cell-wall pectins. *Curr. Biol.* **2008**, *18*, 1943–1948.
14. Peaucelle, A.; Braybrook, S.A.; le Guillou, L.; Bron, E.; Kuhlemeier, C.; Hofte, H. Pectin-induced changes in cell wall mechanics underlie organ initiation in Arabidopsis. *Curr. Biol.* **2011**, *21*, 1720–1726.
15. Wolf, S.; Greiner, S. Growth control by cell wall pectins. *Protoplasma* **2012**, *249*, 169–175.
16. Wolf, S.; Mouille, G.; Pelloux, J. Homogalacturonan methyl-esterification and plant development. *Mol. Plant* **2009**, *2*, 851–860.
17. Willats, W.G.; McCartney, L.; Mackie, W.; Knox, J.P. Pectin: Cell biology and prospects for functional analysis. *Plant Mol. Biol.* **2001**, *47*, 9–27.
18. Hadfield, K.A.; Bennett, A.B. Polygalacturonases: Many genes in search of a function. *Plant Physiol.* **1998**, *117*, 337–343.
19. Torki, M.; Mandaron, P.; Mache, R.; Falconet, D. Characterization of a ubiquitous expressed gene family encoding polygalacturonase in *Arabidopsis thaliana*. *Gene* **2000**, *242*, 427–436.

20. Park, K.C.; Kwon, S.J.; Kim, P.H.; Bureau, T.; Kim, N.S. Gene structure dynamics and divergence of the polygalacturonase gene family of plants and fungus. *Genome* **2008**, *51*, 30–40.
21. Kim, J.; Shiu, S.H.; Thoma, S.; Li, W.H.; Patterson, S.E. Patterns of expansion and expression divergence in the plant polygalacturonase gene family. *Genome Biol.* **2006**, *7*, R87.
22. Park, K.C.; Kwon, S.J.; Kim, N.S. Intron loss mediated structural dynamics and functional differentiation of the polygalacturonase gene family in land plants. *Genes Genomics* **2010**, *32*, 570–577.
23. Markovic, O.; Janecek, S. Pectin degrading glycoside hydrolases of family 28: Sequence-structural features, specificities and evolution. *Protein Eng.* **2001**, *14*, 615–631.
24. Protsenko, M.A.; Buza, N.L.; Krinitsyna, A.A.; Bulantseva, E.A.; Korableva, N.P. Polygalacturonase-inhibiting protein is a structural component of plant cell wall. *Biochemistry (Mosc.)* **2008**, *73*, 1053–1062.
25. Clausen, M.H.; Willats, W.G.T.; Knox, J.P. Synthetic methyl hexagalacturonate hapten inhibitors of antihomogalacturonan monoclonal antibodies LM7, JIM5 and JIM7. *Carbohydr. Res.* **2003**, *338*, 1797–1800.
26. Zhang, Z.Q.; Pierce, M.L.; Mort, A.J. Changes in homogalacturonans and enzymes degrading them during cotton cotyledon expansion. *Phytochemistry* **2007**, *68*, 1094–1103.
27. Abbott, D.W.; Boraston, A.B. The structural basis for exopolygalacturonase activity in a family 28 glycoside hydrolase. *J. Mol. Biol.* **2007**, *368*, 1215–1222.
28. Pressey, R.; Hinton, D.M.; Avants, J.K. Development of Polygalacturonase Activity and Solubilization of Pectin in Peaches during Ripening. *J. Food Sci.* **1971**, *36*, 1070–1073.
29. Smith, C.J.; Watson, C.F.; Morris, P.C.; Bird, C.R.; Seymour, G.B.; Gray, J.E.; Arnold, C.; Tucker, G.A.; Schuch, W.; Harding, S.; *et al.* Inheritance and effect on ripening of antisense polygalacturonase genes in transgenic tomatoes. *Plant Mol. Biol.* **1990**, *14*, 369–379.
30. Smith, C.J.S.; Watson, C.F.; Ray, J.; Bird, C.R.; Morris, P.C.; Schuch, W.; Grierson, D. Antisense RNA inhibition of polygalacturonase gene-expression in transgenic tomatoes. *Nature* **1988**, *334*, 724–726.
31. Atkinson, R.G.; Sutherland, P.W.; Johnston, S.L.; Gunaseelan, K.; Hallett, I.C.; Mitra, D.; Brummell, D.A.; Schroder, R.; Johnston, J.W.; Schaffer, R.J. Down-regulation of *POLYGALACTURONASE1* alters firmness, tensile strength and water loss in apple (*Malus x domestica*) fruit. *BMC Plant Biol.* **2012**, *12*, 129.
32. Pose, S.; Paniagua, C.; Cifuentes, M.; Blanco-Portales, R.; Quesada, M.A.; Mercado, J.A. Insights into the effects of polygalacturonase FaPG1 gene silencing on pectin matrix disassembly, enhanced tissue integrity, and firmness in ripe strawberry fruits. *J. Exp. Bot.* **2013**, *64*, 3803–3815.
33. Quesada, M.A.; Blanco-Portales, R.; Pose, S.; Garcia-Gago, J.A.; Jimenez-Bermudez, S.; Munoz-Serrano, A.; Caballero, J.L.; Pliego-Alfaro, F.; Mercado, J.A.; Munoz-Blanco, J. Antisense down-regulation of the FaPG1 gene reveals an unexpected central role for polygalacturonase in strawberry fruit softening. *Plant Physiol.* **2009**, *150*, 1022–1032.
34. Tucker, G.A.; Schindler, C.B.; Roberts, J.A. Flower abscission in mutant tomato plants. *Planta* **1984**, *160*, 164–167.
35. Gonzalez-Carranza, Z.H.; Elliott, K.A.; Roberts, J.A. Expression of polygalacturonases and evidence to support their role during cell separation processes in *Arabidopsis thaliana*. *J. Exp. Bot.* **2007**, *58*, 3719–3730.

36. Petersen, M.; Sander, L.; Child, R.; van Onckelen, H.; Ulvskov, P.; Borkhardt, B. Isolation and characterisation of a pod dehiscence zone-specific polygalacturonase from Brassica napus. *Plant Mol. Biol.* **1996**, *31*, 517–527.
37. Atkinson, R.G.; Schroder, R.; Hallett, I.C.; Cohen, D.; MacRae, E.A. Overexpression of polygalacturonase in transgenic apple trees leads to a range of novel phenotypes involving changes in cell adhesion. *Plant Physiol.* **2002**, *129*, 122–133.
38. Liu, H.; Ma, Y.; Chen, N.; Guo, S.; Liu, H.; Guo, X.; Chong, K.; Xu, Y. Overexpression of stress-inducible OsBURP16, the beta subunit of polygalacturonase 1, decreases pectin content and cell adhesion and increases abiotic stress sensitivity in rice. *Plant Cell Environ.* **2014**, *37*, 1144–1158.
39. Rhee, S.Y.; Osborne, E.; Poindexter, P.D.; Somerville, C.R. Microspore separation in the quartet 3 mutants of Arabidopsis is impaired by a defect in a developmentally regulated polygalacturonase required for pollen mother cell wall degradation. *Plant Physiol.* **2003**, *133*, 1170–1180.
40. Ogawa, M.; Kay, P.; Wilson, S.; Swain, S.M. ARABIDOPSIS DEHISCENCE ZONE POLYGALACTURONASE1 (ADPG1), ADPG2, and QUARTET2 are Polygalacturonases required for cell separation during reproductive development in Arabidopsis. *Plant Cell* **2009**, *21*, 216–233.
41. Xiao, C.; Somerville, C.; Anderson, C.T. POLYGALACTURONASE INVOLVED IN EXPANSION1 functions in cell elongation and flower development in Arabidopsis. *Plant Cell* **2014**, *26*, 1018–1035.
42. Sitrit, Y.; Hadfield, K.A.; Bennett, A.B.; Bradford, K.J.; Downie, A.B. Expression of a polygalacturonase associated with tomato seed germination. *Plant Physiol.* **1999**, *121*, 419–428.
43. Riov, J. A polygalacturonase from citrus leaf explants: role in abscission. *Plant Physiol.* **1974**, *53*, 312–316.
44. Cao, J. The pectin lyases in *Arabidopsis thaliana*: Evolution, selection and expression profiles. *PLoS One* **2012**, *7*, e46944.
45. Babu, Y.; Musielak, T.; Henschen, A.; Bayer, M. Suspensor length determines developmental progression of the embryo in Arabidopsis. *Plant Physiology* **2013**, *162*, 1448–1458.
46. Irshad, M.; Canut, H.; Borderies, G.; Pont-Lezica, R.; Jamet, E. A new picture of cell wall protein dynamics in elongating cells of *Arabidopsis thaliana*: Confirmed actors and newcomers. *BMC Plant Biol.* **2008**, *8*, 94.
47. Le, B.H.; Cheng, C.; Bui, A.Q.; Wagmaister, J.A.; Henry, K.F.; Pelletier, J.; Kwong, L.; Belmonte, M.; Kirkbride, R.; Horvath, S.; *et al.* Global analysis of gene activity during Arabidopsis seed development and identification of seed-specific transcription factors. *Proc. Natl. Acad. Sci. USA* **2010**, *107*, 8063–8070.
48. Kalaitzis, P.; Solomos, T.; Tucker, M.L. Three different polygalacturonases are expressed in tomato leaf and flower abscission, each with a different temporal expression pattern. *Plant Physiol.* **1997**, *113*, 1303–1308.
49. Tamura, K.; Stecher, G.; Peterson, D.; Filipski, A.; Kumar, S. MEGA6: Molecular evolutionary genetics analysis version 6.0. *Mol. Biol. Evol.* **2013**, *30*, 2725–2729.
50. Somerville, C.; Bauer, S.; Brininstool, G.; Facette, M.; Hamann, T.; Milne, J.; Osborne, E.; Paredez, A.; Persson, S.; Raab, T.; *et al.* Toward a systems approach to understanding plant cell walls. *Science* **2004**, *306*, 2206–2211.

51. Burgert, I.; Keplinger, T. Plant micro- and nanomechanics: Experimental techniques for plant cell-wall analysis. *J. Exp. Bot.* **2013**, *64*, 4635–4649.
52. Knox, J.P.; Linstead, P.J.; King, J.; Cooper, C.; Roberts, K. Pectin esterification is spatially regulated both within cell walls and between developing tissues of root apices. *Planta* **1990**, *181*, 512–521.
53. Shan, Q.; Wang, Y.; Li, J.; Zhang, Y.; Chen, K.; Liang, Z.; Zhang, K.; Liu, J.; Xi, J.J.; Qiu, J.L.; *et al.* Targeted genome modification of crop plants using a CRISPR-Cas system. *Nat. Biotechnol.* **2013**, *31*, 686–688.
54. Feng, Z.; Mao, Y.; Xu, N.; Zhang, B.; Wei, P.; Yang, D.L.; Wang, Z.; Zhang, Z.; Zheng, R.; Yang, L.; *et al.* Multigeneration analysis reveals the inheritance, specificity, and patterns of CRISPR/Cas-induced gene modifications in Arabidopsis. *Proc. Natl. Acad. Sci. USA* **2014**, *111*, 4632–4637.

Multiple Autoregulation of Nodulation (AON) Signals Identified through Split Root Analysis of *Medicago truncatula sunn* and *rdn1* Mutants

Tessema Kassaw [1,†], William Bridges Jr. [2] and Julia Frugoli [3,*]

1 Department of Genetics & Biochemistry, Clemson University, Clemson, SC 29634, USA; E-Mail: Tessema.Kassaw@colostate.edu

2 Department of Mathematical Sciences, Clemson University, Clemson, SC 29634, USA; E-Mail: wbrdgs@g.clemson.edu

3 Department of Genetics & Biochemistry, Clemson University, Clemson, SC 29634, USA

† Currently address: Department of Biology, Colorado State University, Fort Collins, CO 80523, USA.

* Author to whom correspondence should be addressed; E-Mail: jfrugol@clemson.edu

Academic Editor: Rujin Chen

Abstract: Nodulation is energetically costly to the host: legumes balance the nitrogen demand with the energy expense by limiting the number of nodules through long-distance signaling. A split root system was used to investigate systemic autoregulation of nodulation (AON) in *Medicago truncatula* and the role of the AON genes *RDN1* and *SUNN* in the regulatory circuit. Developing nodule primordia did not trigger AON in plants carrying mutations in *RDN1* and *SUNN* genes, while wild type plants had fully induced AON within three days. However, despite lacking an early suppression response, AON mutants suppressed nodulation when roots were inoculated 10 days or more apart, correlated with the maturation of nitrogen fixing nodules. In addition to correlation between nitrogen fixation and suppression of nodulation, suppression by extreme nutrient stress was also observed in all genotypes and may be a component of the observed response due to the conditions of the assay. These results suggest there is more than one systemic regulatory circuit controlling nodulation in *M. truncatula*. While both signals are present in wild type plants, the second signal can only be observed in plants lacking the early repression (AON mutants). *RDN1* and *SUNN* are not essential for response to the later signal.

Keywords: autoregulation of nodulation; *Medicago truncatula*; split-root analysis; nitrogen; *RDN1*; *SUNN*

---

1. Introduction

When legumes set up a symbiosis with bacteria in the genus *Rhizobia*, the plants develop a new organ on the roots (a nodule) to house the bacteria. After entering the plant through complex species-specific recognition and growth events, the bacteria reach the cortical cells of the root, differentiate into bacteriods and fix nitrogen in the low oxygen environment provided by the plant, sharing nitrogen in exchange for carbon skeletons from photosynthesis (for review see [1]). The nitrogenase enzyme used by the bacteria to fix N_2 into useable ammonia is irreversibly inhibited by oxygen and requires the plant to provide this special environment. Although the symbiosis provides nitrogen to the plant, allowing legumes to thrive under conditions other plants cannot, legumes only enter into the symbiosis under nitrate limitation. Suppression of nodulation in the presence of nitrate is thought to occur because the formation and maintenance of the nodule environment is energetically costly to the plant; the biological cost to the plant of nodule establishment, nitrogen fixation and transport is estimated at 12–17 grams of carbon per gram of nitrogen [2]. Likewise, plants that have already entered into the symbiosis will suppress the development of future nodules, even before the original nodules become fully established and provide ammonia [3], suggesting that there are several molecular switches controlling nodule regulation or that the regulation requires coordination of a number of signals.

The suppression of future nodule development is termed autoregulation of nodulation (AON) and it involves long distance signaling from root to shoot and back again; long distance signaling was initially demonstrated in split root inoculation experiments by first inoculating one side of the split root and observing systematic suppression of further nodulation in the other side of the split root [3]. AON is activated early upon Nod factor perception; the elongation zone of the root with emerging root hairs is vulnerable to rhizobial infection and is the portion of the root affected by the autoregulation process [4]. The timing of full induction of AON varies from species to species and is influenced by growth conditions. AON induction in *G. max* has been observed within four days [3] and in *L. japonicus* within five days [5], but reports of shorter times of two days in *Trifolium subterraneum* (clover) [6] and 30 h in *Vicia sativa* [7] have also been published.

Both shoot and root controlled AON defective mutants have been reported in *Lotus japonicus*, *Pisum sativum*, *Glycine max* and *Medicago truncatula*. The first shoot controlled AON gene cloned was *HAR1/SYM29/NARK/SUNN*, encoding a leucine-rich repeat receptor protein kinase like CLAVATA1 (CLV1) of *Arabidopsis thaliana* [8–11]. The protein kinase has an extracellular domain of leucine-rich repeats, a membrane spanning domain and an intracellular protein kinase domain. Mutations in *SUNN* (*super numerary nodules*) result in a five- to tenfold increase in nodule number [11]. Because the mutations exert their effect from the shoot, they are believed to be the receptors for the root-derived CLE signal [12] and/or triggers for the biosynthesis and release of the shoot derived inhibitory signal [13].

Among the genes known to exert their effect from the root is the *RDN1/NOD3* gene (root-determined nodulation) in *M. truncatula* and pea [14]. The *RDN* gene family in *M. truncatula* is orthologous to the *HPAT* gene family in *Arabidopsis*, which has been shown to add arabinose residues to the hydroxyprolines of some CLE peptides [15]. CLE peptides are attractive as the root derived signal, and in *M. truncatula* the expression of two CLE peptides are correlated with nodule development. The *MtCLE12* and *MtCLE13* genes are induced in the root upon nodulation, are expressed in nodule meristems and overexpression reduces nodulation in a SUNN dependent manner [16–19]. RDN1, therefore, presumably, affects the release of the systemic signal received by SUNN. Plants homozygous for either *rdn1-2* or *sunn-4* alleles result in excessive nodulation, with the *sunn-4* mutation giving the most severe effect [14,20].

In wild type plants, successful nodule formation and subsequent nitrogen fixation occur only under nitrogen limiting conditions. In the presence of high concentrations of biologically available nitrogen, plants cease nitrogen fixation and nodule formation is suppressed [21]. In contrast, soybean mutants with defects in AON are reported to be nitrate tolerant, nodulating abundantly even in the presence of nitrate [22,23]. In *M. truncatula*, the plants carrying mutations in *SUNN* or *RDN1* nodulate in the presence of nitrate, but at reduced levels compared to when nitrate is lacking [14]. This suggests that an autoregulation signal and a nitrate signal both act to inhibit nodule formation. Fixed nitrogen is transported in the xylem of nodulators largely as glutamine, asparagine or uriede [24]. The large pool of asparagine that is present in nodules may buffer the transport of nitrogen and thus act to regulate nitrogen fixation via a feedback mechanism [24]. In addition, inhibition of nitrogen fixation can affect the partitioning of carbon and other metabolites from the shoot. Singleton and van Kessel showed that the flux of photosynthate to nitrogen fixing nodules and their associated roots is greater than that to non-nitrogen fixing sections of the root [25], suggesting that nitrogen assimilation, either from N_2 fixation or inorganic sources, has a localized effect on both nodule and root development. Split root systems have also demonstrated that the lack of nitrogen acquisition by a half root system nodulated with non-fixing *rhizobia* triggers a compensation response, enabling the other half root system nodulated with nitrogen fixing partners to compensate for the local nitrogen limitation of the first root [26], which suggests a systemic component as well.

In order to determine if the excess nodule number phenotypes of the *rdn1* and *sunn* mutants involve changes in AON timing, we investigated AON in wild type plants and plants carrying mutant alleles of *RDN1* or *SUNN*. We report an early AON signal evident at three days in wild type but not mutant plants and a second AON signal ten days later observed in the AON mutants. Our findings using measurements of nitrogen fixation, response to supplied nitrogen and inoculation with bacteria unable to fix nitrogen support the assertion that the second AON signal is not associated with early nodulation events but rather with the establishment of nitrogen fixing nodules and although this signal is presumably masked by the CLE12/13-SUNN signaling in wild type plants, it is unaffected and detectable in the *sunn* and *rdn1* AON mutants. Additionally, both wild type and AON mutant plants suppress nodulation under severe nutrient stress, suggesting that the nitrate and stress signals are not affected in AON regulatory mutants.

2. Results and Discussion

2.1. AON Mutants Lack an Early Systemic Response to Previous Nodulation Events, but Suppression Returns after 10–15 Days

As noted in the introduction, the timing of AON has been reported to vary between species. Although we were able to observe autoregulation at four days post inoculation in previous work [27], we refined the experiment to determine the duration of the systemic autoregulatory signal timing in *M. truncatula* using the split-root system in [27]. Plants with two spatially separated equal roots, a tester (Root A) and a responder (Root B) were inoculated with *S. medicae* simultaneously or from two to 15 days apart. We compared the nodulation of Root B to the nodulation of Root A 21 days after the inoculation of root B, reporting the results as a percentage of the total nodules on Root A to account for the higher nodule numbers of AON mutants. When inoculated simultaneously, the responder roots and tester roots of wild type had no significant difference in nodule number (Figure 1a). The value of 150 percent for the mean at Time 0 results from the need to randomly designate one root as Root A when calculating the mean percentage of a simultaneous inoculation and wild type roots inoculated at the same time varied as much as 5–15 nodules/root. Percent nodulation on the responder roots decreased in wild type when the roots were inoculated two days apart, and as the interval between inoculations increased from three days to 15 days, the responder root consistently developed approximately half of the nodules of the first root, a statistically significant difference. Distribution of the nodules on Root A and Root B were similar, consisting of some nodules in the lower parts of the roots and some in developing laterals near the surface. This indicated that AON occurs between two and three days in *M. truncatula* and the degree of suppression remains constant through at least 15 days.

Unlike reports in soybean [28], we did not observe significant suppression under our conditions until day three and nodulation of the root portion receiving the delayed inoculation (Root B) was never completely suppressed by prior inoculation of the other root (Root A); even at a 15-day interval between inoculation of the roots suppression hovered around 50%. The three-day interval for full induction of AON in *M. truncatula* is shorter than previous reports in soybean and *L. japonicus* [3,5], but longer than that those reported for clover and *Vicia sativa* [6,7]. This variation could be due to differences in the split root systems used to determine the timing (see discussion in [27]), a difference between species in the speed of translocation of the signal, morphological and anatomical differences in the species, or the different rhizobial partners for each legume species.

The same analysis was performed using hypernodulation mutants carrying a null allele of the *SUNN* gene, *sunn-4* [11], which contains a stop codon very close to the initial signal sequence or a null allele of the *RDN1* gene, *rdn1-2* [14], which contains an insertion near a splicing site that reduces RNA levels over 400 percent. As expected for mutants isolated for their defects in AON, no significant difference in nodulation percentage between Root A and Root B was observed for inoculations separated by zero to eight days (Figure 1b,c). However, at ten days between inoculations in *rdn1* mutant plants and 12 days in *sunn-4* mutants, we observed suppression of nodulation to wild type suppression percentage levels in Root B at all but one time point. At data points separated 10 days or more between inoculations, the nodules on Root A were pink (a likely indication of active nitrogen fixation because of leghemoglobin synthesis) and thus the suppression coincides with the transition from white nodules to pink nodules, which occurs between eight and 10 days in this system. Nitrogen fixation is a possible explanation for the source of the later regulatory signal. Nitrogen fixation creates a strong localized sink for photosynthate that has effects on both nodulation and root development in uninoculated portion of the roots [25] and application of high inorganic nitrogen affects nodulation [29].

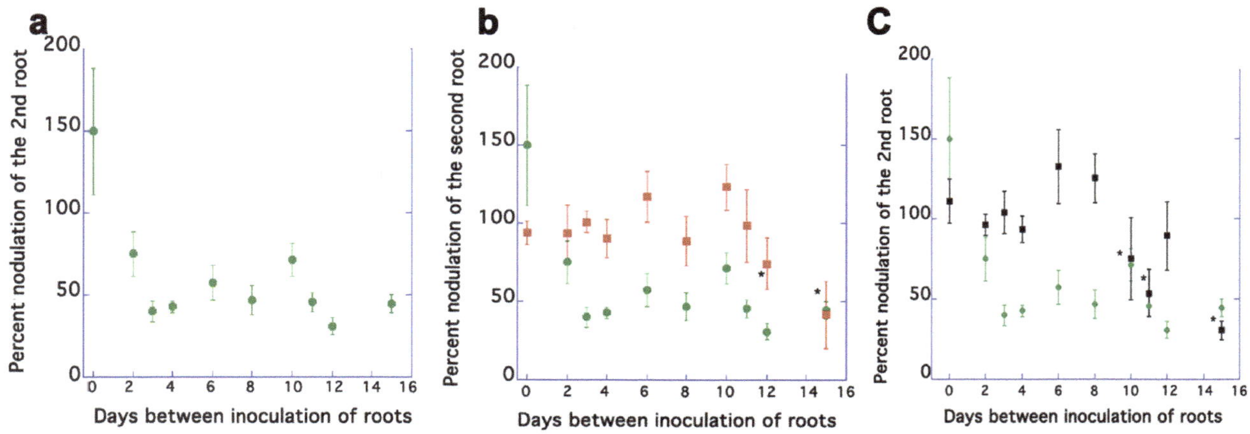

Figure 1. Persistence of the systemic suppression signal in mutant and wild type plants. (**a**) Wild type plants have consistently fewer nodules on the second root when the second root is inoculated 2–15 days after the first. Percent nodulation was calculated by dividing the mean nodule number of the second root by the mean nodule number of the first root. Values for time points 2–15 are statistically significant from time 0. (**b**) Wild type values (green dots) compared to the *sunn-4* allele values (red boxes). Data points marked with * are not statistically different between wild type and mutant (**c**) Wild type values (green dots) compared to the *rdn1-2* allele values (black boxes). Data points marked with * are not statistically different between wild type and mutant. Percent nodulation was calculated by dividing the mean nodule number of the second root by the mean nodule number of the first root. Statistical comparisons among means are based on factorial ANOVA followed by pairwise mean comparisons. Data are shown as means ± SE. ($n = 6$ to 47 plants for each time point per genotype).

2.2. AON Mutants and Wild Type Fix Equivalent Amounts of Nitrogen at Time Points when Systemic Suppression Is Observed

Since we postulated that the mutants were supressing nodulation on the responder root in response to nitrogen fixed by the tester root, to pursue assimilated nitrogen as the signal for the systemic suppression, we compared measures of nitrogen fixation capacity between AON mutants (*rdn1-2* and *sunn-4*) and wild type plants at the time point in question. First, we compared the total number of nodules (Figure 2a), the number of nitrogen fixing nodules (Figure 2b), and the mean nodule fresh weight (Figure 2c) between mutant plants and wild type plants. Both the total and nitrogen fixing nodule numbers were significantly higher in the AON mutants than in the wild type plants, but there was no difference in mean nodule fresh weight, in agreement with greenhouse work on the orthologous mutants in pea [30]. We then measured the rate of nitrogen fixation at 10 and 15 days post inoculation with *S. medicae* in wild type plants and the AON mutants, since despite having more nitrogen fixing nodules, the level of pink color varied between nodules on mutant and wild type plants. However, the level of nitrogen fixation in plants 10 and 15 days after inoculation was the same as wild type in all mutants (Figure 2d), suggesting that *sunn4* and *rdn1-2* mutants compensate for increased nodule number with reduced nitrogen fixation activity per nodule. In support of this, Jeudy *et al.* [31] showed that the systemic regulation of nitrogen fixation activity by the plant is *SUNN* independent; excess capacity in the *sunn-2* mutant allowed it to up regulate nitrogen fixation activity in response to localized nitrogen starvation in a split root system.

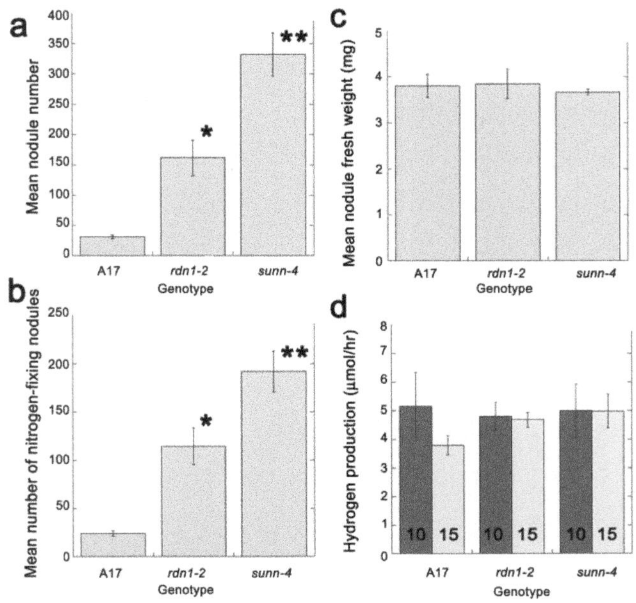

Figure 2. AON mutants and wild type plants fix equivalent amounts of nitrogen. Roots of A17, *rdn1-2* and *sunn-4* plants were evaluated 15 days after inoculation with *S. medicae*. Means ±SE are per genotype (**a**) Mean nodule number (**b**) Mean number of nitrogen fixing (pink) nodules (**c**) Mean nodule fresh weight (ten samples of three nodules each from five to six plants per genotype). (**d**) Hydrogen production (µmol/hr) as a proxy for nitrogen fixation rate (see Experimental Section). Measurements were taken on the same plants at both 10 and 15 days after inoculation with *S. medicae* in two biological replicates. A17 (n = 9), *rdn1-2* (n = 10) and *sunn-4* (n = 9). Statistically significant differences from wild type based on Student's t-test are indicated by * ($p < 0.05$) and ** ($p < 0.01$).

Thus the "nitrogen fixation signal" would be the same strength in wild type and AON mutants. Since both wild type and AON mutant plants reach the same level of nitrogen fixation at 10 days post inoculation, it is possible the systemic repression observed is a response to nitrate acquisition. We note it would not be possible to detect this second suppression in wild type plants in the split root system because wild type plants have already suppressed nodulation on the second root due to CLE12/13-SUNN signaling from developing nodules at the time this second signal is sent.

2.3. Nitrate Reduces Nodulation in Both AON Mutants and Wild Type Plants in this System

In previous work, we demonstrated that in an aeroponic system, supplying *M. truncatula* with nitrate reduced nodule number in both wild type and the AON mutants *sunn* and *rdn1* [14]. To confirm the results applied to our perlite split root system, we tested the systemic nature of the effect in one of the mutants (*sunn-4*) and wild type. Roots were starved for four days (water only) and then the first root was supplied with nitrate for four days in half of the plants while the control plants' first roots continue to receive only water. Four days later, the second root is inoculated with *S. medicae* and nodules on all roots are counted 20 days later. When nodules on the second root are reported as a percent nodulation of the first root to account for the higher nodule number of an AON mutant, both the wild type and *sunn-4* mutants reduce nodule number in the second root to 10%–30% of the water control (Figure 3) and there is no statistical difference in the degree of nodule number reduction. This confirms that the effect is systemic and that defects in AON do not affect the plant's ability to respond to available nitrate. AON mutants are commonly reported as having nitrogen resistant nodulation, in that they nodulate in the presence of nitrate while wild type plants make few to no nodules, but if viewed as a percentage reduction in nodule numbers, our data confirm the plants respond to available nitrate to the same degree as wild type, they just start from a higher number of nodulation events. Thus if nitrate suppression of nodulation is seen as a proportional rather than a binary response, there is no defect in this response in *M. truncatula* AON mutants.

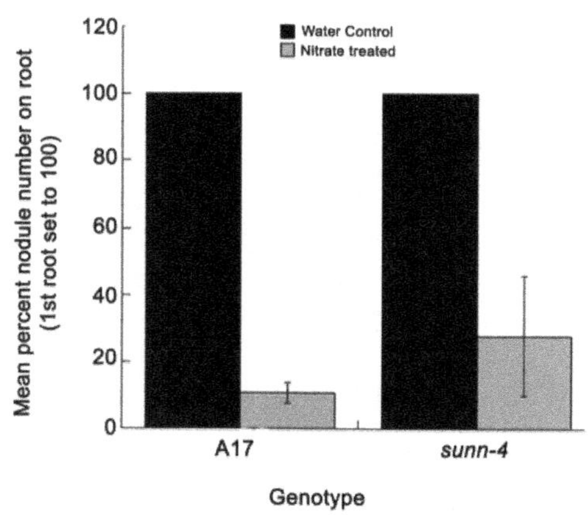

Figure 3. Effect of ammonium nitrate *versus* water treatment of first root on nodulation of second root in a split root system. Split root plants were grown as in [27] and the first root fed with either water or 10 mM NH_4NO_3 for four days before inoculating the second root. Nodules were counted on the second root of A17 (wild type) plants (n = 12) and *sunn-4* plants (n = 7) 20 days after inoculation and expressed as percentage of the water controls. Bars indicate standard error of the percentage means calculated by Taylor series expansion.

2.4. Inoculation with Fix⁻ or Nod⁺/Fix⁻ Rhizobia *Did Not Elicit the AON Response in all Plants*

To further support nitrate as the signal for the second suppression event, we designed a split root experiment in which the first root is inoculated with Fix^- bacteria, which make nodules but do not fix nitrogen, or Nod^- bacteria which do not elicit nodule formation at all. Inoculum was switched from the natural symbiont of *M. truncatula*, *S. medicae*, to the commonly used symbiont of *M. sativa*, *Rhizobium meliloti*, to take advantage of the large number of mutant strains available in *R. meliloti*. We chose three *S. meliloti* strains: wild type *Rm*1021 [32], a Nod$^+$/Fix$^-$ bacterial mutant; *Rm*1312 [33] to test for nitrogen fixation response; and a Nod$^-$/Fix$^-$ bacterial mutant SL44 [34] to test for response to developing meristems (the CLEs). The tester root was inoculated with one of the three strains and the responder root was always inoculated with the wild type strain 20 days after the tester. This later time point was chosen to accommodate the slightly slower pace of nodule development we observed in preliminary experiments when wild type plants were inoculated with *R. meliloti versus S. medicae.*

Inoculating the tester root of wild type and *rdnl* mutant plants with the wild type *Rm*1021 strain resulted in suppression of nodule number on the responder root 20 days later in *rdn1-2* mutant and wild type plants (gray bars in Figure 4a) at levels consistent with those seen with the *S. medicae* strain, indicating that the suppression shown in Figure 1 also occurred with this strain of *rhizobia* in these plants. Interestingly, *sunn-4* mutants decreased nodulation, but not to a level of statistical significance, suggesting the suppression response was muted in this genotype when inoculated with *R. meliloti*. However, inoculation of the tester root with the Nod$^-$/Fix$^-$ SL44 strain, which does not form nodules or nodule meristems, did not suppress nodulation on the responder root in all plants tested (Figure 4c), consistent with the postulate that without the initial signaling that leads to nodule meristems, AON does not occur in any plant. More importantly, inoculation of the tester root with the Nod$^+$/Fix$^-$ bacterialmutant *Rm*1312 resulted in no suppression of nodulation in the responder root of wild type or hypernodulation mutant plants (Figure 4b), suggesting that the ability of the bacteria to fix nitrogen was critical to the regulation observed in wild type plants at 20 days after inoculation, and to the late regulation phenomenon we observed in the mutants. In all plants tested, lack of nitrogen fixation in the first root abolishes regulation in the second root if the second root is inoculated after the 10–12 day time point.

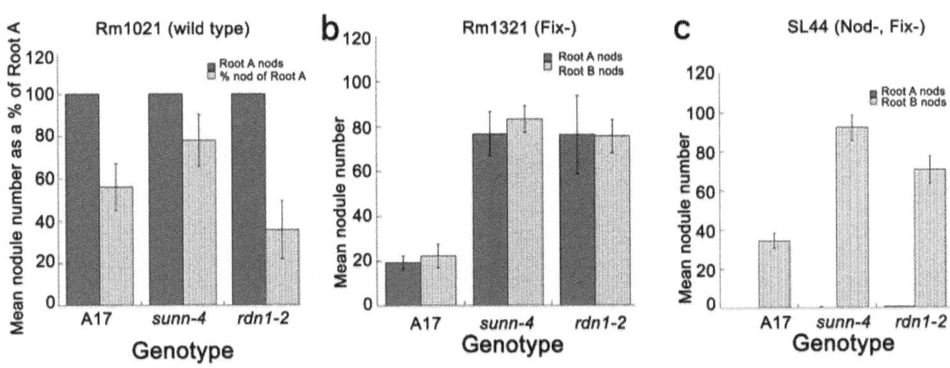

Figure 4. The effect of prior inoculation of the tester root with different strain of *S. meliloti* on subsequent nodulation of the responder root in A17, *sunn-4,* and *rdn1-2*. The responder root was inoculated with *Rm*1021 20 days after the tester root inoculation and nodules were counted 21 days later. (**a**) *Rm*1021 wild type on both roots, expressed as mean nodulation % of the responder root to the tester. Errors are standard error by Taylor expansion. When inoculating with different strains, data is expressed as mean nodule number ± SE on the responder root to mean nodule number ± SE for strains used on the tester root in (**b**) *Rm*1312 (Nod$^+$/Fix$^-$), and (**c**) SL44 (Nod$^-$/Fix$^-$). Error bars for (**b**) and (**c**) are standard error of the mean (n = 5 to 10 plants per genotype per experiment).

2.5. Nutrient Stress Causes Suppression of Nodulation in Wild Type and AON Mutants

Although we could not detect this second signal in wild type plants, we reasoned that if nitrate acquisition was the second systemic signal the AON mutants were responding to, treating the first root with water up until the time at which we observed the suppression (11 days) should abolish the suppression and the second root should hypernodulate in AON mutants and nodulate normally in wild type plants. Using wild type, *sunn-4* and *rdn1-2* plants in the split root system, the first root received only water and the second root was inoculated either after four days (the control mimicking the early conditions in Figure 1) or after 11 days. The value for the first root is set to 100 percent in Figure 5. As expected, after four days of water treatment wild type plants nodulated robustly while the AON mutants displayed significantly higher nodule numbers than wild type plants on the second root ($p = 0.009$ and $p = 0.012$ for *sunn-4* and *rdn1-2,* respectively, Students *t* test). We were surprised to find that after 11 days of water, not only did wild type plants significantly reduce nodule number on the second root ($p < 0.00001$), but *sunn-4* and *rdn1-2* plants did as well, and nodule numbers on the second root of AON mutant split root plants were statistically the same as the second root of wild type plants ($p = 0.0001$). This effect does not rule out the nitrate availability signal as the source of the suppression seen in Figure 1, but it suggests that the stress of being starved for nitrogen and other nutrients also leads to a suppression of nodule number and both these suppression pathways are intact in wild type and AON mutants.

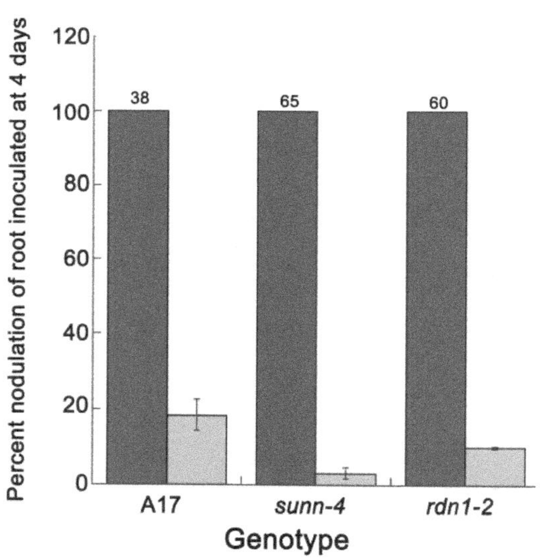

Figure 5. AON mutants and wild type plants suppress nodulation after long periods of water only treatment of the first root. Dark grey bars represent nodulation on the second root after four days of water treatment prior to inoculation and is set to 100% (actual mean on top of bar); light grey indicates mean percent reduction in nodulation on the second root after 11 days treatment of the first root prior to inoculation. Bars indicate standard error of the percentage means calculated by Taylor expansion. $n = 9–16$ plants per genotype per treatment.

To confirm that the plants were indeed stressed for nutrients, we performed C and N analysis on shoot tissue from these same plants and compared them to the plants in Figure 3 that received nitrate for four days. All tissue was harvested on the day of inoculation. After four days of treatment, as expected, plants that received nitrate had almost twice the nitrogen of plants that received only water and there was no difference between genotypes (Figure 6a). Plants which received only water on the first root for 10 days had tissue with significantly less nitrogen than the water treated plants at four days, indicating a higher level of nitrogen starvation across genotypes. However, there was no difference in carbon content between all plants and all treatments (Figure 6b). Thus, nitrogen is the significant driver of the C/N ratio in Figure 6c, which shows all plants changing compositions significantly as nutrients are depleted. As mentioned in the introduction, nodulation is a costly process for the plant [2]. Since these nutrient stressed plants suppressed nodulation, we postulated that the plant reaches a point at which the resources required to mount the nodulation response are not available even though there is a nitrogen deficit that would otherwise promote nodulation. Indeed Voisin, *et al.* [30] showed a strong interaction between N and C metabolism in pea nodule number regulation and variation dependent on growth rates in the week preceding an inoculation. Also, nitrogen is not the only nutrient the "water only" plants are deficient in; phosphorus deficiencies have also been shown to affect the ability to nodulate [35], and these plants are likely low in potassium and some micronutrients as well.

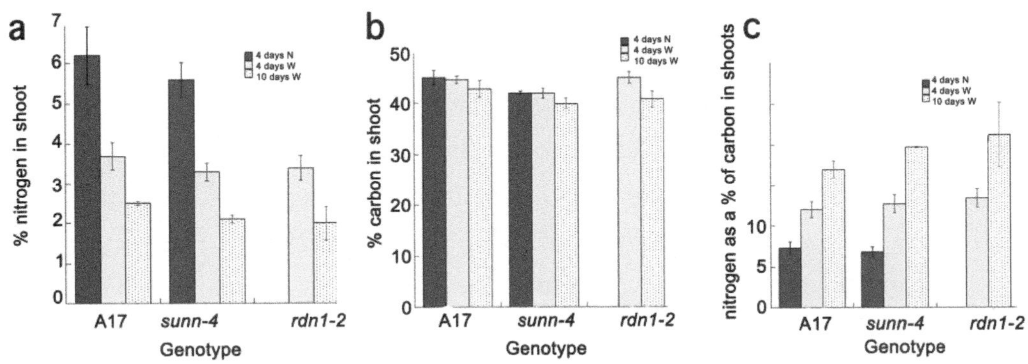

Figure 6. Carbon and nitrogen in shoots as a percentage of dry weight determined by ^{13}C and ^{15}N isotope analysis of organic solids. An equal amount of tissue was pooled from three plants for each genotype and condition at the day of inoculation, dried, and analysis performed in triplicate (**a**). Mean percent nitrogen in leaves from plants in which the first roots was watered with 10 mM NH_4NO_3 for four days from Figure 3 (black bars-data for *rdn1-2* not available) with water for four days from Figure 5 (grey bars) and with water for 10 days from Figure 5 (dotted bars) (**b**). Mean percent carbon from the same material and (**c**) C/N ratio of plants in (a) and (b).

The nutrient stress response observed when inoculating with water is at first glance puzzling: we reasoned that responder roots on plants inoculated with bacteria that could not fix nitrogen (Figure 4) should be as nutrient stressed as those in Figure 5, making the lack of nodule suppression unexpected in the roots responding to inoculation with Fix^- or Nod^- bacteria. Inoculation of *M. sativa* with Nod^- *rhizobia* or even non-interacting Agrobacterium increased nodule development when co-inoculated with wild type bacteria [36] and it may be that the Fix^- or Nod^- inoculations "prime" the plant for future nodulation. While Caetano-Anollés and Bauer did not use a split root system with separate inoculations, our results could suggest a systemic effect, but we recognize this is speculation and further work is required. We note that many environmental variations, even variations in light intensity, can affect nodule number [30], and the experiments in Figure 4 and 5 were performed in the greenhouse in different years, with likely small variations in other environmental inputs besides nutrients, as well as being inoculated with different species of *rhizobia*, which may contribute to the varied results.

3. Experimental Section

3.1. Plant Materials and Growth Conditions

Seeds of *Medicago truncatula*.cv Jemalong A17 and AON defective mutants *rdn1-2* [14] and *sunn-4* [11] were acid scarified and imbibed as described in [20]. Seed was then vernalized in the dark at 4 °C for 2 days on Harrison Modified Farhaeus (HMF) media [37] covered with two half round Whatman filter papers (GE Healthcare, USA), followed by germination in the dark at room temperature for 1 day before being used in the experiments described.

3.2. Split Root Development and Inoculation

Lateral root initiation was stimulated by removing the roots of 5-day-old seedlings at the root-shoot junction. The trimmed plants were placed onto HMF plates sandwiched between two half rounds of Whatman filter paper and split-root systems were established as described previously [27]. Plants were transferred to pots filled with washed, autoclaved perlite and placed in a greenhouse. They were watered daily with a 100-fold dilution of water-soluble 20:10:20 (N:P:K) Peat-Lite Special fertilizer (Scotts Company, OH, USA) for 5 days. After an additional 4 days of nutrient starvation induced by watering with water alone, the plants were used for split-root inoculation experiments with bacteria. The second root (Root B) was inoculated with *rhizobia* 0, 2, 3, 4, 6, 8, 10, 12, 15, or 20 days after the first root (Root A) inoculation and nodules on both roots were counted 21 days after the second root inoculation. Root systems that were not similar in length (within 20% of each other) at the 21-day time point were eliminated from consideration. For simultaneous inoculation (0 time point) one root at random was designated A and the other B.

3.3. Rhizobial Strains and Growth Conditions

To demonstrate the timing of AON, both Root A and Root B were inoculated with *S. medicae* strain ABS7 [38]. To investigate the nature of a second nodule inhibitory signal, the tester root was inoculated with *Rm*1021 [32] or its derivatives *Rm*1312 [33], or SL44 [34] and the responder root with the wild type strain *Rm*1021; nodules on the responder root were counted after 21 days. Root systems that were unequal in size at the 21-day time point were eliminated from consideration.

All strains were grown in liquid TY media (5 g tryptone, 3 g yeast extract, 1.4 g $CaCl_2•2H_2O$, pH 6.5–7 and 15 g agar per liter) containing antibiotics at the following concentrations: tetracycline (15 μg/mL) for *S. medicae* and streptomycin (500 μg/mL) for *S. meliloti* strains. Before inoculation, the *rhizobia* were diluted to 0.2 OD 600 nm with sterile water and 6 mL of the bacterial solution was applied to each root compartment at root collar region (flood inoculation). Nodules were counted using an Olympus SZX12 Dissecting Stereo Microscope 21 days after the second root inoculation after gentle washing of the Perlite from the roots.

3.4. Nitrogen Fixation Measurements

The Q-Box NF1LP package (Qubit Systems Inc, Kingston, Canada) was used for measuring the rate of H_2 production in order to estimate nitrogen fixation. Plants were prepared as described for the split root experiments with the same establishment and nutrient withdrawal regime and then inoculated with *S. medicae*. The flow-through H_2 from N_2-fixing tissues was measured on each plant at 10 days and 15 days post rhizobial inoculation in two independent experiments. The nodulated root systems were sealed with Qubitac^{TM} sealant to avoid air leakage so that only the shoots were exposed to room air. To generate maximum H_2 evolution, roots were flushed with an $Ar{:}O_2$ (80:20) gas combination at the rate of 250 mL/minute for 5 min, so that all electrons measured were utilized for H^+ reduction. The measurement of H_2 evolution as a means of determining nitrogenase activity used the LabPro interface in conjunction with Logger Pro software provided by Vernier (Portland, Oregon) according to the manufacturer's instructions.

3.5. Determination of C and N

For each condition analyzed, three samples consisting each of pooled leaves from three plants (two trifoliates per plant) were dried, crushed, and prepared in tin capsules for C and N composition analysis by light stable isotope ratio mass spectrometry at the Duke Environmental Stable Isotope Laboratory [39]. Each sample weighed between 5.0 to 12.2 mg.

3.6. Statistical Analysis

Statistical analyses were conducted to determine if the true means of the various responses (percent nodulation of the second root, nodule number, number of nitrogen fixing nodules, nodule fresh weight, hydrogen production, nodule number on responder root) differed among the levels of the various experimental factors of interest (days between inoculation of roots, dpi, and genotypes). Analysis of Variance (ANOVA) was used on the split root data (with Type I error rate set at 0.05) to determine if overall significant differences existed among the factor levels. If overall differences were found, pairwise comparison among the least squares means of the factor levels were performed using JMP-9 software (SAS Institute Inc., Cary, IL, USA). For all other analyses, the Student's t test with the Bonferroni correction for multiple comparisons was used to determine the significance of differences from wild types. Calculation of error in mean percentages was done with a Taylor expansion of the variance.

4. Conclusions

We show nodule regulation in *M. truncatula* in our system includes both an early response, related to nodule meristem development (the SUNN/RDN1 pathway) and a late response which does not require SUNN or RDN1 and is correlated with nutrient status, especially nitrate status. While the two signals are redundant in nodulating wild type plants, the response of *rdn1* and *sunn* mutants in the split root system allowed us to dissect out the timing of this second suppression event to coincide with the establishment of nodules actively fixing nitrogen. This timing would allow plants to continue to nodulate if the initial *rhizobia* are not providing nitrogen compounds for the plants or the nodules or roots are damaged and no longer fix an adequate supply.

Both nutrient stress and nitrate are implicated as regulating nodule number outside of the systemic CLE12/13-SUNN signaling pathway, and our experiments provide evidence for at least one additional pathway for nodule suppression that does not require SUNN or RDN1. Determining if the nitrate and/or nutrient stress pathways signal through common components or are different pathways will require additional experimentation with mutants in these other pathways. The recent isolation of mutants in *MtCRA2*, a leucine-rich repeat receptor kinase involved in nitrogen sensing and response, and the discovery of the *Mt*CEP1 peptide involved in nitrogen sensing should facilitate future studies to further understanding [40,41].

Acknowledgments

We would like to thank Melanie Barnett and Sharon Long for the *S. meliloti* strains, and Elise Schnabel for experimental assistance and critical reading of the manuscript. This work was supported by NSF awards #IOS-0641848 and #IOS-1146014. Technical Contribution No. 6132 of the Clemson University Experiment Station.

Author Contributions

T.K. performed the majority of the experiments, W.B. assisted in statistical analysis of the data, and both participated with J.F. in designing the experiments and preparing the manuscript.

Conflicts of Interest

The authors declare no conflict of interest.

References and Notes

1. Oldroyd, G.; Downie, A. Coordinating nodule morphogenesis with rhizobial infection in legumes. *Annu. Rev. Plant Biol.* **2008**, *59*, 519–545.
2. Crawford, N.M.; Kahn, M.L.; Leustek, T.; Long, S.R. Nitrogen and sulfur. In *Biochemistry and Molecular Biology of Plants*; Buchanan, B.B., Gruissem, W., Jones, R.L., Eds.; American Association of Plant Physiologists: Rockville, MD, USA, 2000; pp. 787–849.
3. Kosslak, R.M.; Bohlool, B.B. Suppression of nodule development of one side of a split root system of soybeans caused by prior inoculation of the other side. *Plant Physiol.* **1984**, *75*, 125–130.
4. Bhuvaneswari, T.V.; Bhagwat, A.A.; Bauer, W. Transient susceptibility of root-cells in 4 common legumes to nodulation by rhizobia. *Plant Physiol.* **1981**, *68*, 1144–1149.
5. Suzuki, A.; Hara, H.; Kinoue, T.; Abe, M.; Uchiumi, T.; Kucho, K.; Higashi, S.; Hirsch, A.M.; Arima, S. Split-root study of autoregulation of nodulation in the model legume *Lotus japonicus*. *J. Plant Res.* **2008**, *121*, 245–249.
6. Sargent, L.; Huang, S.Z.; Rolfe, B.G.; Djordjevic, M.A. Split-root assays using *Trifolium subterraneum* show that rhizobium infection induces a systemic response that can inhibit nodulation of another invasive rhizobium strain. *Appl. Environ. Microbiol.* **1987**, *53*, 1611–1619.

7. Van Brussel, A.A.N.; Tak, T.; Boot, K.J.M.; Kijne, J.W. Autoregulation of root nodule formation: Signals of both symbiotic partners studied in a split-root system of *Vicia sativa* subsp. *Nigra*. *Mol. Plant-Microbe Interact.* **2002**, *15*, 341–349.
8. Krusell, L.; Madsen, L.H.; Sato, S.; Aubert, G.; Genua, A.; Szczyglowski, K.; Duc, G.; Kaneko, T.; Tabata, S.; De Bruijn, F.J.; *et al.* Shoot control of root development and nodulation is mediated by a receptor-like kinase. *Nature* **2002**, *420*, 422–426.
9. Nishimura, R.; Hayashi, M.; Wu, G.J.; Kouchi, H.; Imaizumu-Anraku, H.; Murakami, Y.; Kawasaki, S.; Akao, S.; Ohmori, M.; Nagasawa, M.; Pajuelo, E.; Sandal, N.; Stougaard, J. HAR1 mediates systemic regulation of symbiotic organ development. *Nature* **2002**, *420*, 426–429.
10. Searle, I.R.; Men, A.E.; Laniya, T.; Buzas, D.; Iturbe-Ormaetxe, I.; Carroll, B.J.; Gresshoff, P.M. Long-distance signaling in nodulation directed by a clavata1-like receptor kinase. *Science* **2003**, *299*, 109–112.
11. Schnabel, E.; Journet, E.P.; Carvalho-Niebel, F.; Duc, G.; Frugoli, J. The *Medicago truncatula SUNN* gene encoding a *CLV1*-like leucine-rich repeat receptor kinase regulates both nodule number and root length. *Plant Mol. Biol.* **2005**, *58*, 809–822.
12. Okamoto, S.; Shinohara, H.; Mori, T.; Matsubayashi, Y.; Kawaguchi, M. Root-derived cle glycopeptides control nodulation by direct binding to HAR1 receptor kinase. *Nat. Commun.* **2013** *4*, 2191.
13. Ferguson, B.J.; Indrasmunar, A.; Hayashi, S.; Lin, M.H.; Reid, D.E.; Gresshoff, P.M. Molecular analysis of legume nodule development and autoregulation. *J. Integr. Plant Biol.* **2010**, *52*, 61–76.
14. Schnabel, E.; Kassaw, T.; Smith, L.; Marsh, J.; Oldroyd, G.; Long, S.; Frugoli, J. The *ROOT DETERMINED NODULATION1* gene regulates nodule number in roots of *Medicago truncatula* and defines a highly conserved, uncharacterized plant gene family. *Plant Physiol.* **2011**, *157*, 328–340.
15. Ogawa-Ohnishi, M.; Matsushita, W.; Matsubayashi, Y. Identification of three hydroxyproline *o*-arabinosyltransferases in *Arabidopsis thaliana*. *Nat. Chem. Biol.* **2013**, *9*, 726–730.
16. Mortier, V.; Herder, G.; Whitford, R.; van de Velde, W.; Rombauts, S.; D'haeseleer, K.; Hosters, M.; Goormachtig, S. CLE peptides control *Medicago truncatula* nodulation locally and systemically. *Plant Physiol.* **2010**, *153*, 222–237.
17. Lim, C.W.; Lee, Y.W.; Hwang, C.H. Soybean nodule-enhanced CLE peptides in roots act as signals in *Gm*NARK-mediated nodulation suppression. *Plant Cell Physiol.* **2011**, *52*, 1613–1627.
18. Saur, I.M.L.; Oakes, M.; Djordjevic, M.A.; Imin, N. Crosstalk between the nodulation signaling pathway and the autoregulation of nodulation in *Medicago truncatula*. *New Phytol.* **2011**, *190*, 865–874.
19. Mortier, V.; de Wever, E.; Vuylsteke, M.; Holsters, M.; Goormachtig, S. Nodule numbers are governed by interaction between CLE peptides and cytokinin signaling. *Plant J.* **2012**, *70*, 367–376.
20. Schnabel, E.; Mukherjee, A.; Smith, L.; Kassaw, T.; Long, S.; Frugoli, J. The *lss* supernodulation mutant of *Medicago truncatula* reduces expression of the *SUNN* gene. *Plant Physiol.* **2010**, *154*, 1390–1402.
21. Schultze, M.; Kondorosi, A. Regulation of symbiotic root nodule development. *Annu. Rev. Genet.* **1998**, *32*, 33–57.

22. Carroll, B.J.; McNeil, D.L.; Gresshoff, P.M. A super-nodulation and nitrate-tolerant symbiotic (nts) soybean mutant. *Plant Physiol.* **1985**, *78*, 34–40.
23. Caba, J.M.; Recalde, L.; Ligero, F. Nitrate-induced ethylene biosynthesis and the control of nodulation in alfalfa. *Plant Cell Environ.* **1998**, *21*, 87–93.
24. Baker, B.; Zambryski, P.; Staskawicz, B.; Dinesh-Kumar, S.P. Signaling in plant-microbe interactions. *Science* **1997**, *276*, 726–733.
25. Singleton, P.W.; van Kessel, C. Effect of localized nitrogen availability to soybean half-root systems on photosynthate partitioning to roots and nodules. *Plant Physiol.* **1987**, *83*, 552–556.
26. Laguerre, G.; Heulin-Gotty, K.; Brunel, B.; Klonowska, A.; Le Quere, A.; Tillard, P.; Prin, Y.; Cleyet-Marel, J.C.; Lepetit, M. Local and systemic N signaling are involved in *Medicago truncatula* preference for the most efficient *Sinorhizobium* symbiotic partners. *New Phytol.* **2012**, *195*, 437–449.
27. Kassaw, T.; Frugoli, J. Simple and efficient methods to generate split roots and grafted plants useful for long-distance signaling studies in *Medicago truncatula* and other small plants. *Plant Methods* **2012**, *8*, 38.
28. Olsson, J.E.; Nakao, P.; Bohlool, B.B.; Gresshoff, P.M. Lack of systemic supression of nodulation in split root systems of supernodulating soybean (*Glycine max* [l.] *merr.*) mutants. *Plant Physiol.* **1989**, *90*, 1347–1352.
29. Hinson, K. Nodulation responses from nitrogen applied to soybean half-root systems. *Agron. J.* **1975**, *67*, 799–804.
30. Voisin, A.S.; Munier-Jolain, N.G.; Salon, C. The nodulation process is tightly adjusted to plant growth. An analysis using environmentally and genetically induced variation of nodule number and biomass in pea. *Plant Soil.* **2010**, *337*, 399–412.
31. Jeudy, C.; Ruffel, S.; Freixes, S.; Tillard, P.; Santoni, A.L.; Morel, S.; Journet, E.P.; Duc, G.; Gojon, A.; Lepetit, M.; Salon, C. Adaptation of *Medicago truncatula* to nitrogen limitation is modulated via local and systemic nodule developmental responses. *New Phytol.* **2010**, *185*, 817–828.
32. Meade, H.M.; Long, S.R.; Ruvkun, G.B.; Brown, S.E.; Ausubel, F.M. Physical and genetic characterization of symbiotic and auxotrophic mutants of *Rhizobium meliloti* induced by transposon Tn5 mutagenesis. *J. Bacteriol.* **1982**, *149*, 114–122.
33. Ruvkun, G.B.; Long, S.R.; Meade, H.M.; van den Bos, R.C.; Ausubel, F.M. Isrm1: A *Rhizobium meliloti* insertion sequence that transposes preferentially into nitrogen fixation genes. *J. Mol. Appl. Genet.* **1982**, *1*, 405–418.
34. Fisher, R.F.; Egelhoff, T.T.; Mulligan, J.T.; Long, S.R. Specific binding of proteins from *Rhizobium meliloti* cell-free extracts containing NodD to DNA sequences upstream of inducible nodulation genes. *Genes Dev.* **1988**, *2*, 282–293.
35. Tsvetkova, G.E.; Georgiev, G.I. Effect of phosphorus nutrition on the nodulation, nitrogen fixation and nutrient-use efficiency of *Bradyrhizobium japonicum*_soybean (*Glycine max* l. *Merr.*) symbiosis. *Bulg. J. Plant Physiol.* **2003**, 331–335.
36. Caetano-Anollés, G.; Bauer, W.D. Enhanced nodule initiation on alfalfa by wild-type *Rhizobium meliloti* co-inoculated with Nod gene mutants and other bacteria. *Planta* **1988**, *174*, 385–395.
37. Huo, X.; Schnabel, E.; Hughes, K.; Frugoli, J. RNAi phenotypes and the localization of a protein::GUS fusion imply a role for *Medicago truncatula PIN* genes in nodulation. *J. Plant Growth Regul.* **2006**, *25*, 156–165.